Advanced Additive Manufacturing

Edited by Igor V. Shishkovsky

Published in London, United Kingdom

IntechOpen

Supporting open minds since 2005

Advanced Additive Manufacturing
http://dx.doi.org/10.5772/intechopen.95667
Edited by Igor V. Shishkovsky

Contributors
Sze Yi Mak, Ching Hang Bob Yung, Lung Fung Tse, Wing Fung Edmond Yau, Matthias Dahlmeyer, Sebastian Noller, Rajkumar Velu, R. Sathishkumar, A. Saiyathibrahim, Roozbeh (Ross) Salary, Isabel Montealegre-Meléndez, Enrique Ariza Galván, Erich Neubauer, Eva Maria Perez Soriano, Cristina M. Arevalo Mora, Michael Kitzmantel, Park Sangwon, Soundharrajan Vaiyapuri, Jin-Ho Kang, Nileshkumar Dubey, Geetha Manivasagam, Kwi-Dug Yun, Sakthiabirami Kumaresan, Tat-Hean Gan, Bojie Sheng, Jamil Kanfoud, Chad Ulven, Patrick Simpson, Michael Holthaus, Luke Gibbon, Pavel Kuznetsov, Anna Mozhayko, Ivan Shakirov, Vitaliy Bobyr, Mikhail Staritsyn, Anton Zhukov, Konstantin Igorevich Makarenko, Igor V. Shishkovsky, Oleg Dubinin, Shyam Aravamudhan, Smith Woosley, David Muñoz-Rojas, Matthieu Weber, Christophe Vallée, Chiara Crivello, Abderrahime Sekkat, Fidel Toldra-Reig, Mikhael Bechelany, Lukas Stepien, Samira Gruber, Moritz Greifzu, Mirko Riede, Aljoscha Roch

First published in London, United Kingdom, 2022 by IntechOpen
IntechOpen is the global imprint of INTECHOPEN LIMITED, registered in England and Wales, registration number: 11086078, 5 Princes Gate Court, London, SW7 2QJ, United Kingdom
Printed in Croatia

British Library Cataloguing-in-Publication Data
A catalogue record for this book is available from the British Library

Additional hard and PDF copies can be obtained from orders@intechopen.com

Advanced Additive Manufacturing
Edited by Igor V. Shishkovsky
p. cm.
Print ISBN 978-1-83962-820-7
Online ISBN 978-1-83962-821-4
eBook (PDF) ISBN 978-1-83962-822-1

For EU product safety concerns:
IN TECH d.o.o., Prolaz Marije Krucifikse Kozulić 3, 51000 Rijeka, Croatia, info@intechopen.com or visit our website at intechopen.com.

Meet the editor

Igor Shishkovsky is the head of the Additive Manufacturing Lab, Center for Materials Technologies (CMT), Skoltech, Russia, and a known Russian expert in additive manufacturing approaches. In recent years, he has participated in the development of 4D printing methods, topological design and synthesis of unique products from metamaterials, and formulation and implementation of the concept of the digital twin of the laser powder bed fusion process. He is a co-author of more than 200 scientific papers, 14 books/chapters, and 8 patents devoted to additive manufacturing (powder bed fusion, direct energy deposition, 3D laser cladding, etc.) and laser treatment of materials. His current research interests are additive manufacturing of functional gradient parts, 4D printing, and biofabrication of implants and scaffolds.

Contents

Preface

Additive manufacturing (AM), also known as 3D printing, is a technology that produces three-dimensional parts layer by layer from a material. AM has existed at various levels of sophistication for decades. However, it has only recently caught the widespread attention of industries and policymakers thanks to technological breakthroughs and advancements in the past decade that allow full-scale metallic components and structures to be made to high standards. AM is getting more attention and is considered a better choice of fabrication in almost every industry due to its unlimited design freedom, optimization, lightweight, and customization.

In the two sections of this book, readers will find new approaches for biomedical applications and advanced technological solutions. Medicine was and is a source of inspiration and a natural field for AM applications. Therefore, customized medical devices, bone tissue engineering for surgical operations, and/or biocompatible ceramics are promising trends.

The development of new hierarchical metamaterials and multi-materials manufactured by 3D printing will improve the use of these structures in biomedicine and industry and will allow new applications to be found. We suggest that applying the proposed manufacturing techniques in collaboration with topological design will significantly expand and deepen the areas of advanced manufacturing. All such trends will be obvious drivers for an accelerated transition of the world industry to a new technological era.

Igor V. Shishkovsky
Center for Design,
Manufacturing and Materials,
Skolkovo Institute of Science and Technology,
Moscow, Russian Federation

Section 1

Biomedical Applications

Chapter 1

Perspective Chapter: Additive Manufactured Zirconia-Based Bio-Ceramics for Biomedical Applications

Sakthiabirami Kumaresan, Soundharrajan Vaiyapuri, Jin-Ho Kang, Nileshkumar Dubey, Geetha Manivasagam, Kwi-Dug Yun and Sang-Won Park

Abstract

Zirconia was established as one of the chief vital ceramic materials for its superior mechanical permanency and biocompatibility, which make it a popular material for dental and orthopedic applications. This has inspired biomedical engineers to exploit zirconia-based bioceramics for dental restorations and repair of load-bearing bone defects caused by cancer, arthritis, and trauma. Additive manufacturing (AM) is being promoted as a possible technique for mimicking the complex architecture of human tissues, and advancements reported in the recent past make it a suitable choice for clinical applications. AM is a bottom-up approach that can offer a high resolution to 3D printed zirconia-based bioceramics for implants, prostheses, and scaffold manufacturing. Substantial research has been initiated worldwide on a large scale for reformatting and optimizing zirconia bioceramics for biomedical applications to maximize the clinical potential of AM. This book chapter provides a comprehensive summary of zirconia-based bioceramics using AM techniques for biomedical applications and highlights the challenges related to AM of zirconia.

Keywords: additive manufacturing (AM), zirconia, dental restorations, scaffolds, implants, challenges

1. Introduction

The use of biomaterials in the reconstruction of injured body parts and skeletal healing is unavoidable. Diverse biomaterials including ceramics, metals, polymers, hydrogels, and composites are explored and have achieved clinical success as well [1–3]. For bone restoration applications ceramic biomaterials are well recognized by biomaterial engineers and medical experts due to their biocompatibility and osteoconductivity. Each bioceramic has its unique properties, and they can be divided into three categories based on the properties: [1] bioactive ceramics: capable of establishing chemical interaction with the cell surface, [2] bio-inert ceramics: fully unreactive to the living ecosystem, [3] resorbable bioceramics: undergoes in vivo deficiency for phagocytosis or dissolution of the biomaterials in human body fluids [4]. The standard bioactive ceramics used for bone-regeneration applications are bio-glasses and calcium phosphate-based resources, such as beta-tricalcium phosphate, hydroxyapatite, and biphasic calcium phosphate (mixture of beta-tricalcium phosphate and hydroxyapatite). However, alumina and zirconia oxide are the well-established bio-inert ceramics used in classic bone-regeneration applications [5]. Each bioceramics are widely used in the various human parts restoration applications based on the needs and capabilities. Excellent mechanical stability and biocompatibility brand zirconia as a potential dental restoration and bone scaffold material for load-bearing applications [6]. Hence, rigorous efforts were concentrated on zirconia-based ceramics in recent times by medical and research experts for dental and biomedical applications.

1.1 Overview of zirconia

Zirconia is a polycrystalline dioxide ceramic of the transition metal zirconium [3, 7]. It was originally documented in 1789 by Martin Heinrich Klaproth, a German chemist [8]. Zirconia exists in three distinct crystal structures depending on the pressure and temperature: monoclinic, tetragonal, and cubic structures [9]. The monoclinic crystal structure is more constant from room temperature to 1170°C, but it has inferior mechanical properties compared to the other two structures [10]. It is commonly accepted that the monoclinic structure will transform into a tetragonal structure during thermal treatment between 1170°C and 2370°C. This change in crystal structure is accompanied by measurable volume reductions (4–5%) during the cooling period [8]. If the temperature is increased further, the tetragonal structure shrinks to form a cubic structure (between 2370°C and 2680°C, the melting point). During cooling, a noticeable volume expansion of 3–4% was observed, which is attributable to the reversible transformation into the monoclinic crystal structure [9]. During phase transformation, internal stress is induced in the zirconia lattice, which results in crack propagation. To suppress the aforementioned behavior, several metallic oxides or dopants (stabilizing agents such as Y_2O_3, MgO, CaO, and CeO) are added to stabilize the zirconia structure, and the resultant type of zirconia is known as partially stabilized zirconia (PSZ) [11].

The key features of PSZ are their ability to enhance the transformation toughening mechanism, which inhibits/shields the further propagation of cracks. Therefore, PSZ is considered suitable for biomedical applications in orthopedics and dentistry due to its unique toughening behavior. In the late 1970s, zirconia was widely used as an effective substitute material for metals and alumina in biomedical and dental applications. This was due to its long-lasting mechanical behaviors, such as good flexural strength and fracture resistance, admirable biocompatibility, chemical permanency, corrosion resistance, and esthetics [12, 13]. Nevertheless, the aging process of zirconia ceramic is stimulated by low-temperature degradation, which has unfavorable impacts on the mechanical strength of prostheses and subsequent growth of external flaws. The presence of microcracks may compromise the performance in the long term in biological fluids [14].

To date, zirconia-based materials have been used in numerous areas in the engineering (energy and aerospace), medicine (orthopedics), and dental (crowns and implants) fields [15]. Common categories of zirconia-based materials existing on the market for biomedical applications are yttrium tetragonal zirconia polycrystal (Y-TZP), glass-infiltrated zirconia-toughened alumina (ZTA), and magnesia partially stabilized zirconia (Mg-PSZ). The properties of these zirconia-based bioceramics are listed in **Table 1**.

In general, zirconia-based ceramics are manufactured using conventional fabrication techniques, such as injection molding [17], hot and cold isostatic pressing, and slip

Properties	Y-TZP	ZTA	Mg-PSZ
Chemical constituents	Y_2O_3, ZrO_2	Al_2O_3, ZrO_2	MgO, ZrO_2
Crystallinity	Monophasic	Biphasic	Biphasic
Density (g/cm^3)	6.05	5	5
Flexural strength (MPa)	800–1300	750–850	700–800
Hardness (GPa)	10–12	12–15	5–6
Fracture toughness (MPa m$^{1/2}$)	5–10	6–12	8–15

Table 1.
Properties of zirconia-based ceramics [16].

casting [18]. Digital techniques such as computer-aided design (CAD) and computer-aided manufacturing (CAM) are extensively used to fabricate dental restorations [7, 19], as well as in subtractive manufacturing techniques, such as machining and milling. However, these techniques have limitations such as material wastage, difficulties in producing complex structures, being time consuming, and wearing of milling and cutting tools. Recently, additive manufacturing (AM) techniques have been increasingly used for the fabrication of high-potential complex ceramic parts with high precision and at reduced cost [20, 21]. Developments in AM technology for the fabrication of zirconia-based ceramic parts and their applications are discussed in the following section.

1.2 Additive manufacturing techniques

AM is one of the most widely used techniques in recent times, and it is capable of building three-dimensional (3D) complex geometric structures with high dimensional precision and within a short manufacturing time. 3D objects with high levels of complexity and structural architectures are fabricated by stacking up the materials layerwise using simulated design files [22, 23]. AM is also known as 3D printing, solid free-form fabrication, and rapid prototyping. The materials used for AM processes are in the form of powders, liquids, or solids. According to the ISO/ASTM 17296 standard, AM technology is mainly characterized into two types based on the degree of consolidation [24].

1.2.1 Single-step process or direct process

As the name suggests, the combined bulk product is manufactured with a basic/specified geometric shape in a single operation by melting and solidification or multi-pass welding (such as powder bed fusion, selective laser melting (SLM), or directed energy deposition), which is mostly used in metal AM.

1.2.2 Multistep or indirect process

It produces the products in multiple steps. First, the green body parts are constructed with the basic geometric shape by binding the powder particles with help of a polymer or binder. Subsequent steps include shape modification/densification, consolidation of the material, or modification of the material properties (such as binder jetting (BJ) and material extrusion). AM ceramics parts are typically formed using multistep progression [25].

For biomedical and dental applications, the 3D printing process principally comprises the following steps (precisely for clinical applications): 1. procurement of 3D models, 2. designing (CAD), 3. slicing, 4. 3D printing, and 5. postprocessing.

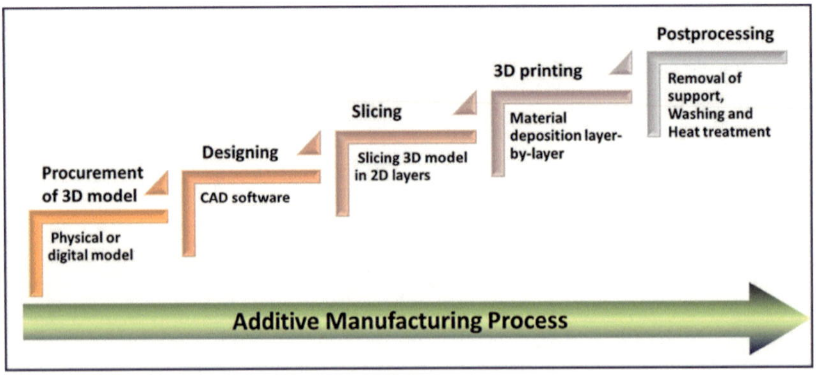

Figure 1.
Illustration of additive manufacturing process.

Briefly, the AM process starts with the sorting of precise medical records (images) of the patients, which are obtained using computed tomography or magnetic resonance imaging. The procured data conforming to digital imaging and communications in medicine standards are transformed into digital models using materialise interactive medical image control system (MIMICS) or 3D DOCTOR software and formed into design files using 3D CAD software. The CAD file is converted to a standard tessellation language (.STL) file, and it is practically sliced to print patterns as per the specific needs of the implant. To acquire the desired print pattern of the products, numerous processing constraints such as printing speed, alignment, printing temperature, layer height, infill, laser condition, and environmental aspects are verified, based on experience or a literature review. The sliced file can be imported into the AM machine for printing/stacking the material in a layer, forming the 3D implant. Finally, the printed parts are exposed to washing, removal of sacrificial layer/support, and heat treatment [25, 26]. The detailed scheme of additive manufacturing process is displayed in **Figure 1**.

The most common AM technologies for the construction of high-strength ceramics are selective laser sintering/melting (SLS/SLM), stereolithography (SLA), digital light processing (DLP), binder jetting (BJ), fused deposition modeling (FDM), and direct ink writing (DIW) [27, 28]. Each AM technology has great commercial potential as well as limitations [29]. Likewise, additively manufactured

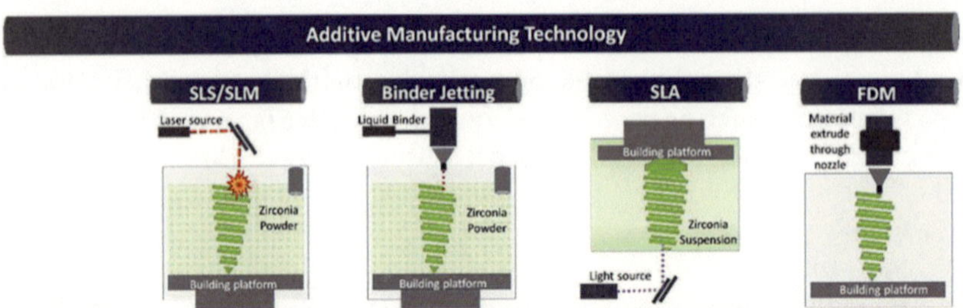

Printing Principle	Powder bed fusion	Powder bed fusion	Vat Photopolymerization	Material extrusion
Material type	Powder	Powder	Slurry	Feedstock/ Ink /Paste
Printing source	Laser irradiation	Liquid binder jetting	Light curing	Nozzle and Pressure
Surface quality	Poor	Good	Good	Average
Mechanical properties	Average	Low	Average	Average
Post processing	Removal of unbound powder and sintering	Removal of unbound powder and sintering	Curing, washing, removal of support, and sintering	Drying and sintering
Accuracy	Poor	Average	Good	Average

Figure 2.
The schematic illustration of different types of AM technologies used for the fabrication of 3D zirconia-based ceramics.

zirconia-based ceramics have inferior mechanical properties due to the persistent porosity and flaw-sensitive properties of zirconia ceramics. Thus, acquiring mechanical properties equivalent to those of ceramics fabricated with more conventional approaches is a big challenge for ceramic AM [30]. However, the technology is still at an early stage, compared with conventional ceramic processing techniques [29]. It is widely recognized that ceramic materials possess a high melting point, high sinterability, and high vulnerability to thermal shock. Therefore, it is challenging to achieve fully consolidated parts, without shortcomings, using AM-based techniques that directly produce sintered objects [25]. To overcome these shortcomings, each AM technology adapts scientific strategies to construct zirconia-based ceramics with high accuracy and quality. In the following section, the formulation strategies of each AM technology are discussed. **Figure 2**, demonstrate the AM technologies used for the fabrication of zirconia parts.

2. Formulation and general properties of zirconia directed to biomedical applications

2.1 Powder bed method

2.1.1 Selective laser sintering/melting (SLS/SLM)

SLS technology uses a high-powered laser beam to sinter/fire the ceramics at an elevated temperature. The laser is aimed at specific areas of the aggregate powdered particles using the distribution to create solid objects [31–33]. The SLM is principally similar to SLS; however, SLM completely melts and fuses the powder particles using a high-powered laser beam to form a solid object [34, 35]. SLS/SLM is an AM technique that uses a laser and is based on the powder bed method that produces 3D solid structures either by sintering or melting the powder materials layerwise following an architecture based on CAD data. (Obtaining high-strength and high-density parts with a laser without debinding/sintering processes can facilitate effective and rapid fabrication, enabling the mass production of ceramic parts (direct AM process) [36, 37].

However, zirconia ceramic is difficult to handle with SLS/SLM, as it has a higher melting point than other bioceramics. In addition, reaching full densification and realizing crack-free final products made of ceramics-based materials using this process are still challenging. Therefore, several studies are investigating the effect of powder properties and processing parameters [21, 35, 38]. Researchers describe the effectiveness of pre-heating the powdered bed, which could improve the mechanical properties of the final ceramic object by reducing the thermal stress, which alleviates crack formation during printing [39, 40]. Most of the zirconia particles use 3–8 mol% yttria-stabilized zirconia (YSZ) to preserve the desired mechanical properties of a tetragonal phase at room temperature. Alternative approaches were also found to be effective in improving the mechanical properties of the zirconia. For example, composites comprising zirconia and alumina are also found to retain the tetragonal phase [40–42]. To improve the mechanical properties of the final zirconia part and prevent cracking, an indirect method in SLS/SLM has been developed and documented [42–44]. Specifically, ceramic powder particles are mixed/coated with a sacrificial polymer binder (which has a lower melting point than the ceramic) and the laser is targeted towards the powder, which melts and fuses the ceramic particles. The fused ceramic particles are then subjected to postprocessing (debinding and sintering) to attain the dense zirconia ceramic scaffolds [42]. The summary of zirconia-based ceramics printing configurations used in SLS/SLM methods is presented in **Table 2**.

Particle size (μm)	Powder composition	Laser & power	Post-processing	Ref.
1–4	Zircar ZYP-30 (10 wt%)	Phenix Systems PM100 (50 W) V = 1250–2000 mm/s	—	[35]
20–70	Alumina toughened zirconia (ATZ) (41.5, 80, 94) wt% ZrO_2, (58.5, 20, 6) wt% Al_2O_3	Nd: YAG laser (150 W) for processing CO_2 laser (1000 W) for pre-heating	—	[45]
22.5–45	7Y-TZP (20–80 wt%)	MCP Realizer SLM 250, Germany	—	[40]
3–50	8Y-TZP + < 2 wt% graphite powder	Phenix ProX 200 Nd:YAG Laser power (W): 78–87	—	[46]
1–5	ATZ of Y-TZP (80 wt%)	CW 200 W Nd-YAG laser (redPOWER, SPI Lasers Ltd., UK) Laser power (W): 34	Post-thermal treatment at 1300°C for 2–10 h	[41]
—	ATZ of Y-TZP (80 wt%)	Realizer SLM 125 equipped with Nd:YAG laser Laser power (W): 90	—	[47]
—	3Y-TZP + 0.5 wt% MgO (magnesium oxide) powder +6.0 wt% epoxy resin	CO_2 laser (λ: 10.6 μm) with power of 100 W Laser power (W): 7	Cold isostatic pressing at 280 MPa	[44]
—	ZrO_2 + nylon 12	Energy density: 0.415 J/mm^2 Laser power: 6.6 W	Cold isostatic pressing at 200 MPa	[43]
—	3Y-TZP + isotactic polypropylene (PP)	CO_2 laser (λ: 10.6 μm) with power of 100 W	Warm isostatic pressing at 64 MPa Sintering in air at 1450°C for 2 h	[42]

Table 2.
Summary of zirconia-based configurations used in SLS/SLM methods [34].

2.1.2 Binder jetting (BJ)

BJ is also based on the powder bed fusion technique, where a binder (binding agent) is selectively deposited to link powder materials. In this technique, a thin layer of ceramic material in powder form is evenly spread over the building platform with the help of a roller [48]. A binding ink is then sprayed onto the ceramic powder particles using the jetting head. The result is the ceramic powders and binders adhering together to form a solid structure. This is repeated multiple times and the layers are printed on top of each other to form the preferred 3D scaffolds. During printing, green ceramic parts are reinforced by boundless powder particles [24]. The BJ process can eliminate the internal residual stresses that evolve during building [1]. Moreover, the postprocessing steps such as the removal of unbound powders and sintering are conducted to consolidate the dense ceramic parts. The effective production of numerous ceramics such as hydroxyapatite, tricalcium phosphate, ZTA, and Al_2O_3 structures with the required porosity using the BJ process have been reported in the literature for biomedical applications [49, 50]. However, obtaining the necessary shrinkage and density in the final product after sintering is still critical. Therefore, many researchers sought to address these issues by integrating nanoparticles into the liquid binder. Recently Huang et al. [32] studied the use of an inorganic colloidal binder (decomposable binder) as a binding agent for the construction of 3Y-ZrO_2 ceramic structures using BJ technology. They selected zirconium basic carbonate as a precursor, and it was dispersed in the colloidal solvent to produce decomposable inorganic colloidal binder because it can be easily decomposed upon sintering and can form zirconia 3D parts with no residue [48]. It was established that the inorganic colloidal binder-based zirconia scaffolds exhibited superior surface quality and density compared to the conventional polymer binder. Conversely, Zhao et al. [32] ***attempted to print zirconia samples using a liquid binder containing zirconia nanoparticles (10 wt%). The density was increased by approximately 86.8%, whereas shrinkage was reduced by approximately 10.6% after sintering the printed parts [51].

2.2 Stereolithography

Among the AM technologies using zirconia, SLA technology is the most well-known and popular method. A photocurable resin comprising photopolymerizable monomers, a photoinitiator, and ceramic particles is molded into a slurry and selectively cured by ultraviolet (UV) radiation in sequential layers to build the 3D object with the desired shape [52]. The geometrical accuracy of the manufactured parts produced using SLA technology is dependent on the laser power, layer thickness, cure depth, and energy dose. The key steps in fabricating ceramic parts with complex geometries and high resolution using SLA are preparing a suitable photocurable ceramic suspension, building the ceramic part, and debinding and sintering [53]. One of the most important factors in this process is the properties of the ceramic suspension. Homogeneous dispersion of zirconia ceramic materials with raw resin is essential for establishing photocurable ceramic resins. The introduction of ceramic materials negatively impacts the properties of raw resin by increasing the viscosity and immobilizing the ceramic/resin suspension. To initiate a matrix around the ceramic materials during photopolymerization, a combination of monomers and oligomers is blended with the ceramic suspension as a binder [54]. It is essential to include a dispersant to prevent agglomerations and retain the resin stability. The dense ceramic parts fabrication is primarily dictated by the volume fraction of the ceramics. Increases in volume fraction improve the final properties of the product (porosity reduction, shrinkage reduction, strength improvement, crack/deformation suppression) [28]. Due to this unique characteristic, SLA-based printers are commercially available in different forms. Hence, design and materials engineers recommend altering the design and printing parameters to the finest quality using state-of-the-art techniques and materials. Many studies have been focused on advancing a suitable photocurable ceramic suspension for the fabrication of zirconia-based ceramic parts (**Table 3**).

2.2.1 Oligomers and monomers

The oligomer (prepolymer) applied to the zirconia in AM methods has a chain structure comprising a medium molecular weight monomer. The oligomer regulates the physical properties of the resin. The reactivity between the monomer and the polymer with a low molecular weight number influences the properties of the cured film through molecular bonding triggered by polymerization. The classification is based on the molecular structure and includes polyester, epoxy, urethane, polyether, and polyacrylic. In general, it is difficult to use the oligomers directly for AM due to their high viscosity [60, 61].

Particle size (μm)	Resin configuration	Solid loading (vol%)	Viscosity (Pa s)	Laser wavelength (nm)	Ref.
0.2	HDDA[a] + TMPTA[a]	55	1.65 at 200 s^{-1}	—	[55]
—	HDDA + IBA[a] + PNPGDA[a]	58	9.02 at 5 s^{-1}	375–425	[56]
0.2	AM[b] + MBAM[b] + Glycerol + Water	40	0.127	—	[57]
0.2	HDDA + PPTTA[a] + PEG[c] + U600[a]	60 (wt%)	—	—	[58]
0.2	HDDA + PEGDA	83 (wt%)	1.23 at 100 s^{-1}	405	[59]

[a]*Acrylate-based monomer.*
[b]*Acrylamide-based monomer.*
[c]*Polyethylene glycol.*

Table 3.
Different formulations and viscosity characteristics for preparation of zirconia suspensions [53].

A monomer is a reactive diluent added to reduce the viscosity of an oligomer. The polymerization can be categorized into two types, namely, a free radical reaction or a cationic reaction [60, 62]. Acrylates and methacrylate are the most used monomers from free radical reactions [62]. Photopolymerization can be stimulated through a free radical initiator, and when the monomer receives a free radical from the initiator, it transfers the free radical to another monomer to form a polymer. The cationic reactive monomers can induce photopolymerization via cationic initiators. Monomers, such as epoxides, vinyl ethers, propenyl ethers, siloxanes, cyclic acetals, and furfurals, are capable of polymerization under a cationic mechanism. Epoxide is the preferred monomer from the cationic reaction groups [63].

2.2.2 Photoinitiator

Monomers and oligomers cannot independently initiate photopolymerization. Therefore, photoinitiators are added to generate reactive species that can trigger the monomers and oligomers. When polymerization is initiated, the reaction proceeds through a chain reaction of double bonds and forms a three-dimensional cross-linked bond together with reactive monomers and oligomers [64, 65]. Free radical photoinitiators added to certain monomers, such as acrylates and methacrylates, absorb UV light to generate free radicals and incite a double bond reaction of the monomers [66]. Cationic initiators can readily react with the binding of certain monomers, such as vinyl ethers and epoxides, because the absorbed UV light produces acids to induce polymerization of the monomers [63].

2.2.3 Dispersant

Dispersants are copolymers with soluble polymer chains and "fixing groups" that impart affinity to the surface of inorganic pigments such as zirconia [53]. The main mechanism in nonaqueous systems with low polarity is steric stabilization. Polymer chains are attached to the pigment surface by adsorption and form a brush-like layer that prevents re-agglomeration due to osmotic and entropy effects. The polymer chains of the dispersant are adsorbed onto the pigment surface to form a layer that prevents re-agglomeration. An effective layer typically ranges from 5 nm to 20 nm, with a particle diameter in the range of 0.05–1 μm. Because the dispersant effects vary with the monomer and oligomer composition, as well as the properties of the ceramic powder, care must be taken regarding the type and content of the dispersant.

2.3 Material extrusion

The compact ceramic raw material supplied to the extruder is difficult to use as an AM material because it has a high tendency of particle aggregation and, thus, increased resistance to flow [67]. Compatible ceramic powder and additives can guarantee permanency for storage and molding through homogeneous particle dispersion after mixing and, thus, facilitate the minimum pressure and viscosity for flow through the printing nozzle [68, 69].

Additionally, there is a need for good bonding and inhibition of separation between the deposited layers during printing [70, 71]. In addition, the included additives must be removed without defects during the post-treatment process [72, 73].

2.3.1 Wax/thermoplastic base

A study on the composition of multicomponent additives for a wax/thermoplastic base is suggested in **Table 4**. In addition to the main additives (such as polyethylene), other components such as wax dispersants and plasticizers are also included to provide strength, elasticity, flexibility, plasticity, and lower viscosity [20].

2.3.2 Water base

In the case of a feedstock in which a large amount of polymer is used as a dispersion medium, defects may occur during debinding after manufacturing. To solve this problem, an aqueous ceramic raw material is used. This water-based ceramic raw material enables the accumulation of zirconia powder with high content and decreases defects during degreasing due to the low content of organic matter.

Processes	Powder	State	Additive materials	Ref.
Wax & thermoplastic base	3 mol% YSZ 300 nm (40 vol%)	Feedstock	Low-density polyethene, paraffin wax, stearic acid	[74]
	3 mol% YSZ 90 nm (47 vol%)	Feedstock	High-density polyethylene, stearic acid, amorphous polyolefin, styrene-ethylene-butylene-styrene copolymer, paraffin wax, extender oil	[75]
	3 mol% YSZ 500 nm (85 wt%)	Feedstock	Ethylene-vinyl acetate copolymer, polyethylene, paraffin wax and stearic acid	[20]
Water base	3 mol% YSZ (45–50 vol%)	Paste	Anionic polyelectrolyte dispersant, hydroxypropyl methylcellulose, polyethyleneimine	[76]
	3 mol% YSZ 500 nm (50 vol%)	Paste	Water, acrylamide, N,N′-methylenebisacrylamide, ammonium citrate	[77]
	3 mol% YSZ 40 nm (60 vol%)	Paste	Ammonium polymethacrylate, methylcellulose, deionized water	[78]

Table 4.
Overview of extrusion processes for zirconia ceramics.

3. Biomedical application of AM zirconia

3.1 Dental applications

The use of zirconia ceramic as a restorative material in the form of dental prostheses started in the early 1980s and gained considerable attention in the dental community, thereafter due to its unique properties (such as excellent esthetics including tooth-like color, high fracture toughness, flexural strength, corrosion resistance, and biocompatibility) [8]. Hence, it has become the best alternative for metal-based dental restorations. Zirconia ceramics have been used in dental applications in the form of dental crowns, dental implants, and fixed partial dentures since 1998 [19]. In general, zirconia restorations are fabricated using digital techniques, including subtractive manufacturing techniques such as CAD/CAM, which is the established method for producing fixed prosthetic restorations [79], where the milling machine is controlled by a computer numeric controlled system. The power-driven milling tools were used to mill/remove the material from a block (presintered or fully sintered ceramic block) to achieve the desired prosthesis background [80]. However, it has certain disadvantages during manufacturing, such as material wastage and wear of milling tools. In addition, its precision is limited, limiting object complexity, tooling equipment dimensions, material properties, among other problems [81]. AM incorporates recent advanced and evolving techniques in digital dentistry, which construct the three-dimensional component by layering the material. It is capable of making cost-effective customized dental prostheses with minimal material consumption and high precision [82]. However, research studies on the 3D printing of zirconia crowns and bridges for dental applications are limited. In addition, various issues such as poor geometrical accuracy, high porosity, and poor margins are unresolved. Recently, several research studies on 3D printing of zirconia ceramics using photopolymerization-based printing (SLA-based technologies) improved the effectiveness and accuracy, making the technique favorable.

3.1.1 Restorative applications

The goal of the dentist is to restore the lost tooth as naturally as possible. The most common material types used in the restorative field are metals and ceramics. However, ceramics possess significant advantages over metal/metal ceramics due to their natural appearance (tooth-like color), which satisfies the esthetic demands, making ceramics the material of choice [83, 84].

YSZ is the most widely used all-ceramic material in dental restorations due to its outstanding material properties [85]. It is used for load-bearing applications, such as dental crowns, bridges, veneers, and implant abutments. YSZ restorations have been used in clinical practice over the past two decades. It is used primarily as a core material for the fabrication of dental prosthesis frameworks. The chipping of ceramic veneers and fracture of the framework, when exposed to continuous masticatory load is often reported [86, 87]. For example, the thermal coefficients of the core material and outer veneer cap (porcelain/lithium disilicate) are different and subjected to different heat treatment temperatures that lead to catastrophic failure. Further, several other factors including surface treatment (airborne-particle abrasion/etching) of the framework and bond strength between the ceramics veneer and zirconia frameworks are consequential [9, 88].

The advancement in zirconia with full-contour monolithic zirconia restorations gained attention to address the aforementioned problems. The fabrication of crowns and bridges using monolithic zirconia is faster and cheaper compared to a manually constructed veneered prosthesis. In recent years, CAD/CAM technology (subtractive) has been used for the fabrication of all-ceramic prostheses and abutments. The very attractive flexural strength and toughness of 3 mol% YSZ make it a classic and suitable material for dental use. Despite the promising properties of zirconia, the optical properties (translucency) are poor, i.e., it is opaque. Therefore, the larger esthetic-related issues initiated the demands for monolithic restorations. Dental material researchers and manufacturers have found several ways to increase translucency characteristics. The light transmission can be improved by either reducing the concentration of aluminum oxide or increasing the concentration of yttrium oxide [15]. For instance, the molar concentration of the yttria is varied (3–5%) to improve the translucency of zirconia with optimal mechanical properties. However, when the concentration of yttria is increased, the material exhibits higher translucency (more esthetics) but also exhibits a reduction in mechanical properties because the structural change into cubic phase becomes dominant. Evidently, the cubic phase does not allow transformation in crystal structure and this leads to a reduction in crack resistance. The "gradient technology" has become the modern advancement in the area of translucent zirconium oxide. A material-specific gradient is introduced into the milling block along with the color gradient (highly chromatic at the cervical region and less chromatic at the incisal region). In particular, the high-strength raw material 3Y-TZP is combined with the highly translucent raw material 5Y-TZP to create a continuous, layer-free color and translucent gradient [89]. The development of AM technology has attracted much attention to the fabrication of zirconia-based restoration with a high potential of making customized dental prothesis with minimal waste (**Figure 3**).

In 2009, Ebert et al. [90] built a zirconia dental crown using the direct inkjet printing method. The printing ceramic suspension was loaded with 27 vol% of zirconia ceramics, with a relative density of 96.9%, flexural strength of 763 MPa, and a fracture toughness of 6.7 MPa $m^{1/2}$. The printed and fired samples showed process-related defects, which were attributed to the clogging of the nozzles during printing that directly affected the mechanical properties. However, the authors demonstrated the potential to print 3D crowns using this technology. Likewise, Özkol et al. [91]

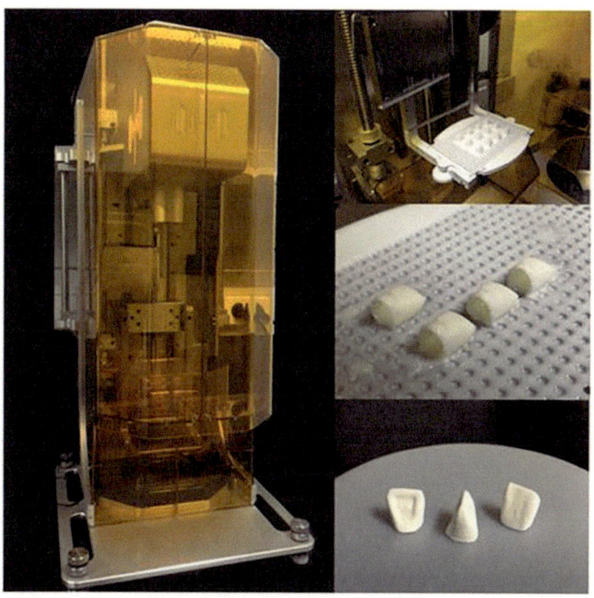

Figure 3.
AM zirconia crowns via DLP technology [56].

Applications	Materials and ceramic content	Fabrication techniques	Density (%) and shrinkage (vol %)	Mechanical properties	Others	Ref.
Dental crown (2009)	YSZ 27 vol%	DIP (from Hewlett Packard)	Density 96.9% and Shrinkage 20%	Flexural strength 763 MPa; Weibull modulus 3.5; Fracture toughness 6.7 MPa $m^{1/2}$	—	[90]
Dental crown (2011)	YSZ 47 vol%	Robo-casting	Shrinkage 30%	—	—	[92]
Dental bridge framework (2012)	YSZ (3Y-TZP) 40 vol%	DIP (from HP deskjet)	Density > 96%	Flexural strength~843 MPa; Weibull modulus 3.6; tensile strength~340 MPa	—	[91]
Dental bridges framework (2013)	ZTA (ZrO_2_80% and Al_2O_3–20%)	SLM	Density~100%	Flexural strength 538 MPa	—	[45]
Dental bridges (2018)	YSZ 40 vol%	SLA (from Shaanxi Hengtong Intelligent Machine Co., Ltd.)	Density 98.58% and shrinkage 20–30%	Flexural strength 200.14 MPa; Vickers hardness 1398 HV	—	[57]
Dental crown (2018)	YSZ 37 vol%	SLA (polymer mold) and gel casting	Density 98.6% and Shrinkage 20.1%	Flexural strength 1170 MPa; Vickers hardness 1383 HV	—	[93]
Dental crown (2019)	YSZ	SLA (from 3DCeram)	—	—	Surface trueness of the 3D printed crown meets the requirement	[94]
Dental crown (2019)	YSZ 45 vol%	SLA (from Porimy 3D Printing Technology Co., Ltd.)	Density-5.83 g/cm^3 and Shrinkage 18.1% in length, 20% in width, and 24.3% in height.	Flexural strength 812 MPa; Weibull strength 866.7 MPa; Weibull modulus 7.44	Cement space 63.40 μm (occlusal area); 135.08 μm (axial area) and 169 μm (marginal area)	[95]
Implant-supported AM crown (2019)	Commercial slurry (3DMixZrO_2)	SLA (from 3DCeram)	—	Fracture resistance 1243 N	—	[96]
Dental crown (2019)	YSZ (3Y-TZP) 48–58 vol%	DLP (from Octave Light R1)	Density 92.79% and Shrinkage—23.81%	Flexural strength 674.74 MPa	Geometrical overgrowth 36.94%	[56]

Applications	Materials and ceramic content	Fabrication techniques	Density (%) and shrinkage (vol %)	Mechanical properties	Others	Ref.
Dental crown (2020)	Commercial slurry (3DMixZrO_2)	SLA (from 3DCeram)	—	—	Marginal and internal discrepancies	[97]
Occlusal veneers (2020)	YSZ 40–60 vol%	Litho-graphy-based ceramics manufacturing process (like DLP) (From Lithoz GmbH)	—	—	Load bearing capacity-Median Fmax values 2026 N	[98]
Dental crown (2020)	YSZ 50–55 vol%	Inkjet	Density 98.5%	Hardness 14.4 GPa; transverse rupture strength 520 MPa	—	[99]
Dental crown (2021)	Commercial slurry SL150	SLA (from Porimy 3D Printing Technology Co., Ltd.)	—	—	Dimensional accuracy 65 μm and marginal adaptation	[100]
Dental crown (2021)	Commercial slurry CSL150 (YSZ) 47 vol%	SLA (from Porimy 3D Printing Technology Co., Ltd.)	—	—	—	[101]
Dental prothesis (bar shaped) (2021)	Commercial slurry (3DMixZrO_2)	SLA (from 3DCeram)	—	Flexural strength 320.32 MPa and 281.12 MPa after aging; fracture resistance 640.64 N and 562.25 after aging	—	[102]
Dental prothesis (bar shaped) (2021)	Commercial slurry (3DMixZrO_2)	SLA (from 3DCeram)	Shrinkage—16.32% in length, 14.25% in width, and 20.33% in height.	—	—	[103]

Table 5.
AM zirconia for dental applications.

attempted to print the zirconia bridge framework using a direct ink printing (DIP) method. The ceramic aqueous ink was prepared with 40 vol% solid content of 3Y-TZP. The printed components were dried and sintered at 1450°C. The relative density of the final product was >96%. Furthermore, finite element analysis was used to determine the stress distribution and the maximum tensile stress of the framework structure. The results of all different loading cases show hot spots on the bottom marginal area of the interdental connectors. The estimated maximum tensile stress values ranged between 250 and 350 MPa. The flexural strength was approximately 843 MPa (**Table 5**).

Lian et al. [57] reported that complex zirconia bridges were produced using the SLA technique with a high shape precision. They prepared a 40 vol% zirconia suspension and the laser scanning speed of 1200 mm/s was optimized for printing. The density and Vickers hardness of the sintered bridges was 98.58% and 1398 HV, respectively. Nevertheless, the flexural strength (200.14) was very low, and it was not good enough for actual dental applications, because of the internal defects formed during the printing process. The authors, therefore, suggested a study of the further optimization of the parameters of the SLA and sintering process. Additionally, in 2019 Wang et al. [94] conducted an in vitro experiment to investigate the surface trueness at different locations (external, intaglio, marginal, and occlusal) of 3D printed zirconia crowns constructed using SLA 3D printing technology.

The point-to-point difference between the scan data (3D printing) and corresponding CAD model data determines the trueness of the fabricated crown. The comparative color maps could demonstrate the accuracy and inaccuracy between the 3D printing and milling techniques. Meanwhile, Li et al. [95] examined the internal and marginal adaptation of 3D printed zirconia crowns and studied the physical and mechanical properties. The authors achieved a consistent flexural strength of 812 MPa and Weibull modulus of 7.44 by using 45 vol% zirconia suspensions. The mechanical strength is sufficient for dental crowns fabrication. While the cement spaces in occlusal (63.4), axial (134.08), and marginal (169.65) areas were not ideal for clinical applications, this can be attributed to light scattering and anisotropic sintering shrinkage.

However, in 2019 Jang et al. [56] investigated the microstructure and physical properties of zirconia products fabricated via DLP technology. The zirconia suspension was prepared using different volume fractions of the ceramic content from 48 vol% to 58 vol%. Cracks were observed on the zirconia specimens, and these cracks increased in number as the zirconia volume fraction decreased. The 3-point bending strength, relative density, and shrinkage of the printed samples were 674.74 MPa, 83.02%, and 23.81%, respectively. The maximum volume fraction possible for 3D printing was 58 vol%.

More recently in 2021, Zandinejad et al. [96] investigated the fracture resistance of AM zirconia crowns cemented to an implant-supported zirconia abutment. They also compared the AM zirconia crowns with milled zirconia, as well as lithium disilicate crowns. A universal testing machine at a crosshead speed of 2 mm/min was used to determine the fracture resistance, and it was verified that the fracture resistance of AM zirconia is equivalent to milled crowns. Nevertheless, intra-oral simulation research on the AM ceramic crowns should be conducted to authorize AM as a real-world technology for the construction of ceramic restorations in clinical dentistry.

3.1.2 Implant application

The popularity of zirconia-based implants is growing enormously as an alternative to alumina and metal-based endosseous implants [104]. Since the late 1980s, zirconia has been used to build surgical implants for the replacement of total hip prostheses in orthopedic surgery [105]. Zirconia-based ceramics have superior mechanical properties and corrosion resistance [106]. Besides, in vitro and in vivo, clinical studies of zirconia implants revealed excellent biocompatibility, osseointegration and a low affinity for bacterial plaque compared to standard metal implants (titanium implants) [107–109]. The utilization of AM technology is beneficial for the fabrication of zirconia-based ceramic dental implants as they can produce customized geometrics and complex structures. The technology can also improve bioactivity without any surface alterations, such as sandblasting, etching or coating [104]. Nevertheless, it is essential that the functional surface quality of zirconia-based implants fabricated from conventional techniques be enhanced to improve mechanical functions such as wear resistance and fatigue. Moreover, the surface treatments can improve bioactive functions, such as cell proliferation, adhesion, bonding strength, and bacterial decolonization [110].

For example, Osman et al. [111] fabricated 3D printed zirconia implants using DLP technology and evaluated the dimensional accuracy, surface topography, and flexural strength (**Table 6**). They showed that custom-designed 3D printed implants revealed satisfactory dimensional precision (root mean square error of 0.1 mm), and the flexural strength (943.2 MPa) is equivalent to that of conservative milled zirconia (800–1000 MPa). The roughness of the surface was found to be 1.59 μm and from the SEM analysis, it was observed that the presence of

Applications	Materials	Fabrication techniques	Mechanical properties	Others	Ref.
Dental implants (2017)	YSZ	DLP (from Delta Co.)	Flexural strength 632.1 MPa; Vickers hardness 14.72 GPa	—	[112]
Root analogue implants (RAI) (2017)	YSZ 27 vol%	DLP (from Admatec)	Weibull modulus 3.5; Fracture toughness 6.7 MPa. $m^{1/2}$	Density 96.9%; Shrinkage 20 vol %	[113]
Dental implants (2017)	YSZ	DLP (from Admatec)	Flexural strength 943 MPa	Dimensional accuracy 0.089 mm and SURFACE roughness 1.59 μm	[111]
Medical implants (cube, cuboidal, and bar shaped) (2019)	ATZ 70 wt%	LCM	Flexural strength 430 MPa	Density 5.45 g/cm^3; accuracy 70–88%	[114]
Hip implant (2019)	YSZ-ZnO (coating)	FDM and gel casting	—	MC3T3-E1 cells; *S. aureus*; *E. coli* and Rabbit hip joint (4 weeks)	[115]
Dental implants (square shaped) (2021)	Commercial slurry (LithaCon 3Y 230; 3DMix ZrO_2; 3D Mix ATZ)	SLA (from 3DCeram and Lithoz GmbH)	Flexural strength 1108.8 MPa (3D Mix ATZ); Weibull modulus 11.1	—	[82]
Dental implants (2021)	ATZ 36–38 vol%	DLP (from Robotfactory)	—	Density 96.8%	[116]

Table 6.
AM zirconia for implant applications.

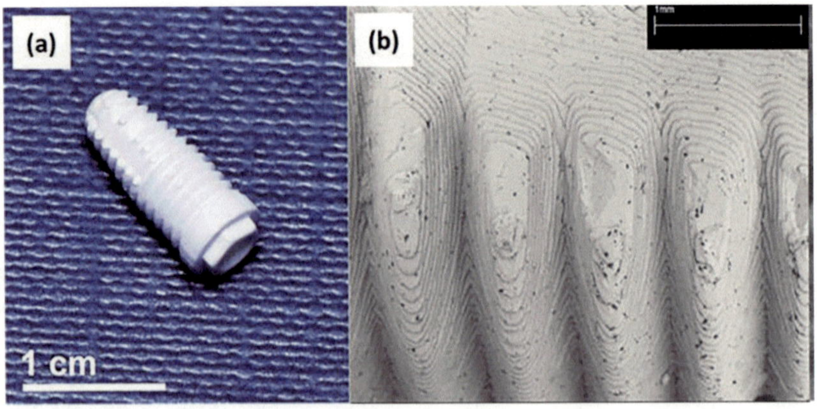

Figure 4.
Dental implant fabricated using DLP-based additive manufacturing technology. (a) ATZ dental implant-green body, (b) micrograph of the lattice structure [116].

microporosities with interconnected pores (196 nm to 3.3 μm) and cracks were visible. These flaws were generated during the sintering process or improper dispersion of ceramic particles into the slurry. To enhance the potential microstructure quality of the printed implants, 3D printing parameters need to be optimized.

However, Nakai et al. [82] inspected the microstructure and flexural strength of zirconia-based ceramics formed using SLA (AM technology) and related to CAD/CAM technology (subtractive technology). In their study, the authors compared the commercially available zirconia-based ceramics products. They were two AM 3Y-TZP (LithaCon 3Y 230 and 3D Mix zirconia) products, and one AM ATZ (3DMix ATZ) product, with conventionally fabricated 3Y-TZP (LAVA plus). The experimental outcomes confirmed that the flexural strength and microstructure of AM zirconia are sufficient and close to that of conventionally (subtractive) manufactured zirconia. AM ATZ exhibited higher flexural strength (1108.8 MPa) than 3Y-TZP. Both 3Y-TZP and ATZ are suitable for dental implants. Moreover, variation in the AM process and the impact of building alignment can alter the mechanical properties of AM zirconia. To promote the practical reliability of AM zirconia implants, the relationship between the surface morphology and bioactivity of zirconia needs to be evaluated in a future study. Recently, Magnani et al. [116] presented the potential capability of DLP printing technology to fabricate the dental implants with a new high-performance ATZ composite material (**Figure 4**).

3.2 Bone-regeneration applications

The clinical success of zirconia bioceramics in the human environment in the form of dental posts, teeth, and crowns in the dentistry field encouraged biomedical researchers to exploit the biological and mechanical properties of zirconia bioceramics for bone-regeneration applications. Accordingly, developing zirconia-based scaffolds with high precision and dimensional stability is vital to satisfy increasingly challenging requirements for bone-regeneration requests. At present, there is a lack of a simple commercial approach to construct 3D zirconia structures, however, the proposal of AM in 3D zirconia scaffold construction shows great potential. Biomedical engineers targeted AM-based technologies for the zirconia scaffold preparations (**Table 7**). Unlike conventional bioceramics, initial attempts to fabricate zirconia bioceramics were mainly concentrated on multi-pass extrusion techniques [44]. The multi-pass extrusion technique is a simple AM technique in which the ethylene-vinyl acetate polymers were blended with zirconia powders to execute extrusion (the extrusion is repeated to construct the scaffold with constant porous core structure). The extrusion proportion, pore-gradient

Materials	Fabrication techniques	Composite/coating materials and infiltration/intermediate layer	Porosity and pore size	Mechanical properties	Biological properties (in vitro and in vivo)	Ref.
YSZ 48–43 vol% (2011)	Multipass extrusion	Intermediate layer HA (α–TCP) –YSZ; coating: HA	77% and 86 μm	Compression strength: 53 MPa	MG-63 cells	[117]
YSZ (2011)	3D Rapid Prototyper (ABS template) followed by slurry impergation	Coating: mesoporous bioglass	63–68% and 500–800 μm	Compression strength: 44.35–123.32 MPa	SBF and BMSC cells	[118]
YSZ 10 vol% (2011)	Sponge replica and electrospinning	Intermediate layer YSZ-BCP; coating; BCP	67.68–69.65%	Compression strength: 4.83–4.97 MPa	MG-63 cells	[119]
ZrO_2 50 vol.% (2012)	Free-form	—	40% and 350 μm	—	Case study (maxilla)	[120]
YSZ 45–40 vol% (2012)	Multipass extrusion	Intermediate layer: HA-YSZ; coating: HA	—	Compression strength: 7–20 MPa	MG-63 cells	[121]
YSZ 46–41 vol% (2012)	Multipass extrusion	Intermediate layer: YSZ-BCP: coating: PCL/BCP	92–78%	Compression strength: 8.27–12.7 MPa	MG-63 cells	[122]
YSZ 70 wt% (2014)	Direct ink writing (DIW)	—	55 and 63%	Compression strength: 8 and 10 MPa	HCT116 cells	[123]
ZrO_2-$CaSiO_3$ (2014)	SLS	Composite: ZrO_2 (10–40 wt%)	70% and 1600 μm	Compression strength: 17.9–44.1 MPa; fracture toughness 1.14–1.66 $MPa.m^{1/2}$	SBF and MG-63 cells	[124]
YSZ-PVP (2016)	Electrospinning	—	—	Modulus 1.11 MPa	HMSC cells	[125]
ZrO_2-β-TCP (2016)	3D Rapid Prototyper (ABS template) followed by impergation	Composite: ZrO_2 (10–50 wt%)	68.5–82.5%	Compression strength: 3–15 MPa; Modulus 184–396 MPa	PBS and MG-63 cells	[126]
ZrO_2-β-TCP (2017)	3D Bioplotter	Composite: ZrO_2 (30 wt%)	60–76.46% and 160–226 μm	Compression strength: 7–12.025 MPa	MG-63 cells	[117]
ZrO_2-PCL 6–30 wt.% (2017)	Electrospinning	—	—	—	3T3 cells	[127]

Materials	Fabrication techniques	Composite/coating materials and infiltration/intermediate layer	Porosity and pore size	Mechanical properties	Biological properties (in vitro and in vivo)	Ref.
YSZ-Al_2O_3 (ZTA) 70 wt% (35.5 vol%) (2018)	Robocasting	Composite: ZTA (YSZ-16 wt.%)	50% and 245 μm	—	HOB cells	[128]
ZrO_2–BCP (2018)	FDM	Composite: ZrO_2 (10 wt%)	350 μm	Compression strength: 0.5 MPa	MG-63 and hMSCs cells	[129]
YSZ 48 vol% (2019)	Robocasting	—	200–500 μm	—	—	[130]
ZrO_2-β- Ca_2SiO_4 (2019)	3D Bioplotter	Composite: ZrO_2 (5, 10, 15 wt%)	~67%	Compression strength: 3.9–6.1 MPa	SBF and BMSC cells; RAT calvarial defect (8 weeks)	[131]
ZrO_2-HA 60 wt% (2019)	DLP	Composite: ZrO_2 (1, 3, 6 wt%)	—	Tensile strength (29.4%); bending strength (23.9%)	BMSC cells	[132]
ZrO_2–PCL (2020)	FDM	Composite: ZrO_2 (5, 10, 20 wt%)	46.2–47% and 459.2–462.7 μm	Compression strength: 5.5–7.9 MPa; Modulus 43–67 MPa	MC3T3-E1 cells	[133]
YSZ (2020)	DLP	Composite: HA (10, 20, 30 wt%)	54.6%	Compression strength: 52.25 MPa; compression strength: after soaking in SBF (25 MPa)	SBF and MC3T3-E1 cells	[134]
YSZ 40 vol% (2020)	FDM and Freeze drying	Intermediate layer: Glass (Infiltration); coating: glass/ Zn-HA (~1 μm) and gelatin/ alginate	40% and 300–450 μm	Compression strength: 68.2–89.8 MPa; Modulus 1.7–2.6 GPa; Strain energy density 1.8–4.2 MJ/m^3	DPCs cells	[74]
YSZ 39.5 vol% (2021)	Direct ink writing (DIW)	Intermediate layer: FA; coating: HA (~20 μm)	61.1–75.3%	Compression strength: 20.8–62.9 MPa	SBF	[135]

Table 7.
AM zirconia for bone tissue regeneration applications.

rate, and microstructure are the critical parameters in controlling the final output of the zirconia scaffolds. More importantly, to increase the biocompatibility of zirconia binary mixtures (ZrO_2/Al_2O_3), fabrication of binary scaffolds with alternating ZrO_2 and Al_2O_3 layers with 3D-interconnected micropores are also demonstrated [136]. However, the multi-pass extrusion designs were not controlled using modern numerical methods. In subsequent years, computerized extrusion-based techniques like 3D Bioplotter and FDM were introduced to precisely design the 3D zirconia scaffolds. Zirconia-based scaffolds (β-Ca_2SiO_4/zirconia scaffolds) fabricated using the 3D-Bioplotter technique were verified to induce bone-regeneration properties in an actual biological atmosphere using a rat model [131]. In FDM, zirconia ceramics are generally blended with polymers such as polycaprolactone to execute a computerized melt mixing process, which can construct a regular grid scaffold [133]. More importantly, biopolymers embedded in zirconia-based scaffolds fabricated using FDM were found to provide additional mechanical support, as well as bioactivity for the zirconia ceramics (**Figure 5**). Compared to the pristine zirconia-based scaffolds (alginate/gelatine), biopolymer embedded zirconia ceramics were found to exhibit the extracellular matrix (ECM) of the bone tissue, which is essential to imitate the biological environment [74]. Subsequently, considerable research efforts were dedicated to formulating zirconia-based scaffolds using the direct ink writing (DIW) or robocasting method (extrusion-based AM-based technique). 3D zirconia scaffolds fabricated with controlled pore openings and thread dimensions using the DIW method were found to possess high porosity (61% and 75%). More importantly, hydroxyapatite/fluorapatite-based coatings on the DIW derived zirconia-based scaffolds were needed to enhance its bioactivity [135]. Photopolymerization-based AM techniques including DLP and SLS were also studied for the fabrication of zirconia-based scaffolds. Specifically, DLP technology was found to have high accuracy and faster processing ability than other AM-based techniques. The ultraviolet light is irradiated on the zirconia suspensions (prepared by optimizing the solid loading of the zirconia powders, organic monomer, potentiators, and dispersant) to articulate the final design. It is important to perform heat treatment in a high-temperature vacuum furnace to avoid internal cracks and imperfections in the heat-treated zirconia scaffolds [132]. Although SLS-based techniques were widely studied for calcium-based bioceramics, the use of SLS techniques to construct zirconia has been limited due to low zirconia concentration. Mostly, zirconia is blended in minimum volume fraction with other bioactive materials like calcium silicates to avoid the

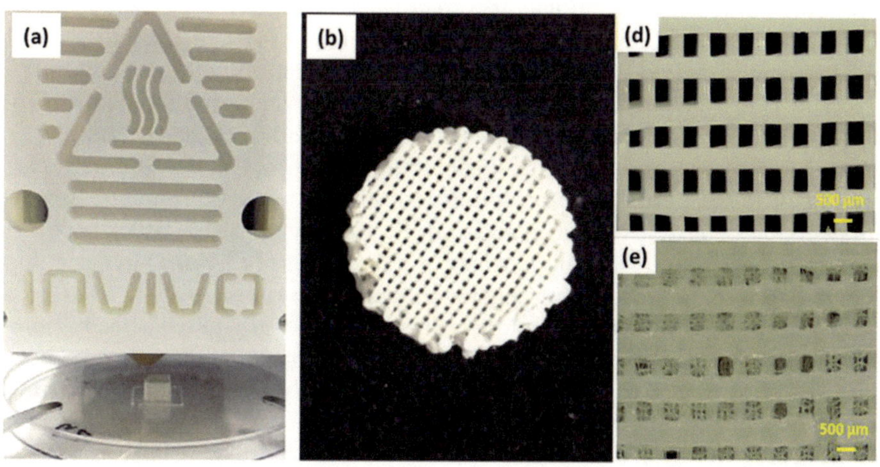

Figure 5.
(a) Scaffold printing using FDM, (b) digital photograph of the printed zirconia scaffold, (d) microscopic images of zirconia scaffold, and (e) polymer embedded zirconia scaffold [74].

unwanted agglomeration-induced material degradation [124]. In addition, to replicate the nano-to-microscale configuration of the ECM of bone tissue, electrospinning of the zirconia-based scaffolds was experimented with. It is believed that the zirconia scaffolds subjected to electrospinning exhibited high endurance to the inbound load from the bone tissue when compared to conventional more fragile scaffolds [125].

4. Outlook and prospective

Zirconia is a classic bioceramic, and its use in the dental and biomedical fields is inevitable. Hence, extensive research efforts have been dedicated to maximizing the potential of AM technologies to formulate the zirconia ceramics into a precise bone or tooth replacement, scaffolds, implants, and crowns. Though, zirconia scaffolds are directly involved in the human environment (in both dental and biomedical fields), the requirements of each field are evidently different. For instance, the zirconia scaffolds should have adequate porosity for bone-regeneration applications and patient-specific design, whereas, zirconia scaffolds for dental restoration and implants need not have a porous structure; instead, they should retain complex shapes with solid/hollow structures. Hence, the scaffold processing via AM also needs to be precise for each application. AM or 3D printing has revolutionized the designing of complex human hard tissues with an excellent surface finish, minimum material wastage, and high fabrication speed compared to conventional techniques. However, AM also suffers from some inherent limitations and challenges. The primary challenges include difficulties in raw material preparation, process control, and immature designs (**Figure 6**). Research advancements achieved by the metal and polymers-based scaffolds via AM-based techniques both in the laboratory and at clinical levels are far ahead when compared to the practically challenging zirconia-based ceramics due to their inherent challenging properties (brittleness, high melting point, and high density). Hence, it is essential to pinpoint the existing challenges in the research investments and activities that restrict the feasibility of AM-based technologies in fabricating zirconia-based ceramics at the laboratory, clinical, and industrial levels.

4.1 Laboratory challenges

Although different types of AM technologies are available for formulating bioceramics, only a few techniques are effective in the fabrication of zirconia parts with minimal imperfections. Despite the large number of AM technologies suitable for processing ceramics, each technique has its individual advantages and limitations. The primary issue for printing starts from the raw material (feedstock/slurry) preparation itself. For example, in extrusion-based techniques, temperature, pressure, nozzle size, and computer-generated design files (scaffold models) can be fed easily to the computer to accomplish the anticipated requirements. However, poor printability, nozzle blockage, and poor flowability of the feedstock have been major bottlenecks (due to the high density and hardness of zirconia) in designing zirconia-based scaffolds for bone-regeneration applications.

Compared to FDM-based techniques, SLA-based techniques have been extensively explored for the fabrication of zirconia-based ceramics due to the excellent surface finish and precision produced by the technology. Commercial SLA printers are now available for zirconia-based ceramics. However, the uneven distribution and particle aggregation of zirconia particles in the slurry suspension upsetting the light scattering properties (cure depth, curing time, and the energy of the UV light source) is a challenging issue. As a result, geometrical overgrowth is unavoidable due to the high refractive index of the zirconia. (SLA-based techniques are mainly

Figure 6.
Major challenges of AM zirconia-based ceramics.

controlled by the light source, refractive index, volume fraction, and particle size.) The most common problem associated with SLA-based techniques for zirconia-based ceramics is the delamination among the layers, which invariably disturbs the physical and mechanical properties of the sintered zirconia parts.

SLS-based techniques, however, can produce scaffolds with high precision, but they are rarely explored for zirconia-based ceramics due to the expensive and complicated control parameters. In particular, the high melting point of zirconia requires pre-heating of the powder bed (>1000°C) to avoid cracks caused by the thermal stress induced by the high-power laser source. Nevertheless, SLS-based techniques have represented a single-step scaffolding process for formulating zirconia scaffolds with full density. The requirement and urgency of developing this technique further for zirconia-based scaffolds are debatable.

For all the above, the major disadvantages of AM-based techniques except direct SLS-based techniques are the low-volume fraction of the zirconia in the feedstock (<60%, in which the polymer occupies the remaining portion). After debinding of polymers, the printed scaffolds can retain only half of the parent zirconia properties, which invariably affects the expected properties of the final sintered zirconia-based scaffolds.

In general, the strength and life of ceramic materials are directly associated with the type and level of residual stress that developed during the AM process. The major issue of any 3D printing system for the fabrication of zirconia parts is the internal (residual) stress, which is formed either during the printing process or during the post-process. The residual stress generated during the post-process includes high-temperature thermal treatment (sintering process) upon cooling or due to the difference in the thermal expansion coefficient (CTE) between the composite material of zirconia/bilayer material [137]. In other words, the mismatch of the CTE of two different materials can induce residual stress (tensile). Correspondingly, it was demonstrated that the selection of slow cooling and firing program of ceramic can potentially reduce the stress, which will also decrease the risk of chipping of porcelain layer in zirconia dental restoration [138]. Moreover, residual stress has a direct effect on the aging process. For instance, the tensile stresses of the zirconia composite can accelerate the aging process of the zirconia parts in body fluids [137]. It can be regulated by the stabilizer material nature and content of the zirconia phase in the composite. The most common diagnostic techniques employed for the residual stress measurements include X-ray diffraction, nanoindentation, Raman spectroscopic analysis, thermal tempering using a two-dimensional (2D) analytical model, and three-dimensional (3D) finite element simulation. However, the magnitude of the residual stress of zirconia parts varies from location to location of geometry. Also, the residual stress distribution is affected by the thickness and geometry of the zirconia parts [139]. Upcoming research should be focused on the residual stress of the AM zirconia parts are need to consider.

4.2 Clinical challenges

Although diverse AM-based research studies claim that zirconia-based scaffolds are practicable, true accomplishments are only determined based on the result of the clinical studies. In this regard, there are many unaddressed areas of applications when applying AM-based zirconia to real-world dental and bone restoration that are unresolved. For example, there are internal defects (cracks, porosity) that are formed during layering/printing or postprinting of the designed zirconia prostheses using AM-based techniques. They could affect the mechanical strength of AM zirconia crowns, bridges, implants, and scaffolds and result in a failure to satisfy the dental and biomedical requirements [57]. Nevertheless, optimum porosity is essential to guide cell adhesion or osteointegration. Hence, the stability among the material properties and biological requests need to be established by optimizing the slurry formulation/feedstock and sintering procedures on whatever AM-based techniques are used. The major challenges of 3D printed dental prostheses for real clinical applications are surface finishing/topography, staircase effects, geometrical overgrowth, and mechanical properties. Specifically, the marginal tolerance requirement (< 0.1 mm) for dental prosthetic applications via AM-based techniques is hard to realize, particularly when material strength and density are also mandatory [54, 140].

Uneven shrinkage is caused by the inbound technical shortage of AM-based techniques. Unresolved accuracy in the z-direction compared to the x and y-directions induces densification of ceramic powders within the layer and related issues (degree of polymerization and layer thickness). Overall, the printing parameters along the z-direction are yet to be optimized in such a way that the dimensional accuracy of the zirconia parts is achieved using AM-based techniques capable of addressing the patient-specific requirements. The technical imperfections in design may lead to plaque accumulation, risk of microleakage, and local inflammation [141, 142]. Thus, the relationship between dimensional precision and clinical adoption is critical to the adoption of any AM-based techniques.

Because the scaffolds need to be in direct contact with biological fluids, parts sterilization is important. Hence, biomedical engineers should be aware of the sterilization requirements while designing zirconia-based parts using AM-based techniques. The scaffolds should not lose their characteristic properties even after sterilization. Limited in vivo studies have been devoted to determining the after-effects of zirconia-based scaffolds on the biological environment. These confirm that the AM-based techniques for zirconia-based ceramics are still in infancy. Hence, biomedical engineers should be conscious of the importance of in vivo studies to realizing the practical applications of zirconia-based scaffolds.

4.3 Industrial (cost and resource) challenges

Leading biomedical implant manufacturing companies including Stryker Corp, ZERAMEX, Straumann ceramic, Nobel Biocare, Zimmer Biomet, Wright Medical,

Globus Medical, and Integra Lifesciences focused on developing, manufacturing, and promoting zirconia-based biomedical implants as a material of choice via additive manufacturing technology. Design flexibility, material productivity, and low-volume production feasibility are the prime factors behind the interest in additive manufacturing technology among the leading companies. However, AM-based technologies facing undeniable difficult challenges to fabricate zirconia-based scaffolds. Though adopting AM-based technologies for zirconia implants needs time and determination, the most important challenge lies in the substantial investment on the principal investment cost for the production floor [143]. Investment in the fabrication of zirconia-based biomedical implants from AM-based technologies is not only about equipment cost. It includes the investments in the AM ecosystem as well, which involves material, software, manpower coaching, postprocessing apparatus, documentation, and merging all facilities capable of mass production. More importantly, capital investment and material resources will be added to the above-stated challenges, which is large enough for a corporation to invest in AM as an aggregate. Hence, long-term cost assessment challenges were ahead for any biomedical implant company to unlock the AM-based technology to process zirconia-based scaffolds for wider marketplaces [143].

5. Conclusions

New technologies often mean new construction techniques and material and resource applications. AM has become a potentially vital technology in fabricating zirconia-based materials for various critical-sized applications, including bone scaffolds and dental crowns, bridges, and implants. As both AM-based technology and zirconia-based materials are in their infancy for scaffold application, it is essential to create awareness and sensitization among researchers. For example, among the AM-based technology, very few 3D printing systems (SLA, SLS, and DLP) are successful in manufacturing zirconia-based ceramics as scaffolds in the lab scale itself. This is inadequate when compared to well-established 3D printing systems for the use of metal and polymer materials, hence there is a prolonged difficulty in the clinical accomplishment of zirconia-based scaffolds. Though the mechanical properties of the zirconia parts achieved via 3D printing are comparable to the conventional zirconia parts, still some inbound issues such as internal defects (crack and porosities) and dimensional accuracies need to be enhanced. Moreover, for the enhanced bioactivity of zirconia parts, precise selection of the bioactive material and surface treatment strategies (coating/composite) are still under search. It has to be declared here that the essential printing parameters, materials preparation, and the development of the printer capability are progressively taken care of by the biomedical experts in the recent reports. Hence, collective efforts need to be dedicated in collaboration with academia, AM-machine developers, and clinical end-users to share their materials and design requirements to achieve the expected goals. The collective scientific outcomes, together with materials engineering and manufacturing technology, are extremely important in actualizing any emerging technology. AM-based technology could be utilized for manufacturing zirconia-based ceramics, which would be a milestone for society if all its current limitations can be systematically and creatively addressed.

Acknowledgements

This work was supported by the National Research Foundation of Korea (NRF). Grant funded by the Korean government (MSIT) (No. 2019R1A2C108945613).

Conflict of interest

The authors declare no conflict of interest.

Notes/thanks/other declarations

We thank Pavithra Kumaresan, Karthik Narayanan, and Hariprasath Sekar for their timely help during editing the manuscript.

Abbreviations

Y-TZP (or) YSZ	yttrium tetragonal zirconia polycrystal (or) yttria stabilised zirconia
ZTA	zirconia toughened alumina
Mg-PSZ (or) MgSZ	magnesia partially stabilized zirconia (or) magnesium stabilized zirconia
Y_2O_3	yttrium oxide
ZrO_2	zirconium dioxide/zirconia
Al_2O_3	aluminium oxide/alumina
MgO	magnesium oxide
ATZ	alumina toughened zirconia
ZnO	zinc oxide
HA	hydroxyapatite
TCP	tricalcium phosphate
BCP	bicalcium phosphate
FA	fluroappatite
CaP	calcium phosphate
PMMA	polymethamethylacralate
PCL	polycaprolactone
CS	chitosan
SF	silk fibrin
POM	polyoxometalates
PLA	polylatic acid
PRP	plasma rich protein
HS	heparin sulfate
CZ	calcium zirconate
Ca_2SiO_4	calcium silicate
PVP	polyvinylpyrrolidone
ABS	acrylonitrile butadiene styrene
Zn-HA	zinc doped hydroxyapatite
DIW	direct ink writing
BJ	binder jetting
FDM	fused deposition modelling
DLP	digital light processing
CAD/CAM	computer aided design/computer aided milling
SLS	selective laser sintering
SLM	selective laser melting
S. aureus	*Streptococcus aureus*
E. coli	*Escherichia coli*
HOS	human osteoscarcoma
SBF	stimulated body fluid

MG63	ostesarcoma cells
MCT3-E1	murine preosteoblast cells
BMSC	bone marrow-derived mesenchymal stem cells
L929	murine fibroblast cells
PBS	phosphate buffered saline
HGF	human gingival fibroblast cells
OB6	murine bone marrow-derived osteoblastic cells
HCT116	human colon carcinoma cells
HOB	human osteoblast cells
hMSC	human mesenchymal stem cells
DPCs	dental pulp cells

Author details

Sakthiabirami Kumaresan[1,2], Soundharrajan Vaiyapuri[3], Jin-Ho Kang[1], Nileshkumar Dubey[4], Geetha Manivasagam[5], Kwi-Dug Yun[1] and Sang-Won Park[1,6]*

1 Department of Prosthodontics, Dental Science Research Institute, School of Dentistry, Chonnam National University, Gwangju, Republic of Korea

2 Chonnam National University Biomedical Evaluation and Research Center, Gwangju, Republic of Korea

3 Department of Materials Science and Engineering, Chonnam National University, Gwangju, Republic of Korea

4 Faculty of Dentistry, National University of Singapore, Singapore

5 Centre for Biomaterials, Cellular and Molecular Theranostics, School of Mechanical Engineering, Vellore Institute of Technology (VIT), Vellore, Tamil Nadu, India

6 RIS Advanced Center for Biomaterials, School of Dentistry, Chonnam National University, Gwangju, Republic of Korea

*Address all correspondence to: psw320@chonnam.ac.kr

References

[1] Ivanova N, Gugleva V, Dobreva M, Pehlivanov I, Stefanov S, Andonova V. We are IntechOpen, the world's leading publisher of Open Access books Built by scientists, for scientists TOP 1%. INTECH. 2016;**i**(tourism):13

[2] Dubey N, Ferreira JA, Malda J, Bhaduri SB, Bottino MC. Extracellular matrix/amorphous magnesium phosphate bioink for 3D bioprinting of craniomaxillofacial bone tissue. ACS Applied Materials & Interfaces. 2020;**12**(21):23752-23763. Available from: https://pubmed.ncbi.nlm.nih.gov/32352748

[3] Saravanan S, Vimalraj S, Thanikaivelan P, Banudevi S, Manivasagam G. A review on injectable chitosan/beta glycerophosphate hydrogels for bone tissue regeneration. International Journal of Biological Macromolecules. 2019;**121**:38-54. Available from: https://www.sciencedirect.com/science/article/pii/S0141813018331829

[4] Best SM, Porter AE, Thian ES, Huang J. Bioceramics: Past, present and for the future. Journal of the European Ceramic Society. 2008;**28**(7):1319-1327. Available from: https://www.sciencedirect.com/science/article/pii/S0955221907005961

[5] Christel P, Meunier A, Dorlot JM, Crolet JM, Witvoet J, Sedel L, et al. Biomechanical compatibility and design of ceramic implants for orthopedic surgery. Annals of the New York Academy of Sciences. 1988;**523**:234-256

[6] Tosiriwatanapong T, Singhatanadgit W. Zirconia-based biomaterials for hard tissue reconstruction. Bone Tissue Regen Insights. 2018;**9**:1179061X18767886. DOI: 10.1177/1179061X18767886

[7] Manicone PF, Rossi Iommetti P, Raffaelli L. An overview of zirconia ceramics: Basic properties and clinical applications. Journal of Dentistry. 2007;**35**(11):819-826

[8] Piconi C, Maccauro G. Zirconia as a ceramic biomaterial. Biomaterials. 1999;**20**:1-25

[9] Saridag S, Tak O, Alniacik G. Basic properties and types of zirconia: An overview. World Journal of Stomatology. 2013;**2**(3):40-47

[10] Sakthiabirami K, Soundharrajan V, Kang J-H, Yang YP, Park S-W. Three-dimensional zirconia-based scaffolds for load-bearing bone-regeneration applications: Prospects and challenges. Materials (Basel). 2021;**14**(12). Available from: https://www.mdpi.com/1996-1944/14/12/3207

[11] Heuer AH, Lange FF, Swain MV, Evans AG. Transformation toughening: An overview. Journal of the American Ceramic Society. 1986;**69**(3):i-iv. DOI: 10.1111/j.1151-2916.1986.tb07400.x

[12] Kelly JR, Denry I. Stabilized zirconia as a structural ceramic: An overview. Dental Materials. 2008;**24**(3):289-298

[13] Ganapathy P, Manivasagam G, Rajamanickam A, Natarajan A. Wear studies on plasma-sprayed Al_2O_3 and 8mole% of Yttrium-stabilized ZrO_2 composite coating on biomedical Ti-6Al-4V alloy for orthopedic joint application. International Journal of Nanomedicine. 2015;**10**(Suppl 1):213-222. Available from: https://pubmed.ncbi.nlm.nih.gov/26491323

[14] Denry I, Kelly JR. Emerging ceramic-based materials for dentistry. Journal of Dental Research. 2014;**93**(12):1235-1242

[15] Volpato CAM, Garbelotto LGD, Fredel MC, Bondioli F. Application of zirconia in dentistry: Biological,

mechanical and optical considerations. In: Advances in Ceramics—Electric and Magnetic Ceramics, Bioceramics, Ceramics and Environment. 2011. Available from: https://www.intechopen.com/chapters/18282

[16] Chen Y-W, Moussi J, Drury JL, Wataha JC. Zirconia in biomedical applications. Expert Review of Medical Devices. 2016;**13**(10):945-963. DOI: 10.1080/17434440.2016.1230017

[17] Gadow R, Kern F. Pressureless sintering of injection molded zirconia toughened alumina nanocomposites. Journal of the Ceramic Society of Japan. 2006;**114**(1335):958-962

[18] Wang S, Yu JY, Li Q, Zheng EY, Duan YG, Qi G. Preparation of gradient ZTA ceramic by centrifugal slip casting method. Advances in Materials Research. 2012;**569**:324-327

[19] Gautam C, Joyner J, Gautam A, Rao J, Vajtai R. Zirconia based dental ceramics: Structure, mechanical properties, biocompatibility and applications. Dalton Transactions. 2016;**45**(48):19194-19215

[20] He Q, Jiang J, Yang X, Zhang L, Zhou Z, Zhong Y, et al. Additive manufacturing of dense zirconia ceramics by fused deposition modeling via screw extrusion. Journal of the European Ceramic Society. 2021;**41**(1):1033-1040. Available from: https://www.sciencedirect.com/science/article/pii/S0955221920307421

[21] Ferrage L, Bertrand G, Lenormand P, Grossin D, Ben-Nissan B. A review of the additive manufacturing (3DP) of bioceramics: Alumina, zirconia (PSZ) and hydroxyapatite. Journal of the Australian Ceramic Society. 2017;**53**(1):11-20. DOI: 10.1007/s41779-016-0003-9

[22] Aytac Z, Dubey N, Daghrery A, Ferreira JA, de Souza Araújo IJ, Castilho M, et al. Innovations in craniofacial bone and periodontal tissue engineering—From electrospinning to converged biofabrication. International Materials Review. 2021;**0**(0):1-38. DOI: 10.1080/09506608.2021.1946236

[23] Kang J-H, Jang K-J, Sakthiabirami K, Oh G-J, Jang J-G, Park C, et al. Fabrication and characterization of 45S5 bioactive glass/thermoplastic composite scaffold by ceramic injection printer. Journal of Nanoscience and Nanotechnology. 2020;**20**(9):5520-5524

[24] Zhang X, Wu X, Shi J. Additive manufacturing of zirconia ceramics: A state-of-the-art review. Journal of Materials Research and Technology. 2020;**9**(4):9029-9048. Available from: https://www.sciencedirect.com/science/article/pii/S2238785420313958

[25] Galante R, Figueiredo-Pina CG, Serro AP. Additive manufacturing of ceramics for dental applications: A review. Dental Materials. 2019;**35**(6):825-846. Available from: https://www.sciencedirect.com/science/article/pii/S0109564118304263

[26] Singh S, Ramakrishna S. Biomedical applications of additive manufacturing: Present and future. Current Opinion in Biomedical Engineering. 2017;**2**:105-115. Available from: https://www.sciencedirect.com/science/article/pii/S2468451117300296

[27] Halloran JW. Freeform fabrication of ceramics. British Ceramic Transactions. 1999;**98**(6):299-303

[28] Deckers J, Vleugels J, Kruth JP. Additive manufacturing of ceramics: A review. Journal of Ceramic Science and Technology. 2014;**5**(4):245-260

[29] Sun J, Chen X, Wade-Zhu J, Binner J, Bai J. A comprehensive study of dense zirconia components fabricated by additive manufacturing. Additive Manufacturing. 2021;**43**:101994.

Available from: https://www.sciencedirect.com/science/article/pii/S2214860421001597

[30] Li W, Ghazanfari A, McMillen D, Leu MC, Hilmas GE, Watts J. Characterization of zirconia specimens fabricated by ceramic on-demand extrusion. Ceramics International. 2018;**44**(11):12245-12252. Available from: https://www.sciencedirect.com/science/article/pii/S0272884218308630

[31] Shishkovsky I, Sherbakov V, Ibatullin I, Volchkov V, Volova L. Nano-size ceramic reinforced 3D biopolymer scaffolds: Tribomechanical testing and stem cell activity. Composite Structures. 2018;**202**:651-659. Available from: https://www.sciencedirect.com/science/article/pii/S026382231734374X

[32] Shishkovsky I, Scherbakov V, Volchkov V, Volova L. Laser-assisted nanoceramics reinforced polymer scaffolds for tissue engineering: Additional heating and stem cells behavior. In: SPIE BiOS Photonics west Proceeding, Dynamics and Fluctuations in Biomedical Photonics XV. 2018;**10493**:104931T. Available from: https://lens.org/083-875- 320-932-820

[33] Shishkovsky I, Nagulin K, Sherbakov V. Study of biocompatible nano oxide ceramics, interstitial in polymer matrix during laser-assisted sintering. International Journal of Advanced Manufacturing Technology. 2015;**78**(1-4):449-455. DOI: 10.1007/s00170-014-6633-6

[34] Grossin D, Montón A, Navarrete-Segado P, Özmen E, Urruth G, Maury F, et al. A review of additive manufacturing of ceramics by powder bed selective laser processing (sintering/melting): Calcium phosphate, silicon carbide, zirconia, alumina, and their composites. Open Ceramics. 2021;**5**:100073. Available from: https://www.sciencedirect.com/science/article/pii/S2666539521000195

[35] Bertrand P, Bayle F, Combe C, Goeuriot P, Smurov I. Ceramic components manufacturing by selective laser sintering. Applied Surface Science. 2007;**254**(4):989-992. Available from: https://www.sciencedirect.com/science/article/pii/S0169433207012603

[36] Shishkovsky IV, Volchkov SE. Ceramics-filled 3D porous biopolymer matrices for tissue-engineering on the stem cell culture: Benchmark testing. In: Bartolo et al., editors. High Value Manufacturing: Advanced Research in Virtual and Rapid Prototyping. Taylor & Francis Group; 2014. pp. 121-126. USBN: 978-1-138-00137-7

[37] Shishkovsky I, Scherbakov V. Selective laser sintering of biopolymers with micro and nano ceramic additives for medicine. Physics Procedia. 2012;**39**:491-499. Open access

[38] Shishkovsky I, Yadroitsev I, Bertrand P, Smurov I. Alumina–zirconium ceramics synthesis by selective laser sintering/melting. Applied Surface Science. 2007;**254**(4):966-970. Available from: https://www.sciencedirect.com/science/article/pii/S0169433207012718

[39] Lee B-T, Kim K-H, Han J-K. Microstructures and material properties of fibrous Al_2O_3–(m-ZrO_2)/t-ZrO_2 composites fabricated by a fibrous monolithic process. Journal of Materials Research. 2004;**19**(11):3234-3241. Available from: https://www.cambridge.org/core/article/microstructures-and-material-properties-of-fibrous-al2o3mzro2tzro2-composites-fabricated-by-a-fibrous-monolithic-process/F489F154FC488B84FCA0607897D22B75

[40] Liu Q, Danlos Y, Song B, Zhang B, Yin S, Liao H. Effect of high-temperature preheating on the selective laser melting of yttria-stabilized zirconia ceramic. Journal of Materials Processing Technology. 2015;**222**:61-74. Available from: https://www.

sciencedirect.com/science/article/pii/S0924013615000862

[41] Verga F, Borlaf M, Conti L, Florio K, Vetterli M, Graule T, et al. Laser-based powder bed fusion of alumina toughened zirconia. Additive Manufacturing. 2020;**31**:100959. Available from: https://www.sciencedirect.com/science/article/pii/S2214860419307870

[42] Shahzad K, Deckers J, Zhang Z, Kruth J-P, Vleugels J. Additive manufacturing of zirconia parts by indirect selective laser sintering. Journal of the European Ceramic Society. 2014;**34**(1):81-89. Available from: https://www.sciencedirect.com/science/article/pii/S0955221913003531

[43] Shi Y, Liu K, Li C, Wei Q, Liu J, Xia S. Additive manufacturing of zirconia parts via selective laser sintering combined with cold isostatic pressing. Check Jixie Gongcheng Xuebao/Chinese Journal of Mechanical Engineering. 2014;**50**(21):118-123

[44] Chen F, Wu J-M, Wu H-Q, Chen Y, Li C-H, Shi Y-S. Microstructure and mechanical properties of 3Y-TZP dental ceramics fabricated by selective laser sintering combined with cold isostatic pressing. International Journal of Lightweight Materials and Manufacture. 2018;**1**(4):239-245. Available from: https://www.sciencedirect.com/science/article/pii/S2588840418300465

[45] Wilkes J, Hagedorn Y, Meiners W, Wissenbach K. Additive manufacturing of ZrO_2-Al_2O_3 ceramic components by selective laser melting. Rapid Prototyping Journal. 2013;**19**(1):51-57. DOI: 10.1108/13552541311292736

[46] Ferrage L, Bertrand G, Lenormand P. Dense yttria-stabilized zirconia obtained by direct selective laser sintering. Additive Manufacturing. 2018;**21**:472-478. Available from: https://www.sciencedirect.com/science/article/pii/S2214860417305602

[47] Koopmann J, Voigt J, Niendorf T. Additive manufacturing of a steel–ceramic multi-material by selective laser melting. Metallurgical and Materials Transactions B. 2019;**50**(2):1042-1051. DOI: 10.1007/s11663-019-01523-1

[48] Huang S, Ye C, Zhao H, Fan Z, Wei Q. Binder jetting yttria stabilised zirconia ceramic with inorganic colloid as a binder. Advances in Applied Ceramics. 2019;**118**(8):458-465. DOI: 10.1080/17436753.2019.1666593

[49] Tarafder S, Balla VK, Davies NM, Bandyopadhyay A, Bose S. Microwave-sintered 3D printed tricalcium phosphate scaffolds for bone tissue engineering. Journal of Tissue Engineering and Regenerative Medicine. 2013;7(8):631-641

[50] Vorndran E, Klarner M, Klammert U, Grover LM, Patel S, Barralet JE, et al. 3D powder printing of β-tricalcium phosphate ceramics using different strategies. Advanced Engineering Materials. 2008;**10**(12):67-71

[51] Zhao H, Ye C, Fan Z, Shi Y. 3D Printing of ZrO_2 Ceramic using Nano-zirconia Suspension as a Binder. 2016;(Icsmim 2015):654-7

[52] Kang J-H, Jang K-J, Sakthiabirami K, Oh G-J, Jang J-G, Park C, et al. Mechanical properties and optical evaluation of scaffolds produced from 45S5 bioactive glass suspensions via stereolithography. Ceramics International. 2020;**46**(2):2481-2488. Available from: https://www.sciencedirect.com/science/article/pii/S0272884219327798

[53] Zakeri S, Vippola M, Levänen E. A comprehensive review of the photopolymerization of ceramic resins used in stereolithography. Additive Manufacturing. 2020;**35**:101177. Available from: https://www.sciencedirect.com/science/article/pii/S2214860420305492

[54] Dehurtevent M, Robberecht L, Hornez J-C, Thuault A, Deveaux E, Béhin P. Stereolithography: A new method for processing dental ceramics by additive computer-aided manufacturing. Dental Materials. 2017;**33**(5):477-485

[55] Zhang K, He R, Xie C, Wang G, Ding G, Wang M, et al. Photosensitive ZrO_2 suspensions for stereolithography. Ceramics International. 2019;**45**(9): 12189-12195. Available from: http://www.sciencedirect.com/science/article/pii/S0272884219306625

[56] Jang K-J, Kang J-H, Fisher JG, Park S-W. Effect of the volume fraction of zirconia suspensions on the microstructure and physical properties of products produced by additive manufacturing. Dental Materials. 2019;**35**(5):e97-e106. Available from: https://www.sciencedirect.com/science/article/pii/S0109564118312119

[57] Lian Q, Sui W, Wu X, Yang F, Yang S. Additive manufacturing of ZrO_2 ceramic dental bridges by stereolithography. Rapid Prototyping Journal. 2018;**24**(1):114-119. DOI: 10.1108/RPJ-09-2016-0144

[58] He R, Liu W, Wu Z, An D, Huang M, Wu H, et al. Fabrication of complex-shaped zirconia ceramic parts via a DLP-stereolithography-based 3D printing method. Ceramics International. 2018;**44**(3):3412-3416. Available from: https://www.sciencedirect.com/science/article/pii/S0272884217325932

[59] Sun J, Binner J, Bai J. 3D printing of zirconia via digital light processing: Optimization of slurry and debinding process. Journal of the European Ceramic Society. 2020;**40**(15):5837-5844. Available from: https://www.sciencedirect.com/science/article/pii/S0955221920304465

[60] Khudyakov IV. Fast photopolymerization of acrylate coatings: Achievements and problems. Progress in Organic Coatings. 2018;**121**:151-159. Available from: https://www.sciencedirect.com/science/article/pii/S0300944017310329

[61] DiSanto P. Book reviews. Journal of Transplant Coordination. 1996;**6**(4):219-220

[62] Allen NS. Photoinitiators for UV and visible curing of coatings: Mechanisms and properties. Journal of Photochemistry and Photobiology A: Chemistry. 1996;**100**(1):101-107. Available from: https://www.sciencedirect.com/science/article/pii/S1010603096044267

[63] Allonas X. Photopolymerization, cationic. Encyclopedia of Polymer Science and Technology. 2019:1-30

[64] Eren TN, Okte N, Morlet-Savary F, Fouassier JP, Lalevee J, Avci D. One-component thioxanthone-based polymeric photoinitiators. Journal of Polymer Science, Part A: Polymer Chemistry. 2016;**54**(20):3370-3378

[65] Hafkamp T, van Baars G, de Jager B, Etman P. A feasibility study on process monitoring and control in vat photopolymerization of ceramics. Mechatronics. 2018;**56**:220-241. Available from: https://www.sciencedirect.com/science/article/pii/S095741581830028X

[66] Bae CJ, Ramachandran A, Chung K, Park S. Ceramic stereolithography: Additive manufacturing for 3D complex ceramic structures. Journal of the Korean Ceramic Society. 2017;**54**(6):470-477

[67] McNulty TF, Mohammadi F, Bandyopadhyay A, Shanefield DJ, Danforth SC, Safari A. Development of a binder formulation for fused deposition of ceramics. Rapid Prototyping Journal. 1998;**4**(4):144-150. DOI: 10.1108/13552549810239012

[68] Gonzalez-Gutierrez J, Cano S, Schuschnigg S, Kukla C, Sapkota J, Holzer C. Additive manufacturing of metallic and ceramic components by the material extrusion of highly-filled polymers: A review and future perspectives. Materials (Basel). 2018;**11**(5):2-36. Available from: https://www.mdpi.com/1996-1944/11/5/840

[69] Venkataraman N, Rangarajan S, Matthewson MJ, Harper B, Safari A, Danforth SC, et al. Feedstock material property-process relationships in fused deposition of ceramics (FDC). Rapid Prototyping Journal. 2000;**6**(4):244-253. DOI: 10.1108/13552540010373344

[70] Spoerk M, Gonzalez-Gutierrez J, Sapkota J, Schuschnigg S, Holzer C. Effect of the printing bed temperature on the adhesion of parts produced by fused filament fabrication. Plastics, Rubber and Composites. 2018;**47**(1):17-24

[71] Carneiro OS, Silva AF, Gomes R. Fused deposition modeling with polypropylene. Materials and Design. 2015;**83**:768-776. Available from: https://www.sciencedirect.com/science/article/pii/S0264127515004037

[72] Khaliq MH, Gomes R, Fernandes C, Nóbrega J, Carneiro OS, Ferrás LL. On the use of high viscosity polymers in the fused filament fabrication process. Rapid Prototyping Journal. 2017;**23**(4):727-735. DOI: 10.1108/RPJ-02-2016-0027

[73] Banerjee S, Joens CJ. 7—Debinding and sintering of metal injection molding (MIM) components. In: Heaney DF, editor. Handbook of Metal Injection Molding, Woodhead Publishing Series in Metals and Surface Engineering. Woodhead Publishing; 2012. pp. 133-180. Available from: https://www.sciencedirect.com/science/article/pii/B9780857090669500078

[74] Sakthiabirami K, Kang J-H, Jang J-G, Soundharrajan V, Lim H-P, Yun K-D, et al. Hybrid porous zirconia scaffolds fabricated using additive manufacturing for bone tissue engineering applications. Materials Science and Engineering: C. 2021;**123**:111950. Available from: https://www.sciencedirect.com/science/article/pii/S0928493121000898

[75] Cano S, Gonzalez-Gutierrez J, Sapkota J, Spoerk M, Arbeiter F, Schuschnigg S, et al. Additive manufacturing of zirconia parts by fused filament fabrication and solvent debinding: Selection of binder formulation. Additive Manufacturing. 2019;**26**:117-128. Available from: https://www.sciencedirect.com/science/article/pii/S2214860418309904

[76] Gaddam A, Brazete DS, Neto AS, Nan B, Ferreira JMF. Three-dimensional printing of zirconia scaffolds for load bearing applications: Study of the optimal fabrication conditions. Journal of the American Ceramic Society. 2021;**104**(9):4368-4380. Available from: https://ceramics.onlinelibrary.wiley.com/doi/abs/10.1111/jace.17874

[77] Shao H, Zhao D, Lin T, He J, Wu J. 3D gel-printing of zirconia ceramic parts. Ceramics International. 2017;**43**(16):13938-13942

[78] Yu T, Zhang Z, Liu Q, Kuliiev R, Orlovskaya N, Wu D. Extrusion-based additive manufacturing of yttria-partially-stabilized zirconia ceramics. Ceramics International. 2020;**46**(4):5020-5027. Available from: https://www.sciencedirect.com/science/article/pii/S0272884219331190

[79] Grech J, Antunes E. Zirconia in dental prosthetics: A literature review. Journal of Materials Research and Technology. 2019;**8**(5):4956-4964. Available from: https://www.sciencedirect.com/science/article/pii/S2238785419300419

[80] Strub JR, Rekow ED, Witkowski S. Computer-aided design and fabrication

of dental restorations: Current systems and future possibilities. Journal of the American Dental Association (1939). 2006;**137**(9):1289-1296

[81] Denry I, Kelly JR. State of the art of zirconia for dental applications. Dental Materials. 2008;**24**(3):299-307

[82] Nakai H, Inokoshi M, Nozaki K, Komatsu K, Kamijo S, Liu H, et al. Additively manufactured zirconia for dental applications. Materials (Basel). 2021;**14**(13):3694

[83] Rosenblum MA, Schulman A. A review of all-ceramic restorations. Journal of the American Dental Association (1939). 1997;**128**(3):297-307

[84] Darmawan BA, Fisher JG, Trung DT, Sakthiabirami K, Park S-W. Two-step sintering of partially stabilized zirconia for applications in ceramic crowns. Materials (Basel). 2020;**13**(8):1857. Available from: https://www.mdpi.com/1996-1944/13/8/1857

[85] Guazzato M, Albakry M, Ringer SP, Swain MV. Strength, fracture toughness and microstructure of a selection of all-ceramic materials. Part II. Zirconia-based dental ceramics. Dental Materials. 2004;**20**(5):449-456

[86] Heintze SD, Rousson V. Survival of zirconia- and metal-supported fixed dental prostheses: A systematic review. The International Journal of Prosthodontics. 2010;**23**(6):493-502

[87] Beuer F, Stimmelmayr M, Gernet W, Edelhoff D, Güth J-F, Naumann M. Prospective study of zirconia-based restorations: 3-year clinical results. Quintessence International. 2010;**41**(8): 631-637

[88] Fischer J, Stawarczyk B, Hämmerle CHF. Flexural strength of veneering ceramics for zirconia. Journal of Dentistry. 2008;**36**(5):316-321

[89] Schweiger J, Bomze D, Schwentenwein M. 3D printing of zirconia–What is the future? Current Oral Health Reports. 2019;**6**(4):339-343. DOI: 10.1007/s40496-019-00243-4

[90] Ebert J, Ozkol E, Zeichner A, Uibel K, Weiss O, Koops U, et al. Direct inkjet printing of dental prostheses made of zirconia. Journal of Dental Research. 2009;**88**(7):673-676

[91] Özkol E, Zhang W, Ebert J, Telle R. Potentials of the "Direct inkjet printing" method for manufacturing 3Y-TZP based dental restorations. Journal of the European Ceramic Society. 2012;**32**(10): 2193-2201. Available from: https://www.sciencedirect.com/science/article/pii/S0955221912001380

[92] Silva NRFA, Witek L, Coelho PG, Thompson VP, Rekow ED, Smay J. Additive CAD/CAM process for dental prostheses. Journal of Prosthodontics. 2011;**20**(2):93-96. Available from: https://onlinelibrary.wiley.com/doi/abs/10.1111/j.1532-849X.2010.00623.x

[93] Liu K, Zhang K, Bourell DL, Chen F, Sun H, Shi Y, et al. Gelcasting of zirconia-based all-ceramic teeth combined with stereolithography. Ceramics International. 2018;**44**(17): 21556-21563. Available from: https://www.sciencedirect.com/science/article/pii/S0272884218322958

[94] Wang W, Yu H, Liu Y, Jiang X, Gao B. Trueness analysis of zirconia crowns fabricated with 3-dimensional printing. The Journal of Prosthetic Dentistry. 2019;**121**(2):285-291

[95] Li R, Wang Y, Hu M, Wang Y, Xv Y, Liu Y, et al. Strength and adaptation of stereolithography-fabricated zirconia dental crowns: An in vitro study. The International Journal of Prosthodontics. 2019;**32**(5):439-443

[96] Zandinejad A, Revilla-león M, Methani MM, Khanlar LN. The Fracture

Resistance of Additively Manufactured Monolithic Zirconia vs . Bi-Layered Alumina Toughened Zirconia Crowns when Cemented to Zirconia Abutments. Evaluating the Potential of 3D Printing of Ceramic Crowns: An In Vitro Study. Dentistry Journal. 2021;**9**(10):115

[97] Revilla-León M, Methani MM, Morton D, Zandinejad A. Internal and marginal discrepancies associated with stereolithography (SLA) additively manufactured zirconia crowns. The Journal of Prosthetic Dentistry. 2020;**124**(6):730-737. Available from: https://www.sciencedirect.com/science/article/pii/S0022391319306109

[98] Ioannidis A, Bomze D, Hämmerle CHF, Hüsler J, Birrer O, Mühlemann S. Load-bearing capacity of CAD/CAM 3D-printed zirconia, CAD/CAM milled zirconia, and heat-pressed lithium disilicate ultra-thin occlusal veneers on molars. Dental Materials. 2020;**36**(4):e109-e116. Available from: https://www.sciencedirect.com/science/article/pii/S0109564120300166

[99] Shi Y, Wang W. 3D inkjet printing of the zirconia ceramic implanted teeth. Materials Letters. 2020;**261**:127131. Available from: https://www.sciencedirect.com/science/article/pii/S0167577X1931763X

[100] Wang W, Sun J. Dimensional accuracy and clinical adaptation of ceramic crowns fabricated with the stereolithography technique. The Journal of Prosthetic Dentistry. 2021;**125**(4):657-663. Available from: https://www.sciencedirect.com/science/article/pii/S0022391320302134

[101] Li R, Chen H, Wang Y, Sun Y. Performance of stereolithography and milling in fabricating monolithic zirconia crowns with different finish line designs. Journal of the Mechanical Behavior of Biomedical Materials. 2021;**115**:104255. Available from: https://www.sciencedirect.com/science/article/pii/S1751616120307931

[102] Revilla-León M, Al-Haj Husain N, Ceballos L, Özcan M. Flexural strength and Weibull characteristics of stereolithography additive manufactured versus milled zirconia. The Journal of Prosthetic Dentistry. 2021;**125**(4):685-690. Available from: https://www.sciencedirect.com/science/article/pii/S0022391320300871

[103] Revilla-León M, Mostafavi D, Methani MM, Zandinejad A. Manufacturing accuracy and volumetric changes of stereolithography additively manufactured zirconia with different porosities. Journal of Prosthetic Dentistry. 2021. Available from: https://www.sciencedirect.com/science/article/pii/S0022391320305047

[104] Osman RB, Swain MV. A critical review of dental implant materials with an emphasis on titanium versus zirconia. Materials (Basel). 2015;**8**(3):932-958

[105] Abd El-Ghany OS, Sherief AH. Zirconia based ceramics, some clinical and biological aspects: Review. Future Dental Journal. 2016;**2**(2):55-64. Available from: http://www.sciencedirect.com/science/article/pii/S2314718016300398

[106] Van der Zel J. Zirconia Ceramic in Dental CAD/CAM: How CAD/CAM Technology Enables Zirconia to Replace Metal in Restorative Dentistry. Journal of Dental Technology. 2007;**2**:17-24

[107] Depprich R, Zipprich H, Ommerborn M, Mahn E, Lammers L, Handschel J, et al. Osseointegration of zirconia implants: An SEM observation of the bone-implant interface. Head & Face Medicine. 2008;**4**:25

[108] Schultze-Mosgau S, Schliephake H, Radespiel-Tröger M, Neukam FW. Osseointegration of endodontic endosseous cones: Zirconium oxide vs

titanium. Oral Surgery, Oral Medicine, Oral Pathology, Oral Radiology, and Endodontics. 2000;**89**(1):91-98

[109] Scarano A, Di Carlo F, Quaranta M, Piattelli A. Bone response to zirconia ceramic implants: An experimental study in rabbits. The Journal of Oral Implantology. 2003;**29**(1):8-12

[110] Sakthiabirami K, Vu VT, Kim JW, Kang JH, Jang KJ, Oh GJ, et al. Tailoring interfacial interaction through glass fusion in glass/zinc-hydroxyapatite composite coatings on glass-infiltrated zirconia. Ceramics International. 2018;**44**(14):16181-16190. Available from: http://www.sciencedirect.com/science/article/pii/S0272884218313063

[111] Osman RB, van der Veen AJ, Huiberts D, Wismeijer D, Alharbi N. 3D-printing zirconia implants; a dream or a reality? An in-vitro study evaluating the dimensional accuracy, surface topography and mechanical properties of printed zirconia implant and discs. Journal of the Mechanical Behavior of Biomedical Materials. 2017;**75**:521-528

[112] Cheng Y-C, Lin D-H, Jiang C-P, Lin Y-M. Dental implant customization using numerical optimization design and 3-dimensional printing fabrication of zirconia ceramic. International Journal for Numerical Methods in Biomedical Engineering. 2017;**33**(5):e2820

[113] Anssari Moin D, Hassan B, Wismeijer D. A novel approach for custom three-dimensional printing of a zirconia root analogue implant by digital light processing. Clinical Oral Implants Research. 2017;**28**(6):668-670. DOI: 10.1111/clr.12859

[114] Schwarzer E, Holtzhausen S, Scheithauer U, Ortmann C, Oberbach T, Moritz T, et al. Process development for additive manufacturing of functionally graded alumina toughened zirconia components intended for medical implant application. Journal of the European Ceramic Society. 2019;**39**(2):522-530. Available from: https://www.sciencedirect.com/science/article/pii/S095522191830548X

[115] Zhu Y, Liu K, Deng J, Ye J, Ai F, Ouyang H, et al. 3D printed zirconia ceramic hip joint with precise structure and broad-spectrum antibacterial properties. International Journal of Nanomedicine. 2019;**14**:5977-5987

[116] Magnani G, Fabbri P, Leoni E, Salernitano E, Mazzanti F. New perspectives on zirconia composites as biomaterials. Journal of Composites Science. 2021;**5**(9):244. Available from: https://www.mdpi.com/2504-477X/5/9/244

[117] Sapkal PS, Kuthe AM, Mathankar S, Deshmukh AA. 3D bio-plotted tricalcium phosphate/zirconia composite scaffolds to heal large size bone defects. Molecular & Cellular Biomechanics. 2017;**14**(2):125-136. Available from: https://www.proquest.com/scholarly-journals/3d-bio-plotted-tricalcium-phosphate-zirconia/docview/2397225089/se-2

[118] Lin F, Yan C, Zheng W, Fan W, Adam C, Oloyede AK. Preparation of mesoporous bioglass coated zirconia scaffold for bone tissue engineering. Advances in Materials Research. 2012;**365**:209-215

[119] Kim Y-H, Lee B-T. Novel approach to the fabrication of an artificial small bone using a combination of sponge replica and electrospinning methods. Science and Technology of Advanced Materials. 2011;**12**(3):35002. DOI: 10.1088/1468-6996/12/3/035002

[120] Hadjicharalambous C, Buyakov A, Buyakova S, Kulkov S, Chatzinikolaidou M. Porous alumina, zirconia and alumina/zirconia for bone repair: Fabrication, mechanical and in

vitro biological response. Biomedical Materials. 2015;**10**(2):025012

[121] Jang D-W, Nguyen T-H, Sarkar SK, Lee B-T. Microwave sintering and in vitro study of defect-free stable porous multilayered HAp–ZrO2 artificial bone scaffold. Science and Technology of Advanced Materials. 2012;**13**(3):35009. DOI: 10.1088/1468-6996/13/3/035009

[122] Mondal D, So-Ra S, Sarkar SK, Min YK, Yang HM, Lee BT. Fabrication of multilayer ZrO_2–biphasic calcium phosphate–poly-caprolactone unidirectional channeled scaffold for bone tissue formation. Journal of Biomaterials Applications. 2012;**28**(3): 462-472

[123] Li Y, Li L, Li B. Direct write printing of three-dimensional ZrO_2 biological scaffolds. Materials and Design. 2015;**72**:16-20. Available from: http://www.sciencedirect.com/science/article/pii/S0261306915000643

[124] Shuai C, Feng P, Yang B, Cao Y, Min A, Peng S. Effect of nano-zirconia on the mechanical and biological properties of calcium silicate scaffolds. International Journal of Applied Ceramic Technology. 2015;**12**(6): 1148-1156

[125] Cadafalch Gazquez G, Chen H, Veldhuis SA, Solmaz A, Mota C, Boukamp BA, et al. Flexible yttrium-stabilized zirconia nanofibers offer bioactive cues for osteogenic differentiation of human mesenchymal stromal cells. ACS Nano. 2016;**10**(6): 5789-5799. DOI: 10.1021/acsnano.5b08005

[126] Sapkal PS, Kuthe AM, Kashyap RS, Nayak AR, Kuthe SA, Kawle AP. Indirect casting of patient-specific tricalcium phosphate zirconia scaffolds for bone tissue regeneration using rapid prototyping methodology. Journal of Porous Materials. 2017;**24**(4):1013-1023. DOI: 10.1007/s10934-016-0341-6

[127] Thakare VG, Joshi PA, Godse RR, Bhatkar VB, Wadegaokar PA, Omanwar SK. Fabrication of polycaprolactone/zirconia nanofiber scaffolds using electrospinning technique. Journal of Polymer Research. 2017;**24**(12):232. DOI: 10.1007/s10965-017-1388-z

[128] Stanciuc A-M, Sprecher CM, Adrien J, Roiban LI, Alini M, Gremillard L, et al. Robocast zirconia-toughened alumina scaffolds: Processing, structural characterisation and interaction with human primary osteoblasts. Journal of the European Ceramic Society. 2018;**38**(3):845-853. Available from: http://www.sciencedirect.com/science/article/pii/S0955221917305708

[129] Sa M-W, Nguyen B-NB, Moriarty RA, Kamalitdinov T, Fisher JP, Kim JY. Fabrication and evaluation of 3D printed BCP scaffolds reinforced with ZrO2 for bone tissue applications. Biotechnology & Bioengineering. 2018;**115**(4):989-999. DOI: https://doi.org/10.1002/bit.26514

[130] Brazete D, Neto AS, Ferreira JMF. Optimization of zirconia inks to fabricate 3D porous scaffolds by robocasting. LEK Technology. 2019; **49**(1):5-10

[131] Fu SY, Yu B, Ding HF, Shi GD, Zhu YF. Zirconia incorporation in 3D printed β-Ca_2SiO_4 scaffolds on their physicochemical and biological property. Wuji Cailiao Xuebao/Journal of Inorganic Materials. 2019;**34**(4):444-454

[132] Zhang J, Huang D, Liu S, Dong X, Li Y, Zhang H, et al. Zirconia toughened hydroxyapatite biocomposite formed by a DLP 3D printing process for potential bone tissue engineering. Materials Science and Engineering: C. 2019;**105**: 110054. Available from: http://www.sciencedirect.com/science/article/pii/S0928493119310537

[133] Wang Q, Ma Z, Wang Y, Zhong L, Xie W. Fabrication and characterization of 3D printed biocomposite scaffolds based on PCL and zirconia nanoparticles. Bio-Design and Manufacturing. 2021;**4**(1):60-71. DOI: 10.1007/s42242-020-00095-3

[134] Cao Y, Shi T, Jiao C, Liang H, Chen R, Tian Z, et al. Fabrication and properties of zirconia/hydroxyapatite composite scaffold based on digital light processing. Ceramics International. 2020;**46**(2):2300-2308. Available from: http://www.sciencedirect.com/science/article/pii/S0272884219327567

[135] Kocyło E, Franchin G, Colombo P, Chmielarz A, Potoczek M. Hydroxyapatite-coated ZrO_2 scaffolds with a fluorapatite intermediate layer produced by direct ink writing. Journal of the European Ceramic Society. 2021;**41**(1):920-928. Available from: http://www.sciencedirect.com/science/article/pii/S0955221920306634

[136] Lee BT, Kang IC, Cho SH, Song HY. Fabrication of a continuously oriented porous Al_2O_3 body and its in vitro study. Journal of the American Ceramic Society. 2005;**88**(8):2262-2266

[137] Wei C, Montagnac G, Reynard B, Le Roux N, Gremillard L. Interplay between internal stresses and matrix stiffness influences hydrothermal ageing behaviour of zirconia-toughened-alumina. Acta Materialia. 2020;**185**:55-65. Available from: https://www.sciencedirect.com/science/article/pii/S1359645419308018

[138] Reginato VF, Kemmoku DT, Caldas RA, Bacchi A, Pfeifer CS, Consani RLX. Characterization of residual stresses in veneering ceramics for prostheses with zirconia framework. Brazilian Dental Journal. 2018;**29**(4):347-353

[139] Zhang Y, Allahkarami M, Hanan JC. Measuring residual stress in ceramic zirconia–porcelain dental crowns by nanoindentation. Journal of the Mechanical Behavior of Biomedical Materials. 2012;**6**:120-127. Available from: https://www.sciencedirect.com/science/article/pii/S1751616111002931

[140] Jang K-J, Kang J-H, Sakthiabirami K, Lim H-P, Yun K-D, Yim E-K, et al. Evaluation of cure depth and geometrical overgrowth depending on zirconia volume fraction using digital light processing. Journal of Nanoscience and Nanotechnology. 2018;**19**(4):2154-2157

[141] Kosyfaki P, del Pilar Pinilla Martín M, Strub JR. Relationship between crowns and the periodontium: A literature update. Quintessence International. 2010;**41**(2):109-126

[142] Contrepois M, Soenen A, Bartala M, Laviole O. Marginal adaptation of ceramic crowns: A systematic review. The Journal of Prosthetic Dentistry. 2013;**110**(6):447-454.e10

[143] 10 of the Biggest Challenges in Scaling Additive Manufacturing for Production in 2020 [Expert Roundup]—AMFG. Available from: https://amfg.ai/2019/10/08/10-of-the-biggest-challenges-in-scaling-additive-manufacturing-for-production-expert-roundup/

Chapter 2

Perspective Chapter: Additive Manufacturing in Customized Medical Device

Ching Hang Bob Yung, Lung Fung Tse, Wing Fung Edmond Yau and Sze Yi Mak

Abstract

The long-established application of rapid prototyping in additive manufacturing (AM) has inspired a revolution in the medical industry into a new era, in which the clinical-driven development of the customized medical device is enabled. This transformation could only be sustainable if clinical concerns could be well addressed. In this work, we propose a workflow that addresses critical clinical concerns such as translation from medical needs to product innovation, anatomical conformation and execution, and validation. This method has demonstrated outstanding advantages over the traditional manufacturing approach in terms of form, function, precision, and clinical flexibility. We further propose a protocol for the validation of biocompatibility, material, and mechanical properties. Finally, we lay out a roadmap for AM-driven customized medical device innovation based on our experiences in Hong Kong, addressing problems of certification, qualification, characterization of three dimensional (3D) printed implants according to medical demands.

Keywords: hybrid additive manufacturing, customized medical device, anatomical conformation, personalized medicine

1. Introduction

Throughout the ages, medicine, by inherent definition, has always been focused on the treatment of persons and individuals. Whilst the pursuit of scientific progress has inexorably propelled this process toward a systematic and harmonized approach to treatments [1], the advent of personalized medical solutions has begun to reintegrate the personalized and idiosyncratic element to the therapeutic action [2, 3]. This has erupted into a vast and expansive medical discipline in the current day, ranging from diagnostic testing [4] to tailored drug treatments [5], to the customized medical devices that will be focal here. The orthopedic customized medical device has become one of the more mundane and immediately practical manifestations of personalized medicine.

Tunneling on customized prosthetic implants such as those in use in dental, maxillofacial, and orthopedic disciplines, the dichotomy between the conventional manufacturing technologies and additive manufacturing (AM) become apparent; where subtractive manufacturing and its kin excel in excellent control of

IntechOpen

repeatability, scalability, surface finishing, and product proportions, the piecewise variation seen in personalized medical solutions levels the playing field significantly, so much so that topologies unique to additive manufacturing (AM) processes such as freeform, anatomically compliant geometries and bioinductive honeycomb porous structures are allowed to shine through [6].

That being said, there are still substantial barriers between the current state of additive manufacturing and ancillary technologies, and mature, well-characterized medical applications [7]. Medical device development has and will for the foreseeable future be driven by clinical needs, and as medical device customization continues to progress, this personalized approach brings medical professionals ever closer to the engineering-based approaches used during the design and manufacture of medical devices [8]. Operating in completely disparate paradigms, efficient bridging of this chasm will be imperative going forward [9].

Based in Hong Kong, the authors have been working toward the realization of streamlined AM utilization in the manufacture of customized medical devices over the past 5 years. Experience and involvement in the formulation of customized medical devices ranging from surgical guides and instruments to long-term orthopedic implants have culminated in a relatively refined and progressively formulaic modus operandi. Putting forth a structured workflow and robust manufacturing process validation protocols, we look to initiate discussion in the space by this proof-of-concept, not in terms of technical operational detail but the constitution of the proposed system and its potency and soundness.

2. Additive manufacturing-assisted fabrication of the medical device

A typical workflow of preparing a customized medical device consists of four stages, namely, anatomic modeling, surgical planning and design, additive manufacturing, and postprocessing, as shown in **Figure 1**. This workflow has been testified and applied to fabricating 11 personalized surgical instruments in Hong Kong [10].

2.1 Anatomic modeling

Once clinical needs are identified, anatomical modeling is constructed based on the patient's anatomy. Generic processes utilized during anatomical modeling are displayed in **Figure 2**. Computed tomography (CT) DICOM data is read as 2D grayscale pixel arrays arranged in a series of planes (**Figure 2A**). Desired anatomical structures are isolated on each individual array through intensity thresholding, artifacts, noise, and distortions are minimized by using image processing tools (**Figure 2B**). Series of 2D slice pixel arrays are interpolated and converted into a three dimensional (3D) computer-aided design (CAD)-friendly format (**Figure 2C**). These models and other patient information are the basis for surgical planning and design.

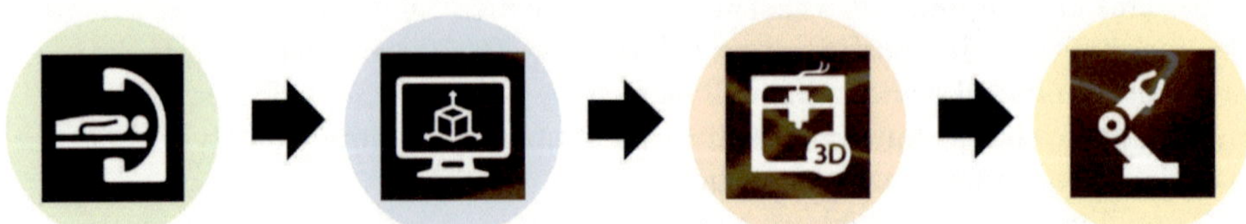

Figure 1.
A typical workflow of preparing a customized medical device. Stage 1: anatomic modeling. Stage 2: surgical planning and design. Stage 3: additive manufacturing. Stage 4: postprocessing.

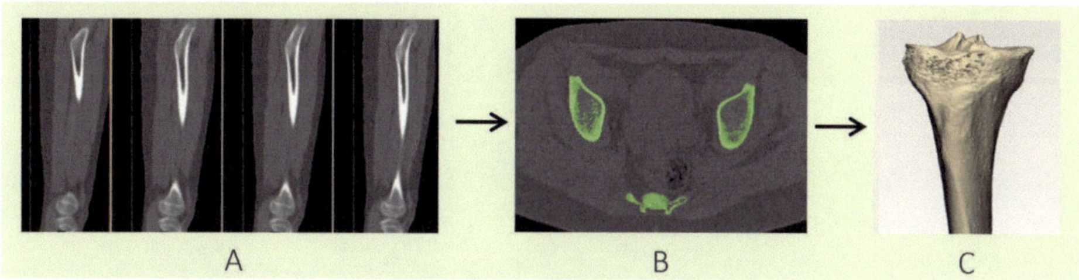

Figure 2.
Stage 1: anatomic modeling, (A) scanning, (B) segmentation, and (C) construction of 3D model. Stage 2: surgical planning and design.

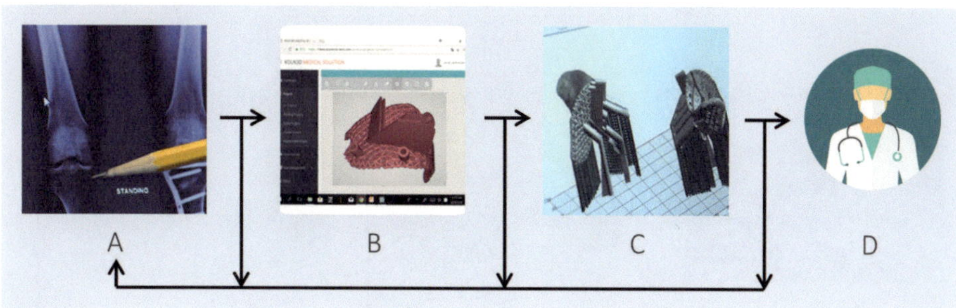

Figure 3.
Stage 2: surgical planning and design, (A) surgical planning, (B) CAD modeling, and (C) CAM modeling.

2.2 Surgical planning and design

The surgical planning and design stage include an iterative process of (**Figure 3A**) surgical planning, (**Figure 3B**) CAD modeling, and (**Figure 3C**) computer-aided manufacturing (CAM) modeling, **Figure 3**. It requires immense communication between surgeons and engineers. The success of the design strongly depends on the level of details as well as the effectiveness of the communication of inputs from both parties. An example of surgical planning and design of a patient-specific instrument is presented in **Figure 4**.

2.3 Additive manufacturing

Proceeding from design to manufacturing, one AM method commonly used for a metal medical device is direct metal laser sintering (DMLS) under the powder bed fusion category. In a typical DMLS setup (LUMEX Avance 25, Matsuura), additive manufacturing is achieved by repeated procedures of (**Figure 5A**) recoating and (**Figure 5B**) laser sintering. The hybrid AM approach incorporates an additional procedure of (**Figure 5C**) computer numerical control (CNC) machining whenever several layers are built [11]. Here, we demonstrate the DMLS method by using cobalt-chromium alloy (**Figure 5**). Spherical powder of size ranges from 25 μm to 40 μm (Koln3DCobaltChrome, Sandvik) is recoating onto the powder bed by a flat blade swiping sideways. The layer thickness is set at 0.4–0.5 mm whereas the laser power is set in the range of 100–400 W.

We examine the morphology and elemental composition of cobalt-chromium alloy before and after sintering. The morphology of cobalt-chromium alloy powder observed under field emission scanning electron microscope (FESEM) is shown in **Figure 6a**. The powder size ranges from 25 μm to 40 μm. After laser sintering,

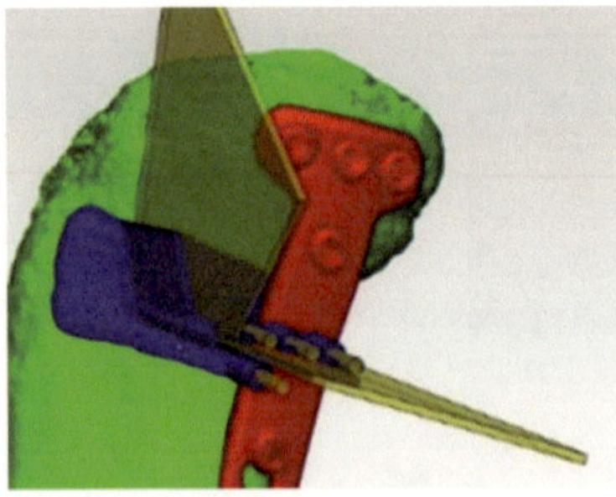

Figure 4.
Surgical planning elements highlighted: segmented patient CT data (green) is combined with surgeon input (transparent yellow) culminating in customized surgical instrument (blue) and standardized implant (red).

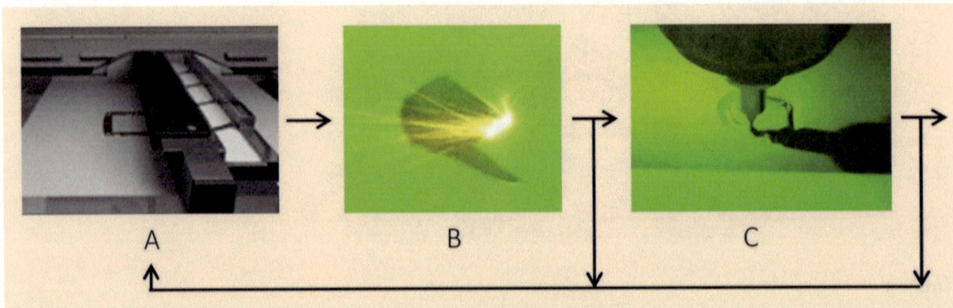

Figure 5.
Stage 3: additive manufacturing, (A) recoating, (B) laser sintering, and (C) CNC machining.

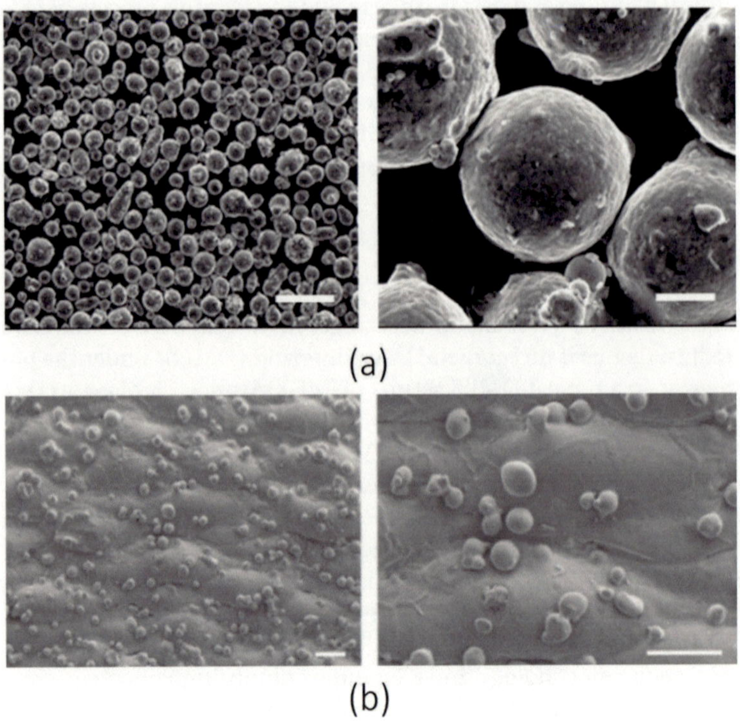

Figure 6.
FESEM images. (a) Cobalt chromium alloy powder in size range of 25–40 μm. Scale bars are 100 μm (left) and 10 μm (right). (b) Surface of a part made by sintering of cobalt-chromium alloy. Scale bars are 100 μm.

unsintered powder remains on the surface of the built part, as shown in **Figure 6b**. To improve the surface finishing of the AM product, postprocessing is required. Some AM parts undergo heat treatment to improve mechanical properties such as ductility and hardness.

Field emission scanning electron microscope/energy dispersive X-ray analysis (FESEM/EDX) was performed using an FEI Quanta 400 FEG MK2 electron microscope and an AMETEK EDAX (PV776068-ME) X-ray analyzer to investigate the composition of the samples. The back-scattered electrons (BSE) images are formed by scanning the sample with a high-energy beam of primary electrons. The primary electrons interact with the sample and generate low-energy secondary electrons and back-scattered electrons, these electrons are collected, and the surface topography of the sample can be constructed. In addition to low-energy secondary electrons, X-rays are also generated by the interaction of the primary electrons and the sample. The characteristic of X-ray emission can give qualitative elemental information of the sample. In the present case, a standardless ZAF algorithm was used for quantification.

Elemental composition measurement is performed on the powder and the sintered part. The sintered samples used in this experiment undergo heat treatment processes. Main elements, such as cobalt (Co), chromium (Cr), and molybdenum (Mo), etc., are measured by EDX. We do not observe any significant changes in the elemental composition of cobalt-chromium alloy before and after the sintering process, which are in the form of powder and sintered parts, respectively, as shown in **Figure 7**. The result shows that the discrepancy in elemental composition varies by within ±2 wt%.

2.4 Postprocessing

Completing the AM process, postprocessing is performed for (**Figure 8A**) support removal and (**Figure 8B**) polishing according to the specific clinical needs, as illustrated in **Figure 8**. The postprocessing procedures are namely product-based plate detachment, support material removal, surface machining, and surface polishing. Even though a high degree of design complexity is enabled by AM

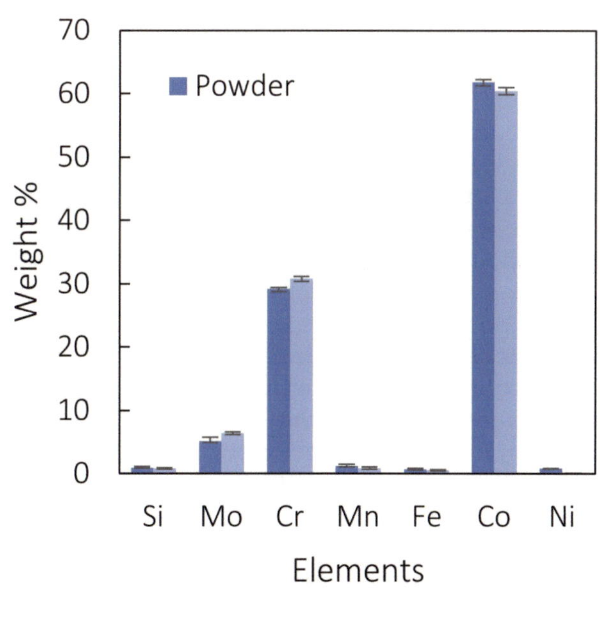

Figure 7.
Elemental composition of (a) cobalt-chromium alloy powder and (b) surface of the corresponding sintered part. Error bar is one standard deviation of five measurements.

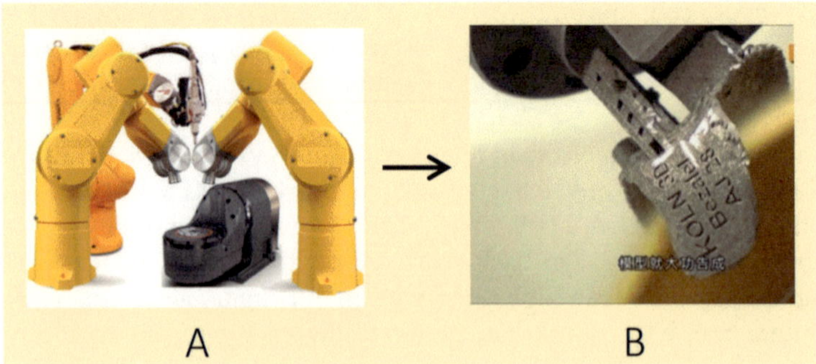

Figure 8.
Stage 4: postprocessing, (A) support removal, and (B) polishing.

technology, the low efficiency in postprocessing is a prevailing limiting factor in the entire process. To date, these postprocessing procedures are commonly conducted manually and relatively time-consuming depending on the complexity of the AM product. Recently, robotic control is introduced to automate the process and is gaining popularity in the manufacturing industry [12]. This technology is highly appealing to the medical industry for it possesses many advantages over manual operation such as higher accuracy and repeatability [13]. Full automation of robotic postprocessing systems is on its way to transforming the medical industry.

2.5 Case study

Here, we present a case study of a teenager with chondral lesions on the posterior medial quadrant of the talar dome. The treatment was performed with the aid of medial malleolar osteotomy surgical jig (**Figure 9**). Our proposed workflow for the preparation of additive manufacturing-assisted fabrication of medical devices has been adopted.

A teenager patient admitted with severe ankle pain when walking was diagnosed with abnormalities on the posterior medial quadrant of the talar dome. The suspected cause of chondral lesion is vascularization defect in subchondral talar bone. The treatment approach is laid out by (1) medial malleolus removal with the surgical jig to expose chondral lesion, (2) removal of defective chondral tissue, (3) articular surface repair, and (4) reattachment of the medial malleolus.

With patient and regulatory approval, the medical device was prepared subsequently. In the stage of anatomical modeling, a CT scan of the ankle with slice thickness 0.625 mm and slice resolution 0.5 mm was performed and the DICOM data of the talocrural joint were segmented and converted to surface mesh body. In the stage of surgical planning and device design, the chondral defect was first located. Cutting planes and fixation screw trajectories computationally were simulated and determined. Subsequently, guides for cutting planes and screw trajectories were designed, followed by patient-matched surface design according to anatomical landmarks and features coalesce to form the final design. Proceeding to the stage of additive manufacturing, the device was fabricated using a DMLS 3D printer with CoCrMo alloy (ASTM F-75 grade) powder. The process was validated with biocompatibility, mechanical, and compositional tests with critical dimensions verified. In the final stage of postprocessing, hot isostatic pressing (HIP) was performed to resolve residual stresses. After support removal, the device was ready for cleaning and packaging. The additive manufacturing-assisted fabrication of medial malleolar osteotomy surgical jig was completed.

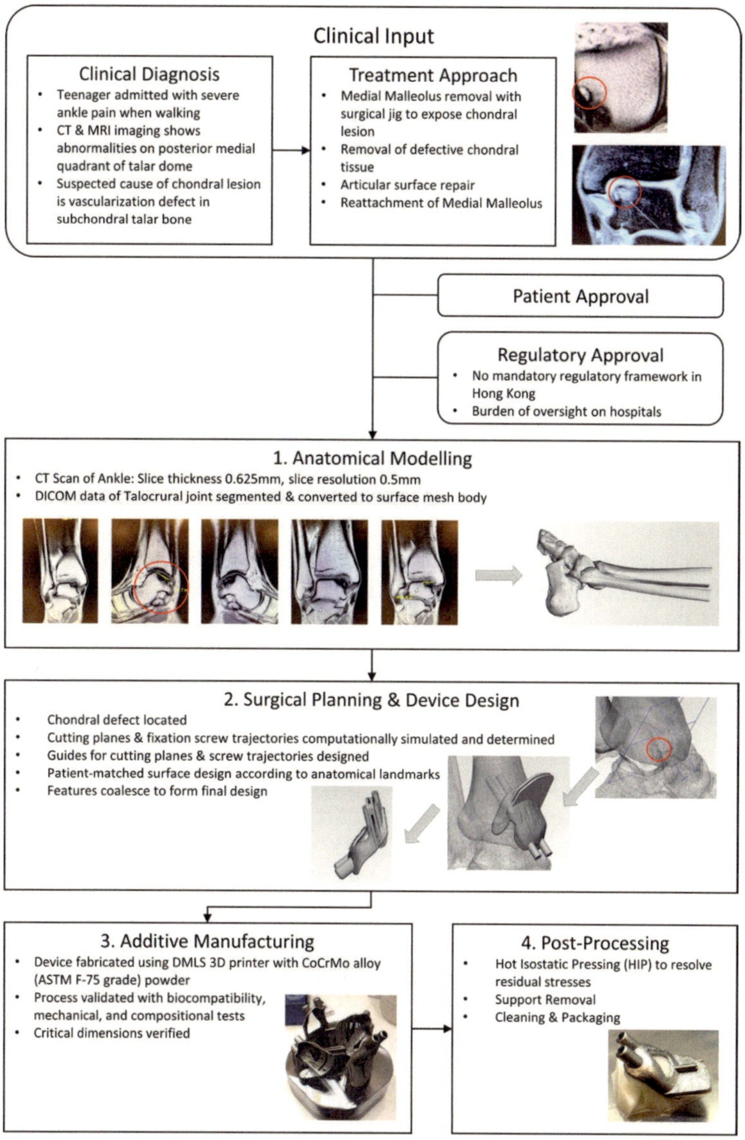

Figure 9.
Case study: medial malleolar osteotomy surgical jig.

3. Critical clinical concerns in AM technology

3.1 Error of medical AM manufacturing

AM-assisted fabrication of the medical device is not an automatic process. The clinical data including anatomy and functions of the body parts need to be analyzed and segmentation of the relevant parts on the data source is of paramount importance to the beginning of every AM process.

Upon arriving at a diagnosis, with confirmation through medical imaging, the anatomical and functional data will be transformed from DICOM data to stereolithography (STL) or CAD data formats interpretable by 3D printers. Engineers and clinicians will then corroboratively engage the design and customization effort. Given this approach, there are common pitfalls to take note of and avoid.

3.1.1 Pathologies around the joint

In the case of a tumor around the hip joint, combined use of CT scan and magnetic resonance imaging (MRI) can help to create accurate models for surgical planning by coregistration of two sets of DICOM. The bony margins can be defined on a CT scan, whereas the soft tissue component of the tumor and sites of tumor invasion, periosteal elevation, and/or edema within the bone are best defined on MRI. A single model created using the information from both modalities allows the surgeon to plan resection and reconstruction utilizing all available information.

3.1.2 Quality of bones in the very young and very old

The differences in pixel density (on CT scan) or signal intensity (on MRI) between immature bones and cartilage and between osteoporotic bones and osteophytes and diseased tissues can be subtle. This is particularly challenging when dealing with complex pelvic pathology. It would need manual input to delineate osteophytes and osteopenic areas and pathology areas. This means that the surgeon, radiologist, or engineer will have to manually identify, at least in part, the various anatomic structures so they can be printed as separate objects. Segmented images designated for printing patient-specific models for surgical planning should be carefully reviewed before the AM manufacturing process and validated after removal of the disease body part.

The above list of conditions is not exhaustive which can cause a fundamental error to the final product with AM manufacturing process. An accurate translation from medical needs to product innovation is safeguarded by the effective communication and information exchange between various parties involved, as discussed in Section 2.

The understanding of various materials for the additive manufacturing process impacts the functionality of the final product. Training, technical competence, and experience utilizing medical software and software for 3D printers determine the quality of the object architecture. Clinical concerns in AM technology fall into the following categories—translation from medical needs to product innovation, anatomical conformation and execution, and validation.

The medical implants for the replacement of defective parts of the human anatomy can be validated with various means. One way is to scan the resected body part and overlay it with the source data of AM [14]. This will provide a quantitative measurement of the accuracy of pre- and postmanufacturing data.

3.2 Geometric conformity of AM medical devices

To investigate the geometric conformity of the AM parts, five distinct metallic medical devices, namely jig 1, 2, and implant 1, 2, 3, have been made from the AM approach and 3D scanned. Turntable mode on Shining 3D Einscan Pro HD 3D scanner has been used to scan robotically milled product coated with AESUB blue scanning spray. The resultant model has then been matched and analyzed with corresponding functions in the materialize MIS 24 software suite. A clinically critical zone is defined for which the geometrical accuracy is examined. Overlaying the clinical critical surface area of the design model and the scanned model, a point-to-point spatial distance is measured for each point within the clinically critical zone. The average discrepancy and the discrepancy histogram are presented in **Table 1**. Jig 1 yields the maximum average discrepancy of 0.13 ± 0.35 mm. For a typical computed tomography (CT) scan, the spatial and axial resolutions are 0.5 mm and 0.6 mm, which set the spatial accuracy requirement of the customized medical device. The result suggests that the AM-assisted fabrication approach attains the satisfactory special accuracy required clinically.

	Jig 1	Jig 2	Implant 1	Implant 2	Implant 3
Spatial discrepancy (mm)	0.13 ± 0.35	0.02 ± 0.15	0.07 ± 0.21	−0.01 ± 0.12	0.10 ± 0.12
Graphical representation			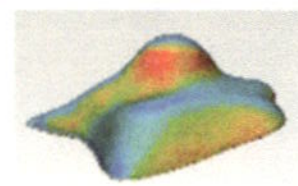	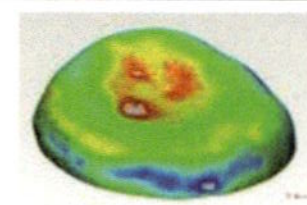	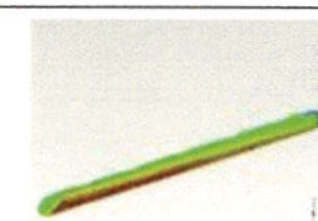
Discrepancy histogram	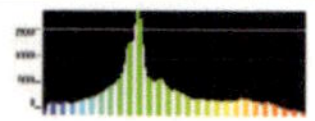	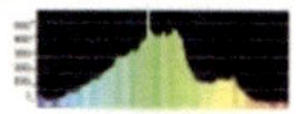			

Table 1.
Robotically machined 3D-printed products were 3D scanned and compared computationally with the base CAD model. Results of the analysis are displayed.

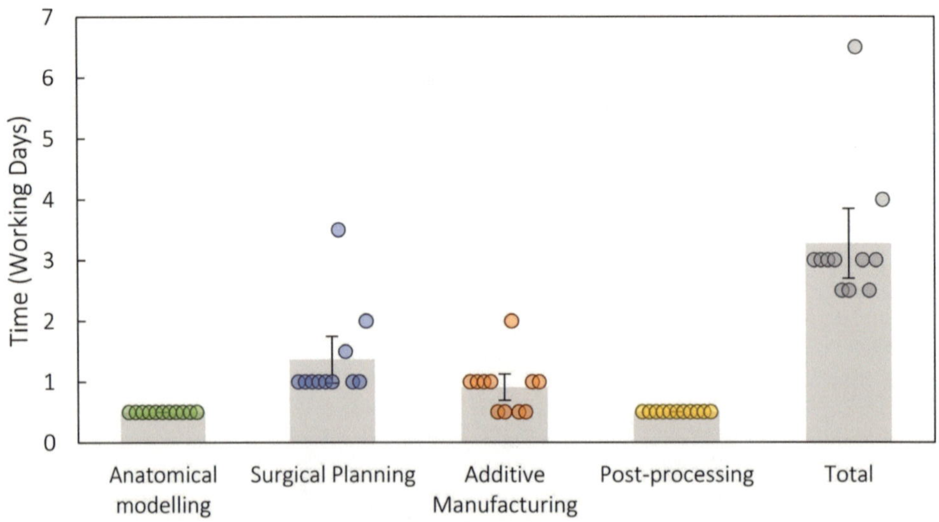

Figure 10.
Time consumed per each of the four stages. Statistics of 11 cases. Error bars indicate the standard deviation.

3.3 Time consumed in each stage of AM-assisted fabrication

Apart from spatial resolution, the time consumed in AM-assisted fabrication is investigated. About 11 patient-specific instruments undertaking the proposed four-stage workflow being made, including five for hallux valgus osteotomy, four for high tibial osteotomy, one each for proximal femur osteotomy, and calcaneal osteotomy. A statistic of the time taken per each stage is measured, as shown in **Figure 10**. Among the cases reported, the average total time taken to complete the four stages is 3.3 ± 0.6 working days. The first and last stages, that is anatomical modeling and postprocessing, are relatively routine and typically take 0.5 days to complete. Contrarily, the time is taken for surgical planning and design (Stage 2) varies. Since Stage 2 involves an iterative process requiring communication between surgeons and engineers, the time involved is not only dependent on the complexity of the design but also the effectiveness of communication between different parties. As a result, time spent on Stage 2 has been found to be the lengthiest and the most variant. The time consumed in additive manufacturing (Stage 3) is dependent on the size and resolution of the AM part. Overall, the short time frame (in the order of a few days) to fabricate metallic patient-specific instruments by AM approach has brought numerous opportunities to cater for nonemergent clinical applications, such as osteotomy as demonstrated in this work.

4. Protocol for validation

While validation is a procedurally quintessential part of a fabrication that ensures process compatibility with intended product applications, it is especially imperative when dealing with medical devices that inherently possess the risks involved with interacting directly with human physiology. To this end, regulatory bodies such as the US Food and Drug Administration (FDA), the Chinese National Medical Products Administration (NMPA), and the decentralized regulatory authorities under the European Commission have developed medical device classification systems that categorize devices in accordance with the risk, their respective intended applications bring with them. Whilst there are nuances between the numerous classifications, medical devices are typically assessed according to their intended purpose or use, with the duration, invasiveness, reusability, sterility, and activeness being commonly scrutinized aspects that are used as indicators of the associated risk. Among medical devices, this is particularly heightened when concerning prosthetic implants that are invasive in nature and are often implicated in extremely prolonged physiological exposure during and after surgical procedures. Knowing this, validation is an exceedingly risk-dependent procedure, and as such one should always keep in mind and analyze the intended purpose or use of a product, what the associated explicit and implied product requirements are, and if and how relevant AM processes could potentially impact the conformity to these requirements.

The traditional pharmaceutical process validation structure of installation qualification (IQ), operational qualification (OQ), and performance qualification (PQ) is generally an effective methodology when transplanted into the context of AM processes. Originating from a similarly health-related industry, this does not significantly vary when being applied to the fabrication of medical devices, though there are indeed adjustments to accommodate for the aforementioned variation in associated risk inherent to the nature of the products in question. Defining the scope and breadth of the process validation will grant it greater clarity in the considerations to be made and the extent of the validation activities to be performed. While we will continue mostly focusing on hybrid additive manufacture processes (CNC-DMLS), be mindful that the type and nature of the AM processes concerned will affect the complexity of respective validation activities.

Installation qualification (IQ) is the ascertainment, through the documentation of objective evidence acquired through predefined verification methods, that all relevant equipment and machinery, whether primary, auxiliary, or ancillary, has been installed in accordance with predetermined requirements or recommendations. In practice in terms of hybrid AM processes, this typically involves infrastructural checks on items and ancillary systems such as electrical supply, compressed air supply, inert gas supply, and chiller, information that is customarily provided by most equipment manufacturers. Equipment manufacturers will also commonly have supportive services surrounding these activities in the form of complete user documentation documents as well as site acceptance tests (SAT) to qualify the commissioning of the equipment. Calibration of all measurement devices used throughout verification processes occurring during the entire validation is also usually included as part of installation qualification. IQ validation activities for the production of medical devices do not deviate from these elements, though one small detail to take note of is that the biologically oriented verification processes taking place subsequently in OQ and PQ validations might not have calibration available as it is traditionally understood. Some examples of this include the chemical assays used to determine physiological chemical characteristics and the histological examinations for assessing biological reactions toward materials. Whilst these evaluative processes are typically performed by accredited laboratories, verifications are done in-house should always have additional device accuracy verifications performed.

Operational qualification (OQ) is the process that results in the establishment of equipment operational parameters, limits, conditions, and requirements that optimally are expected to result in products meeting and product specifications. Performance qualification (PQ) builds on the findings of OQ validation, effectively stress testing the manufacturing process under simulated worst-case scenarios to ensure product specifications are met regardless, or those potential deviations and their respective rates of occurrence are acknowledged. Both OQ and PQ typically comprise a series of tests and verifications, as well as the documentation of all pertinent methodologies, results, evidence, and conclusions. Operational conditions and parameters, their control, veracity, and repeatability, are first verified to safeguard the authenticity of the test environment. In the case of hybrid AM processes, this can range from laser control parameters such as laser power and path overlap to sintering chamber environmental conditions such as oxygen concentration, build plate temperature, and enclosure temperature. This includes verification of metal powders used, as well as processes used to handle said metal powders and their respective control parameters. Once all highlighted aspects of operational control have been verified, operational limits and conditions can then be established and tests can be conducted on products produced by using operational parameters across this range, with acceptance criteria enacted based on product specifications. Through data collection and trend analysis, one can deduce and provide justification for the establishment of optimal operational parameters and conditions during product manufacture, and thus concludes OQ validation and moving on to PQ validation. In terms of hybrid AM processes, since there is not much variation when considering maximal system throughput and worst-case scenarios, PQ is often simply performed through periodically testing products manufactured at maximum printing load.

Nonmedical applications of hybrid AM processes will generally inspect for mechanical aspects such as strength and malleability, material properties such as product composition, grain structure, presence of impurities, as well as explicit elements of product specification such as product form and critical dimensions. While international standards concerning appraisal methodology of these quantities are well established from beyond the medical field, the uniqueness of medical devices applications often warrants their own testing methodology. For example, whilst bending strength and stiffness are well-characterized quantities in their own right, ISO 9585 and ASTM F382 both describe methodologies that are specific to bone plates.

In the case of medical devices, critical evaluative processes, standardized methodologies, and highly specific parameters for biocompatibility conformance have been outlined in the document series ISO 10993. In particular, Part 1 of ISO 10993 systematically outlines by flowchart all considerations necessitated by regulatory bodies when assessing the risk associated with a medical device, echoing factors previously mentioned surrounding intended use or purpose such as duration

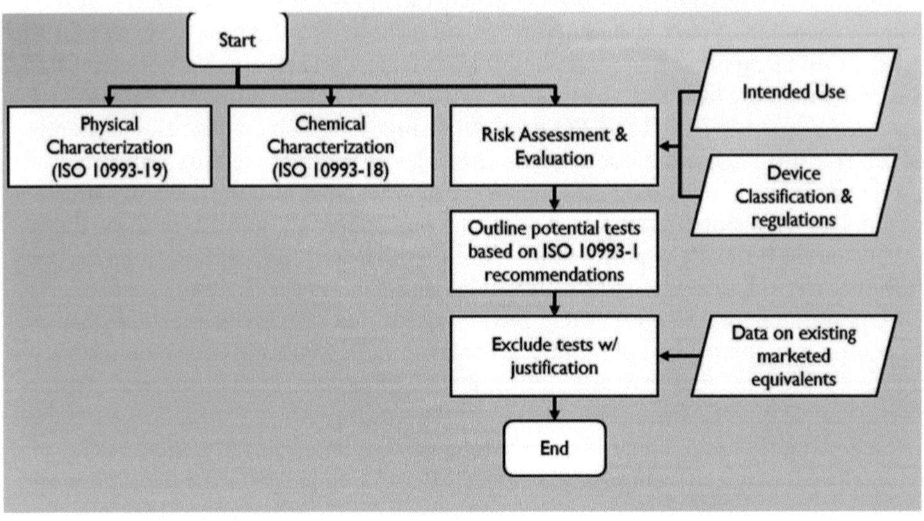

Figure 11.
Flowchart of selection of biocompatibility tests.

of use and invasiveness, but also taking into account the nature of physiological surfaces contacted as well as the characteristics of the device itself. Depending on the outcome of the evaluation, a range of endpoints of biological evaluation are recommended, indicating the types of tests required to demonstrate an affirmative biological evaluation (**Figure 11**).

Critical to the case of hybrid AM processes is the stipulation that evaluation is only necessary if there is no available preexisting biocompatibility data regarding materials involved in the manufacture of the product. Whilst customized medical devices are still somewhat of a novel therapeutic solution, functionally identical or similar products have long been in use and have been extremely well-characterized biocompatibility. The same can be said for material composition, where traditionally subtractively manufactured equivalents are well defined in terms of biocompatibility. Since hybrid AM process fabricated products are homogenous and typically established metal alloys, combined with animal guidelines detailed in ISO 10993-2, it can be expected that barring medical devices where material properties are completely novel with no biocompatibility data, most hybrid AM products will only require physical and chemical characterization, as well as cytotoxicity. *In vitro* cytotoxicity, one of the other more baseline biocompatibility evaluations required is established in Part 5 of ISO 10993, utilizing cell culture assays to gauge the viability of cell growth in the vicinity of the product, and serves as a catch-all gatekeeping evaluation that preliminarily judges feasibility of biological product applications. The procedures for chemical and physical characterization have been detailed in Parts 18 and 19 of ISO 10993, respectively, entailing extensive procedural and testing standards and references by which to carry out testing. Both put emphasis on the risk-based approach of the characterization, deriving required tests from the inherent risk of application and intended use. In the case of customized medical devices produced by hybrid AM processes, chemical characterization typically comprises of immersion and corrosion tests, detailed in Part 15 of ISO 10993, that seek to detect potentially hazardous metal ions, as well as extraction tests, found in Part 16 of ISO 10993, that identify organic compounds likely biologically disruptive in nature. The physical characterization will heavily depend on the material in use, as well as the intended use of the product, but generally, hybrid AM customized medical products will make use of scanning electron microscopes (SEMs) to identify product material and demonstrate equivalence with known materials currently in use [15].

Device type	Biological contact	Length of contact	ISO standard for tests required						
			ISO 10993-18	ISO 10993-19	ISO 10993-5	ISO 10993-10	ISO 10993-11	ISO 10993-6	ISO 10993-3
Implant	Tissue/bone	<24 h	×	×	×	×			
		24 h<, <30 days	×	×	×	×	o	×	o
		>30 days	×	×	×	×	o	×	o
External brace	Intact skin	<24 h	×	×	×	×			
		24 h<, <30 days	×	×	×	×			
		>30 days	×	×	×	×			

×: tests in ISO standard are recommended if no preexisting marketed equivalents exist.
o: particular tests in ISO standard are recommended if no preexisting marketed equivalents exist.

Table 2.
Biocompatibility tests required for 2 types of example medical devices.

On the flip side, if the application and/or material in use is truly novel with insufficient precedent data for biocompatibility evaluation, a wide range of costly *in vitro* and *in vivo* tests will be required to satisfy the requirements of ISO 10993. Assuming the case of orthopedic implants, in addition to previously mentioned physical, chemical, and cytotoxicity characterizations, sensitization and irritation tests (ISO 10993—Part 10), pyrogenicity and systemic, subacute, subchronic, and chronic toxicity tests (ISO 10993—Part 11), implantation tests (ISO 10993—Part 6), as well as genotoxicity and carcinogenicity testing, will be required. Fortunately, this is not a common occurrence for hybrid AM products where precedent and equivalence are the norms (**Table 2**) [15].

5. A roadmap for AM-driven customized medical device innovation in Hong Kong

As personalized medical solutions popularize across the globe, there has been a scramble from regulatory bodies in bringing patient-conforming medical devices under the scope of preexisting regulative structure. Given the uniqueness of certain anatomical features and the significant interpersonal variation that exists, by definition, these medical devices will have noticeable differences on a piece-by-piece basis. This results in increased difficulty in quality control and subsequently, regulation since these devices intrinsically are never completely identical and thus one cannot demonstrate conformance to regulatory requirements through a sample device in the present when every following iteration of the device will inherently be different, with variation often being guided by the anatomical features of patients. To this end, regulators have taken similar, risk-based approaches in incorporating these elements into their respective frameworks. The discussion will focus on the globally preeminent regulatory bodies of USFDA and the national regulatory bodies under the European Commission, as well as the locally relevant NMPA of China.

Personalized medical devices are generally split into three groups of products by regulatory bodies, based on the resemblance of their respective intended uses and manufacturing models with conventional, mass-produced medical devices that regulatory bodies devise their systems around. Here, we must part with the terminology of "Customized Medical Device" in favor of more precise language. The first group is referred to as adaptable medical devices (AMD) and are personalized medical devices that are mass manufactured as a series of compatible components and assemblies, only to be tailored to the patient's unique requirements at the point of care, by medical professionals in accordance with their medical judgment as well as device guidelines. They are essentially mass-produced products with an element of personalization in their intended use and are usually treated as such by regulatory bodies, simply following standard device classification protocols to determine regulatory requirements. In fact, the European Medical Device Regulation (EU MDR), FDA, and NMPA all do not implement additional regulatory procedures with respect to this type of device. Some AM-produced products make use of this ease of control and regulation, where the device is comprised of standardized, mass-produced parts as well as a relatively small AM-produced component that conforms to a subsection of patient anatomy that typically sees greater variation and hence requires personalization [16].

The second group is described as patient-matched medical devices (PMD) and are characterized by a largely identical manufacturing process, as well as a design envelope that encapsulates and bounds potential design features, including patient-specific features, and their potential variation. Significant portions of AM-produced medical devices will fall under this category, with products that have slightly adaptive if not harmonized intended uses, indications, contraindications, and design envelopes, that are essentially functionally identical but are designed to cater to differing individual anatomies. The classification of PMDs is an adaptation that regulatory bodies have utilized in incorporating personalized medical devices, and by proxy AM products, into a simpler regulatory framework, where these devices are generally viewed in a similar fashion as conventional medical devices but may require additional documentation, justifications, and design controls to compensate for the increased risk brought on by potential variations in design. The NMPA, EU MDR, and FDA do not have any additional regulatory requirements with respect to "Patient-Matched Medical Devices," however, the FDA has developed and issued a guidance document titled "Technical Considerations for Additive Manufactured Medical Devices" in 2017, such that whilst there is no additional explicit requirement, there is a baseline of expectation when putting forth AM products that the FDA assumes [17].

The final category is the most original and authentic to the idea of personalized medicine, that being custom-made medical devices (CMD). This refers to devices that, at the request or prescription of a medical professional, are truly made for only one particular individual and is generally targeted toward extremely rare conditions

	CMD	PMD	AMD
Intended use	Intended for use only for a particular individual (including medical professionals), to address specific feature or condition of the said target individual	Intended for use on the specific patient, with certain features matched with said patient's anatomical data, done so according to the design envelope.	Intended for general use, with device personalized during the application, according to assembly instructions
Prescription	Mandatory	Not mandatory	Not mandatory
Design responsibility	Doctor, under manufacturer consultation	Manufacturer, under doctor consultation	Manufacturer
Production	Unique, tailored	Repeated validated process, possibly in batches	Mass produced
EU MDR regulatory requirements	Statement (Annex XIII) + conformity assessment (Annex IX/Annex XI)—Class III implantable devices only	Conformity assessment (Annex IX/Annex XI)	Conformity assessment (Annex IX/Annex XI)
US FDA regulatory requirements	Custom device exemption (FD&C act section 520(b) + requirements) + QS regulation (CFR title 21)	PMA (Class III/Class II), 510(k) approval (Class I/Class II)	PMA (Class III/Class II), 510(k) approval (Class I/Class II)
CN NMPA regulatory requirements	Notification to regulatory body for initial cases undergo standard market clearance/approval as soon as possible	Standard market clearance/approval: Pre-market notification (Class I) Pre-market registration (Class II+)	Standard market clearance/approval: Pre-market notification (Class I) Pre-market registration (Class II+)

All information is highly generalized and will deviate from product to product.

Table 3.
Personalized medical device classification and regulatory requirements.

where it is unfeasible to market the device or high specific conditions where no singular adaptive design envelope could realistically cover all facets of device. As a result of this distinctive product design and intended use, the manufacturing process is often unique in its entirety. Curiously the definition extends from devices tailored toward patients to devices catering to the medical professionals treating the patient. Owing to its inherently flexible nature, AM processes are often involved in the production of these devices, foremost being custom-made orthopedic implants. To date most regulatory bodies have implemented special exemptions and requirements to allow the use of these devices, conceding that truly customized devices will have to be evaluated and accounted for outside the general regulatory system. Annex IX [18] within the EU MDR has very clear requirements for CMDs, that a specific statement shall be prepared for all CMDs expressing key information, as well as other ancillary requirements. The EU MDR also states that Class III implantable "Custom-Made Medical Devices" shall additionally be subject to the typical conformity assessment. Meanwhile, the NMPA regulates CMDs less stringently, where "Custom-Made Medical Devices" are allowed for use after notification with key information is sent and acknowledged by regulatory authorities, with the precondition that the CMDs is to undergo standard market clearance and approval as soon as clinical data and feedback following the initial utilizations allow for the registration of the CMD (**Table 3**) [19, 20].

6. Conclusions

The demand for a customized medical device is at an all-time high. Thanks to the accuracy and variety of form and function to attain intended biomechanical function with adequate biocompatibility, AM-assisted fabrication has profound advantages in clinical flexibility. This work has demonstrated with examples a framework of AM-assisted fabrication of metallic medical devices serving intended clinical needs within a suitable time frame. The AM-assisted fabrication platform established is potentially utilizable with synthesized biomaterials and pharmaceutics [21]. Opportunities are gravitating to surgeons and researchers navigating to efficacious outcomes in clinical applications.

Acknowledgements

SYM acknowledges funding support from sponsored research RS 200263 at the University of Hong Kong.

Author details

Ching Hang Bob Yung[1], Lung Fung Tse[1], Wing Fung Edmond Yau[1] and Sze Yi Mak[1,2*]

1 Koln 3D Technology (Medical) Limited, Hong Kong

2 The University of Hong Kong, Hong Kong

*Address all correspondence to: smak@hku.hk

References

[1] Whitcomb DC. What is personalized medicine and what should it replace? Nature Reviews Gastroenterology & Hepatology. 2012;**9**(7):418-424

[2] Dzau VJ, Ginsburg GS, Van Nuys K, Agus D, Goldman D. Aligning incentives to fulfil the promise of personalised medicine. Lancet. 2015;**385**(9982): 2118-2119

[3] Hamburg MA, Collins FS. The path to personalized medicine. New England Journal of Medicine. 2010;**363**(4):301-304

[4] Bates S. Progress towards personalized medicine. Drug Discovery Today. 2010;**15**(3):115-120

[5] Schork NJ. Personalized medicine: Time for one-person trials. Nature. 2015;**520**(7549):609-611

[6] Chim YN, Chow SKH, Mak SY, Li MMC, Yung BCH, Yau EWF, et al. 3D-printed cobalt-chromium porous metal implants showed enhanced bone-implant interface and bone in-growth in a rabbit epiphyseal bone defect model. Bone Reports. 2020;**13**:100375

[7] Li C, Pisignano D, Zhao Y, Xue J. Advances in medical applications of additive manufacturing. Engineering. 2020;**6**(11):1222-1231

[8] Javaid M, Haleem A. Additive manufacturing applications in medical cases: A literature based review. Alexandria Journal of Medicine. 2018;**54**(4):411-422

[9] Tuomi J, Paloheimo K-S, Vehviläinen J, Björkstrand R, Salmi M, Huotilainen E, et al. A novel classification and online platform for planning and documentation of medical applications of additive manufacturing. Surgical Innovation. 2014;**21**(6):553-559

[10] Lau LCM, Chui ECS, Fan JCH, Man GCW, Hung YW, Ho KKW, et al. Patient-specific instrumentation (PSI) referencing high tibial osteotomy technological transfer and education: Protocol for a double-blind, randomised controlled trial (Protected HTO trial). BMJ Open. 2021;**11**(2):e041129

[11] Mak SY, Tam KL, Yung CHB, Yau WFE. Hybrid metal 3D printing for selective polished surface. Materials Science Forum. 2021;**1027**:136-140

[12] Yau EWF, Szeto WH, Mak SY, FSY W, Chuah KB, To CP, editors. Design and development of a tri robot machining cell for medical product manufacture. In: 2021 5th International Conference on Robotics and Automation Sciences (ICRAS); 11-13 June 2021. Beijing, China: IEEE; 2021

[13] Szeto WH, Wong FSY, Yau EWF, Mak SY, Chuah KB. Kinematic modelling of a tri robot machining cell. In: 2021 6th International Conference on Control and Robotics Engineering (ICCRE); 16-8 April 2021. Wuhan, China: IEEE; 2021

[14] Wong KC, Kumta SM, Geel NV, Demol J. One-step reconstruction with a 3D-printed, biomechanically evaluated custom implant after complex pelvic tumor resection. Computer Aided Surgery. 2015;**20**(1):14-23

[15] International Organization for Standardization. ISO 10993-1:2018. Biological Evaluation of Medical Devices—Part 1: Evaluation and Testing within a Risk Amangement Process. Geneva: ISO; 2018

[16] IMDRF Personalized Medical Devices. Personalized Medical Devices—Regulatory Pathways [Internet]. 2020. Available from: http://www.imdrf.org/docs/imdrf/final/technical/imdrf-tech-200318-pmd-rp-n58.pdf

[17] U.S. Department of Health and Human Services, Food and Drug Administration, Center for Devices and Radiological Health, Center for Biologics Evaluation and Research. Technical Considerations for Additive Manufactured Medical Devices. 2017. Available from: https://www.fda.gov/regulatory-information/search-fda-guidance-documents/technical-considerations-additive-manufactured-medical-devices

[18] Council Regulation (EC) 2017/745 of 5 April 2017 on medical devices, amending Directive 2001/83/EC, Regulation (EC) No 178/2002 and Regulation (EC) No 1223/2009 and repealing Council Directives 90/385/EEC and 93/42/EEC

[19] National Medical Products Administration, National Health Commission of the People's Republic of China. (定制式医疗器械监督管理规定试行). National Health Commission of the People's Republic of China; 2019. Available from: https://www.nmpa.gov.cn/directory/web/nmpa/xxgk/ggtg/qtggtg/20190704160701585.html

[20] State Administration for Market Regulation. 医疗器械注册与备案管理办法. State Administration for Market Regulation; 2021. Available from: http://gkml.samr.gov.cn/nsjg/fgs/202108/t20210831_334228.html

[21] Guzzi EA, Tibbitt MW. Additive manufacturing of precision biomaterials. Advanced Materials. 2020;**32**(13):1901994

Chapter 3

Perspective Chapter: Advanced Manufacturing for Bone Tissue Engineering and Regenerative Medicine

Roozbeh (Ross) Salary

Abstract

This book chapter delineates advanced additive manufacturing processes used in clinical practice for high-resolution fabrication of mechanically-robust and dimensionally-accurate bone tissue scaffolds with a focus on pneumatic micro-extrusion, fused deposition modeling, polymer jet printing, and digital light processing. The main components as well as the underlying physics behind each process are explained. Furthermore, this chapter is integrated with a review of literature; the aim is to show how these additive manufacturing processes are potentially utilized in clinical practice for bone tissue engineering. This chapter serves as an introductory platform toward advanced studies and/or research works in the area of bone regenerative medicine. Finally, this chapter will be helpful to engineering and medical students as well as researchers from academia and industry.

Keywords: bone tissue engineering, regenerative medicine, advanced manufacturing

1. Introduction

1.1 Objective and scope

The objective of this chapter is to introduce AM processes, which have been utilized in clinical practice for the fabrication of mechanically-robust and dimensionally-accurate bone tissue scaffolds for the treatment of osseous fractures, defects, and diseases (such as osteoporosis, bone tumor resection, and orthopedic trauma). Particularly, this chapter concentrates on PME, FDM, PJP, as well as DLP with a special focus on the PME-AM process. **Figure 1** illustrates additively-manufactured biocompatible bone tissue scaffolds and constructs having porous internal structures. Please note that the phantom as well as the femur bone were composed of a medical-grade composite material [1, 2], while the cubic microporous scaffold was composed of PCL [3–5].

The aforementioned AM processes enable high-resolution, non-contact, and multi-material deposition of functional bio-inks, polymer materials, as well as composite materials for tissue engineering applications. In spite of their benefits and potential applications, the AM processes are intrinsically complex. The process complexity, to a great extent, stems from not only complex physical phenomena (such as phase change and non-Newtonian material deposition), but also dynamic material-process interactions. In addition, there are a broad range of design factors and process

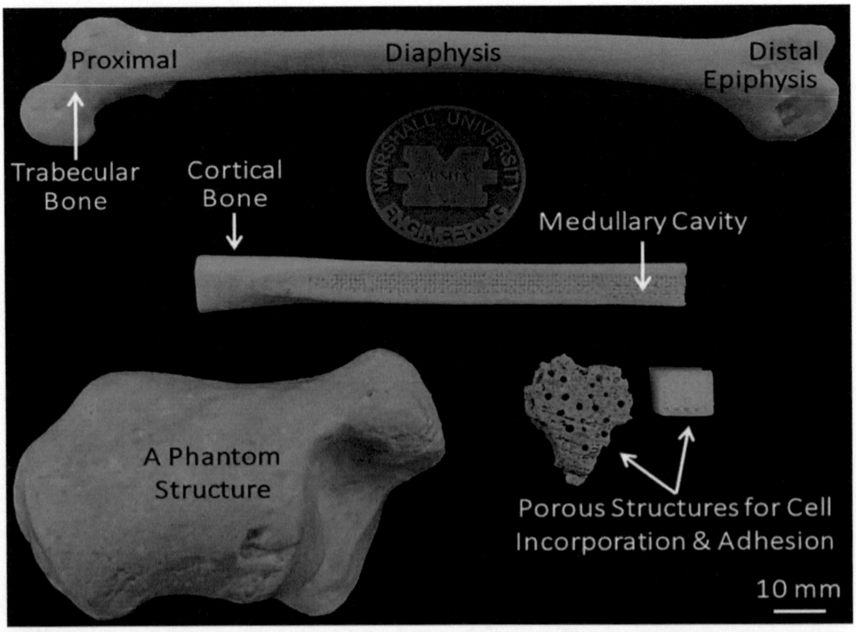

Figure 1.
Biocompatible bone scaffolds and constructs with microporous internal structures, which allow for not only cell incorporation and adhesion, but also diffuse proliferation for clinical practice (Source: [1–5]).

parameters (such as porosity, surface roughness, scaffold topology, nozzle diameter, material viscosity, as well as material deposition pressure, flow rate, and temperature) that contribute to the complexity of the AM processes. Consequently, investigation of the influence of the significant design and process parameters (in addition to their interactions) on the mechanical, biomedical, and morphological properties of the fabricated bone structures would be inevitable [1–10]. In the absence of such knowledge, orthopedic surgeons and clinicians will be unable to efficiently treat osseous fractures in the presence of constraints, such as sex, age, bone density, and immune system rejection. Hence, the aim of this chapter is to introduce the AM processes and highlight their significant process parameters. The PME, FDM, PJP, and DLP processes will be reviewed in Sections 2.1–2.4, respectively. A review of other AM processes (i.e., powder bed fusion as well as binder jetting) used in clinical practice is given in Section 2.5. Finally, the conclusions are presented in Section 3.

2. AM processes for bone tissue engineering

2.1 Pneumatic micro-extrusion (PME)

PME is a material extrusion AM process [11], which has emerged as a robust high-resolution method for the fabrication of a wide range of biological tissues, scaffolds, and structures. Advanced PME systems, for example, Cellink BIO X (Boston, MA, USA), have a layer resolution and positioning precision of 100 and 10 μm, respectively.

As demonstrated in **Figure 2**, the PME process utilizes a high-pressure gas flow (typically air, supplied by a compressor) as a medium of transport and deposition. A polymer material (typically in powder form) is loaded into the deposition head's cartridge (also known as barrel) and subsequently heated above the polymer's melting temperature; this results in formation of a non-Newtonian molten

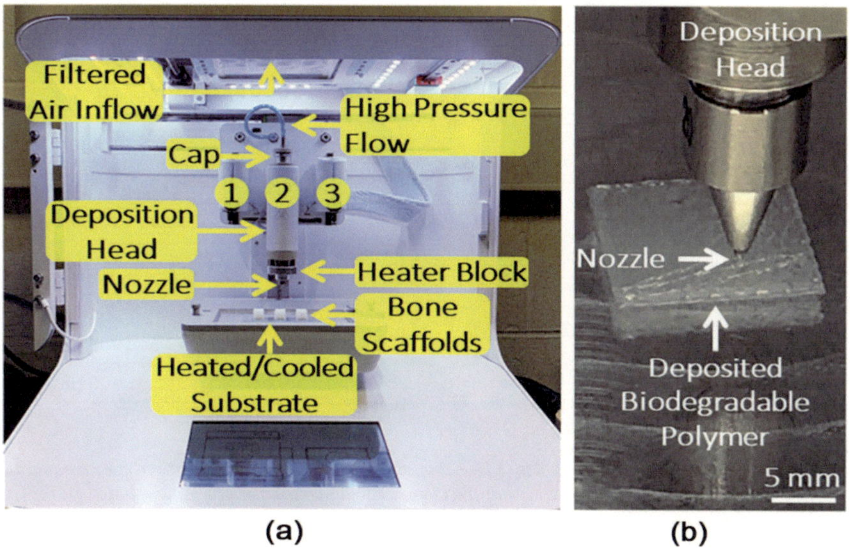

Figure 2.
(a) The material deposition chamber as well as the main components of the PME AM process; and (b) pneumatic micro-deposition of a polymer material (PCL) on a heated glass substrate using a converging microcapillary nozzle (Source: [3–5]).

polymer flow prior to deposition [3–5]. An internal/external air compressor or a pressure source (not shown), provides a steady pressure flow into the cartridge. Having high thermal conductivity, the cartridge allows for rapid melting of the loaded polymer material. The molten polymer is, subsequently, deposited on a heated/cooled free surface via a converging microcapillary nozzle with the aid of the pressurized gas flow. A fan with a filter at the top of the chamber not only delivers a clean air flow, but also aids in maintaining a fixed level (rate) of polymer solidification (and thus layer adhesion). The surface temperature is kept below the melting temperature of the polymer material. Spoerk et al. [12] observed that an optimal bed temperature would be critical for proper layer adhesion and thus accurate material deposition.

Klemstine et al. [3] investigated the mechanical properties of biocompatible and biodegradable triply periodic minimal surface-based bone scaffolds, composed of PCL and fabricated using the PME-AM process. Having a molecular weight (Mn) as well as a density of approximately 50,000 and 1.145 g/mL (at 25°C), respectively, PCL is a semi-crystalline, hydrophobic polyester-based polymer, derived from caprolactone monomer using ring-opening polymerization. It has a glass transition temperature of −60°C and a melting temperature in the range of 59–64°C. In addition, PCL has a tensile strength and elasticity modulus (indicative of stiffness) of 16 MPa and 0.4 GPa, respectively [13]. The PME fabrication of the bone scaffolds was on the basis of a set of optimal process parameters, detailed in **Table 1**.

In a research work, Yu et al. [4] investigated the effects of influential scaffold design factors and process parameters—as listed in **Table 2**, including layer thickness, layer width, infill density, print speed, flow pressure, deposition head temperature, and infill pattern—on the dimensional accuracy as well as the mechanical properties of PME-fabricated PCL bone scaffolds. The assessment of the scaffold dimensional accuracy was based on not only a digital image processing platform established in the MATLAB environment, but also physical measurements (used to corroborate the veracity of the image-based assessment). The underlying algorithms embedded in the image-processing platform are discussed in detail in [14, 15].

Parameter	Type	Level [unit]
Layer height (thickness)	Design	200 [μm]
Infill pattern	Design	Concentric
Nozzle size	Machine	200 [μm]
Bed temperature	Machine	10 [°C]
Print speed	Machine	2.5 [mm/s]
Deposition head temperature	Machine	180 [°C]
Deposition flow pressure	Machine	300 [kPa]
Pre-flow delay	Machine	900 [ms]
Post-flow delay	Machine	400 [ms]

Table 1.
The optimal PME process parameters used by Klemstine et al. for the additive fabrication of biocompatible bone scaffolds composed of PCL. Please note that the bone scaffold fabrication process was based on the Cellink BIO X 3D-bioprinting system (Source: [3]).

Parameter	Type	Level [unit]
Variables		
Layer height/thickness	Design	125–200 [μm]
Layer/line width	Design	125–200 [μm]
Print speed	Machine	0.30–0.45 [mm/s]
Infill density	Design	0.20–0.35
Deposition head temperature (DHT)	Machine	100–125 [°C]
Flow pressure	Machine	520–560 [kPa]
Infill pattern	Design	• Honeycomb; • Rectilinear; • Concentric; • Cubic; • Gyroid.
Fixed parameters		
Filling/raster angle	Design	90 [°]
Number of shells	Design	2
Scaffold diameter	Design	10 [mm]
Scaffold height	Design	3 [mm]
Nozzle size	Machine	200 [μm]
Bed temperature	Machine	45 [°C]
Fan speed	Machine	100 [%]

Table 2.
The experimental design table established by Yu et al. to systematically study the influence of seven design and process parameters on the functional properties of PCL bone scaffolds fabricated using the PME process. Please note that the bone scaffold fabrication process was based on the Cellink INKREDIBLE^{+} *3D-bioprinting system (Source: [4]).*

Please note that the image-processing platform, in addition, allows for *in situ* characterization, monitoring, and ultimately control of the PME process.

Figure 3 illustrates the influence of the deposition head temperature (as an example of the PME process parameters) on the morphology as well as the microstructure of the fabricated PCL bone scaffolds. It is implied from the figure that the scaffold diameter increases as a result of an increase in the deposition head

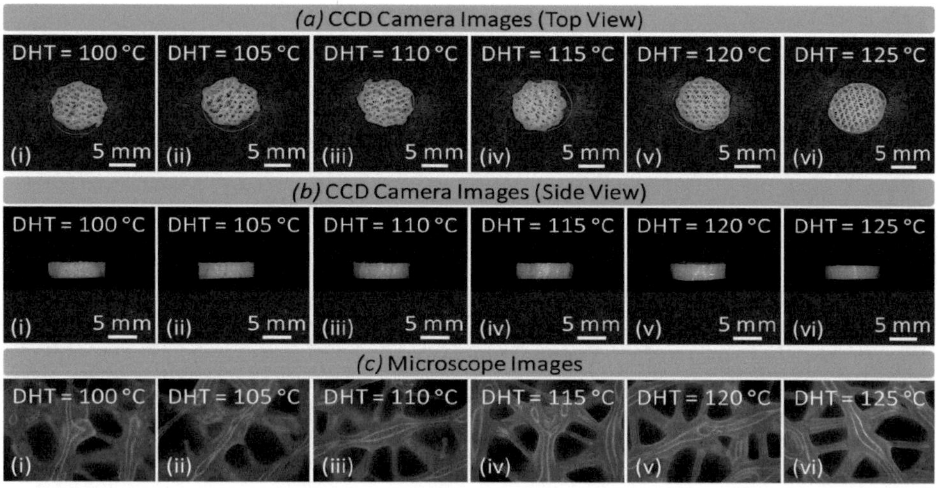

Figure 3.
The influence of deposition head temperature (DHT) at six levels of 100, 105, 110, 115, 120, and 125°C on: (a, b) the morphology, and (c) the microstructure of PME-fabricated PCL bone scaffolds (Source: [4]).

temperature. This phenomenon, largely, stems from the fact that an increase in the deposition head temperature leads to a decrease in the polymer viscosity and consequently, an increase in material deposition rate (resulting in formation of larger bone scaffolds in diameter). Besides, Yu et al. [4] observed that the scaffold stiffness increased when the deposition head temperature increased from 100 to 125 [°C]; this trend can be due to the increased amount of deposited mass per scaffold structure as a result of a decline in the polymer viscosity (when the deposition head temperature increases).

Yeow et al. [4, 8] developed a 3D computational fluid dynamics (CFD) model with the aim to investigate the underlying non-Newtonian fluid dynamics of material transport and deposition in the PME process, formulated as a transient multi-phase flow problem. Demonstrated in **Figure 4(a)**, the geometry of the CFD model consisted of a cartridge, a connector, as well as a microcapillary nozzle (having a diameter of 200 μm). It turned out that approximately five layers of inflation would be sufficient to obtain accurate solution near the wall boundaries. Shown in **Figure 4(b)**, four boundary conditions were defined for the CFD model, including flow pressure inlet, stationery wall, volume fraction, and pressure outlet. Both the energy equation and the laminar viscous model were included in the CFD model (in addition to the continuity and momentum equations) respectively to account for the effects of viscous heating and to capture the effects of viscosity.

It was observed that the transport of molten PCL through the micro-capillary nozzle (under a flow pressure of 550 kPa) would be a viscous flow having a Reynolds number (Re) of $\ll 1$, implied from **Figure 4(c)**. This PME material deposition regime is unlike that of the other additive manufacturing processes, such as AJP, where material deposition is intrinsically turbulent [14, 16, 17].

The following research works exemplify the use of PME in clinical practice. In a research work, Du et al. [18] demonstrated additive fabrication of mesoporous bioactive glass/silk fibroin composite scaffolds with high osteogenic ability using the PME process. The functional properties of the fabricated scaffolds were characterized on the basis of porosity, compressive strength, degradation, biocompatibility, as well as apatite forming ability. The results of an animal study showed that the mesoporous bioactive glass/silk fibroin composite scaffolds (loaded with human bone marrow mesenchymal stem cells) had not only more significant osteogenic

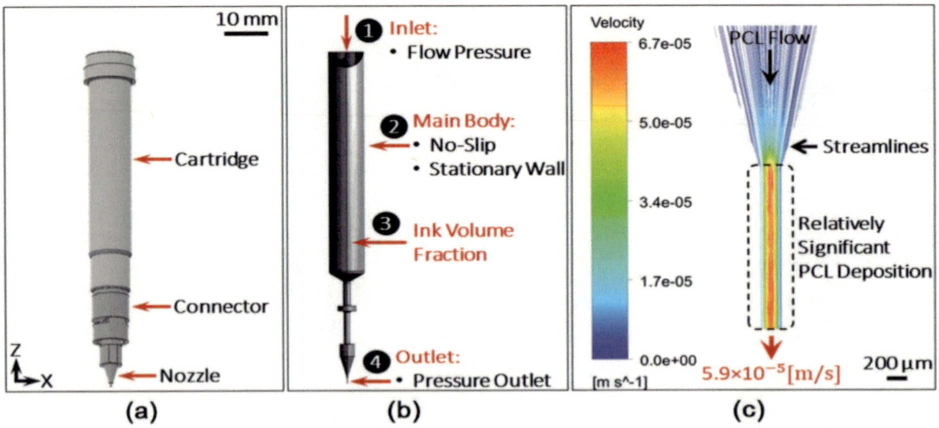

Figure 4.
(a) The main components of the PME deposition head assembly; (b) the boundary conditions defined for the 3D-CFD model; and (c) simulation of the velocity field in the PME deposition head under a flow pressure of 550 kPa (Source: [4, 8]).

potential, but also superior compressive strength and biocompatibility than mesoporous bioactive glass/polycaprolactone (PCL) scaffolds.

In addition, Du et al. [19] demonstrated 3D-fabrication of pearl/calcium sulfate composite scaffolds—characterized with high osteogenic ability, uniform interconnected macropores, high porosity, as well as improved mechanical properties—using PME integrated with a hydration process. The fabricated scaffolds not only showed satisfactory apatite-forming ability, but also stimulated the proliferation as well as differentiation of rat bone mesenchymal stem cells. The osteogenic potential of the scaffolds was assessed based on micro-computed tomography (μ-CT) imaging and histological analysis.

Park et al. [20] demonstrate a novel technique, combining 3D printing with spatial-temporal deposition and control of growth factors, with the aim to prevascularize bone tissues. Having osteogenic and vasculogenic potential, human dental pulp stem cells were deposited using the PME process together with bone morphogenetic protein-2 as well as vascular endothelial growth factor. An animal study was conducted where both micro-vessel formation and angiogenesis were observed.

Cui et al. [21] investigated 3D-fabrication of polyion complex hydrogel-based scaffolds incorporating multi-walled carbon nanotubes using PME for bone regeneration. The hypothesis of the work was that the addition of multi-walled carbon nanotubes would enhance bone repair efficiency. It was observed that the fabricated scaffolds not only were biocompatible with rat bone marrow-derived mesenchymal stem cells, but also had a high degree of osteogenic differentiation, mineralized matrix formation, and osteogenesis upregulation.

2.2 Fused deposition modeling (FDM)

Similar to PME, FDM is a material extrusion AM process, which has been extensively utilized in tissue engineering applications. As illustrated in **Figure 5**, in the FDM process, a polymer/composite material, typically in the form of a filament, is fed into a heat block (maintained at a temperature above the melting temperature of the polymer material) with the aid of a stepper motor; this leads to formation of a molten polymer flow. A non-Newtonian fluid, the molten polymer is, subsequently, passed through a converging microcapillary nozzle (made up of, e.g., hardened steel or brass) in order to increase the linear momentum of the molten polymer flow

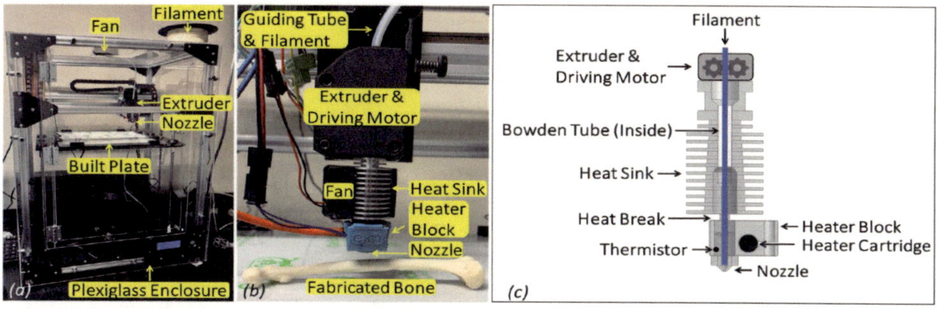

Figure 5.
An FDM 3D-printing system, equipped with a Plexiglas enclosure and a fan installed to create a thermally uniform environment for steady-state material deposition. (a, b) Real pictures, and (c) a schematic diagram of the FDM deposition head assembly (Source: [1, 2, 22]).

prior to deposition on a heated or unheated free surface. Finally, the built plate (also known as platen) is automatically translated downward (controlled as a function of layer height), and the next layer is deposited on top of the previous layer. In fact, the layer height controls the amount of overlap between two subsequent deposited vertical layers. Please note that layer-to-layer bonding is influenced by not only the extrusion temperature, but also the layer height.

Chaffins et al. [1] investigated the mechanical properties of FDM-fabricated medical-grade bone scaffolds composed of a biocompatible composite material (with low moisture absorption) containing polyamide, polyolefin, and cellulose fibers. The composite material has an elongation at break, elasticity modulus (stiffness), and ultimate strength of 4%, 850 MPa, and 23 MPa, respectively. The FDM fabrication of the bone scaffolds was on the basis of a set of optimal process parameters, detailed in **Table 3**. An equilibration time of 3 hours was taken with the aim to ensure thermal equilibrium (regulated by the chamber fan) and as a result, steady-state material deposition in the FDM process. Please note that the fan speed affects the rate of material solidification after deposition. With the aid of a slicer software program, that is, Cura (Ultimaker, Utrecht, the Netherlands), the 3D CAD models of the bone scaffolds were converted into a G-code, and consequently a tool-path was created prior to fabrication.

The following research works exemplify the use of FDM in clinical practice. In a research work by Lai et al. [23], porous bone scaffolds with biomimetic structure were fabricated, based on a novel composite material composed of magnesium, PLGA, and β-tricalcium phosphate for the treatment of bone defects. A steroid-associated osteonecrosis rabbit model was established to assess the biosafety as well as the osteogenic and angiogenic properties of the fabricated scaffolds. It turned out that the scaffolds led to an increase in blood perfusion in addition to vessel ingrowth after surgery (approximately in 4–8 weeks), observed respectively with the aid of dynamic contrast-enhanced magnetic resonance imaging and micro-computed tomography (μ-CT)-based angiography. Furthermore, the fabricated scaffolds led to significant bone formation with enhanced functional properties.

Deng et al. [24] demonstrated additive fabrication of bi-lineage constructive scaffolds (composed of manganese-doped β-tricalcium phosphate) for bone regeneration. The physicochemical properties and bi-lineage bioactivity of the fabricated scaffolds as well as the mechanism of stimulating osteochondral regeneration were characterized. It was observed that the addition of manganese to β-tricalcium phosphate not only reduced the lattice parameters and

Parameter	Type	Level [unit]
Scaffold porosity (CAD-based)	Design	60 [%]
Number of shells	Design	1
Scaffold dimensions	Design	15 × 15 × 15 [mm]
Layer height (thickness)	Design	200 [μm]
Layer (line) width	Design	200 [μm]
Infill density	Design	100 [%]
Nozzle size	Machine	400 [μm]
Bed temperature	Machine	95 [°C]
Fan speed	Machine	83 [%] [10 Volts]
Print speed	Machine	15 [mm/s]
Deposition head temperature	Machine	235 [°C]
Flow (feed) rate	Machine	100 [%]
Steady state chamber temperature	Machine	37 [°C]
Build plate adhesion type	Machine	Brim

Table 3.
The optimal FDM process parameters used by Chaffins et al. for the additive fabrication of bone scaffolds composed of a biocompatible composite material. Please note that the bone scaffold fabrication process was based on the $FT5-R_2$ *3D-printing system (Folger Tech, Milford, NH, USA) (Source: [1]).*

crystallization temperature, but also enhanced the density and the compressive strength of the fabricated scaffolds. The results of an animal study, in addition, showed that the ionic products from manganese-doped β-tricalcium phosphate improved the proliferation and promoted the differentiation of chondrocytes and rabbit mesenchymal stem cells. Furthermore, the results showed that the fabricated scaffolds significantly improved the regeneration of subchondral bone tissues, transplanted *in vivo*.

Hassanajili et al. [25] demonstrated characterization of polylactic-acid/polycaprolactone/hydroxyapatite composite scaffolds, fabricated using a combined fabrication process where material extrusion was utilized for the fabrication of a negative mold (composed of poly(vinyl alcohol), soluble in water) for casting integrated with freeze drying/particle leaching method. Liquid replacement technique was utilized to measure scaffold porosity. Cell adhesion, scaffold cytotoxicity, cell viability, and mineral deposition (indicative of osteoinductive capacity) together with modulus of elasticity, porosity, and pore size were measured to characterize the functional properties of the fabricated bone scaffolds.

Oladapo et al. [26] investigated the functional characteristics of biomimetic bone scaffolds, composed of poly lactic acid (PLA) matrix reinforced with carbohydrate particles and fabricated using the FDM process. In fact, the presence of carbohydrate particles allows for ion or ionic group substitutions and enhances the kinetics of absorption and ultimately, the mechanical properties of fabricated scaffolds. The bioactivity, surface roughness, apparent porosity, as well as mechanical properties of the fabricated scaffolds were analyzed. It was observed that there was a significant, proportional relationship between the carbohydrate content and surface roughness. In addition, the presence of carbohydrate particles led to a decline in scaffold stiffness and compressive strength (when compared with pure PLA).

2.3 Polymer jet printing (PJP)

Demonstrated in **Figure 6**, the PJP process operates on the basis of simultaneous deposition (jetting) of build as well as support materials (composed of liquid photopolymers) on a free surface [7]. Both build and support materials are radiation-curable. Having a resolution of, for example, 600 and 1200 dots per inch (DIP), inkjet heads are utilized for the deposition of liquid photopolymers onto a build platform. Subsequently, the deposited photopolymers are immediately cured *in situ* using a UV light source; this mechanism allows for fabrication of layers on top of each other. The photopolymer materials undergo a chemical transformation and become solid upon irradiation of the UV light [11].

Weese et al. [7] investigated the effects of four influential PJP process parameters, detailed in **Table 4**, on the mechanical properties of fabricated femur bone structures. The PJP process parameters include: (i) *print direction*, controlling the sequence of material deposition and layer formation; (ii) *resolution factor*, indicative of the resolution of material deposition; (iii) *UV light intensity*, controlling the intensity of the UV light source; and (iv) *deposition head temperature*.

The additive fabrication of the femur bone structures was based on a PJP 3D-printing system (Objet30 Pro, Stratasys Ltd., Eden Prairie, MN, USA), having an accuracy of 100 μm. The 3D-printing system, in addition, allows for deposition of materials with a layer thickness of as small as 28 μm. A slicer software program, that is, Objet Studio (Stratasys Ltd., Eden Prairie, MN, USA), was similarly used to convent 3D CAD models into a G-code and create a tool-path. The build photopolymer material used for the fabrication of the femur bone structures had a polymerized density, water absorption, and glass transition temperature (T_g) of 1.17–1.18 [g/cm^3], 1–1.5%, and 53°C, respectively. Furthermore, the build material had a

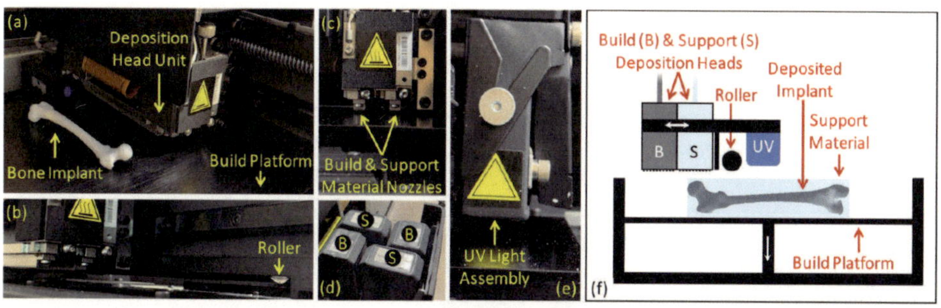

Figure 6.
(a-e) Real pictures of the main components of the PJP process including: (a) the deposition head assembly; (b) the roller; (c) the material deposition nozzles; (d) the build and support material cartridges shown by the letters B and S, respectively; and (e) the UV light assembly. (f) A schematic diagram of the PJP process. Please note that the deposition head assembly includes build as well as support material nozzles (Source: [7]).

Parameter	Type	Level [unit]
Print direction	Machine	[Unidirectional, bidirectional]
Resolution factor	Machine	[0.25, 2.00]
UV light intensity factor	Machine	[0.34, 1.35]
Deposition head temperature factor	Machine	[0.65, 1.95]

Table 4.
The PJP process parameters used by Weese et al. for the additive fabrication of femur bone structures (Source: [7]).

stiffness as well as a tensile strength of 2–3 GPa and 50–65 MPa, respectively. Unlike the build material, the support material was soluble in a solution of 2% sodium hydroxide (NaOH) and 1% sodium metasilicate (Na_2SiO_3).

The following research work exemplifies the use of PJP in clinical practice. In a research work by Libonati et al. [27], synthetic composite structures (composed of acrylic-based photopolymers and having a pattern inspired by the microstructure of cortical bone) were designed and characterized. The composite structures were fabricated using the PJP process, and their functional performance in terms of fracture behavior was characterized by mechanical testing. It was observed that the cortical bone-inspired design would potentially enhance toughness amplification and would be essential for balance with material strength. In addition, the PJ-fabricated composite structures showed similar cortical bone-related failure mechanisms, including crack deflection, crack branching, constrained microcracking, as well as fibril bridging.

2.4 Digital light processing (DLP)

DLP is a vat-photopolymerization AM process. It is, to some extent, similar to the PJP process, where radiation-curable resins become a solid upon exposure to UV light through a process called *photopolymerization*. **Figure 7** illustrates the main components of the DLP process. Once the first layer has been cured, the build platform is translated automatically upward, and the next later is cured on top of the previous cured later. The DLP process has been utilized for the high-resolution fabrication of 3D structures with complex internal geometries such as bone and dental implants. In a research work by Raines et al. [28], dental implants with complex internal structures were fabricated using the DLP process. The fabricated dental implants were composed of a biocompatible resin. The DLP process consists of several design and process parameters, as listed in **Table 5**. 3D-Sprint (3D Systems, Rock Hill, SC, USA) was the slicer software program of choice. Finally, the DLP 3D-printing system used in their work was FabPro 1000 (3D Systems, Rock Hill, SC, USA).

2.5 A review of other AM processes used in clinical practice

2.5.1 Powder bed fusion

Pei et al. [29] demonstrated an integrated method for the biomechanical design as well as fabrication of bionic bone tissue implants using SLS additive

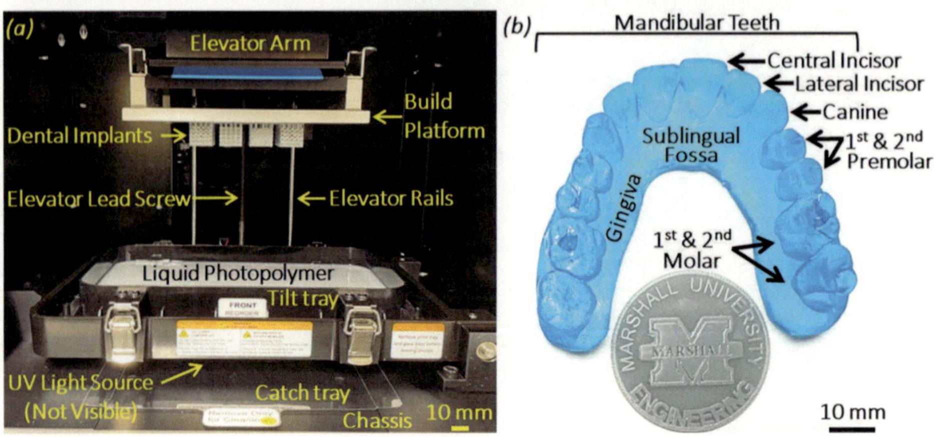

Figure 7.
(a) The main components of the DLP additive manufacturing process. (b) An X-ray microCT-based, biocompatible dental jaw model, fabricated using the DLP-AM process (Source: [28]).

Parameter	Type	Level [unit]
Scaffold porosity	Design	60 [%]
Number of shells	Design	1
Scaffold dimensions	Design	15 × 15 × 15 [mm]
Layer thickness	Design	100 [μm]
Cure depth	Machine	115 [μm]
Print resolution	Machine	454 × 454 dpi
Layer resolution	Machine	100 [μm]

Table 5.
The DLP process parameters used by Raines et al. for the fabrication of dental structures, composed of a biocompatible liquid photopolymer (Source: [28]).

manufacturing process with the aim to repair the femoral head. Composed of titanium, the fabricated implants were based on interconnected diamond-lattice pore units that would prevent stress shielding. FEA was utilized for design optimization and numerical characterization of the mechanical properties of the bone implants. On the basis of an animal study, it was observed that pore-unit parameters would significantly influence implant porosity, pore size distribution, and mechanical strength.

Zhang et al. [30] designed and fabricated porous scaffolds (composed of Ti6Al4V alloy and having diamond-lattice pore units with customized shape and tunable mechanical properties) using SLS process for bone tissue regeneration and ultimately femoral-head repair. The weak points within the structure of the scaffolds were analyzed using a FEA model. Struts diameters, pore size, as well as porosity were identified as critical implant design parameters. The biocompatibility and osteogenic potential of the fabricated scaffolds were assessed *in vivo* based on an animal study.

Similarly, Zhao et al. [31] investigated the mechanical properties of porous titanium alloy scaffolds fabricated using SLS process for bone tissue reconstruction. Various scaffolds (including Diamond, Gyroid, Orthogonal, Truss, and Cube) were designed based on parameterization modeling. The mechanical characteristics of the designs as well as the SLS-fabricated scaffolds were assessed numerically using FEA and experimentally using mechanical testing, respectively. It was observed that the fabricated scaffolds were dimensionally accurate characterized with an error of <3% (when comparted with their reference CAD models).

2.5.2 Binder jetting

In a research work by Inzana et al. [32], composite bone scaffolds (composed of calcium phosphate and collagen) were fabricated using binder jet additive manufacturing process. Collagen was dissolved into a phosphoric acid-based binder solution to synthesize a collagen-incorporated calcium phosphate composite. The addition of collagen, to a great extent, improved the flexural strength of the fabricated scaffolds as well as cell viability. In addition, it was observed that the properties of the binder solution would play a significant role in the cytocompatibility, osteoconductivity, and mechanical strength of the fabricated bone scaffolds. Furthermore, to achieve reliable material deposition, the viscosity and surface tension of the collagen solution were reduced with the aid of physiologic heat treatment and Tween 80, respectively. To assess the healing potential, the fabricated scaffolds were implanted into a critically-sized murine femoral defect for 9 weeks; it tuned out that the implants not only were osteoconductive, but also led to new bone growth.

Zhou et al. [33] investigated the characteristics of synthesized composite materials composed of calcium phosphate powders (based on hydroxyapatite and β-tricalcium phosphate) as well as calcium sulfate powders for the fabrication of bone scaffolds using binder jet printing process (with a water-based binder). The printability of the composite materials was assessed in terms of powder bed packing, binder deposition regimes, particle size, calcium phosphate:calcium sulfate ratio, powder-binder wettability, and the strength of the fabricated scaffolds. It was observed that the compressive strength of the scaffolds increased as the ratio of calcium phosphate:calcium sulfate increased. In addition, the use of fine powders (i.e., ≤ 20 μm in diameter) led to slow binder penetration rate, large penetration depth, low wetting ratio, and insignificant green-state strength. Besides, the beta tricalcium phosphate-based powders had lower wetting ratio as well as green-state strength than the hydroxyapatite-based powders.

Bergmann et al. [34] demonstrated additive fabrication of customized bone substitute implants composed of bioresorbable β-tricalcium phosphate as well as bioresorbable and adjustable bioactive glass fabricated using binder jet printing process for swift maxillofacial or craniofacial defect repair. The fabricated scaffolds had a bending strength of approximately 15 MPa. In addition, the results of an X -ray powder diffraction (XRD) analysis revealed the presence of both calcium sodium phosphate and calcium silicate (that are biocompatible and biodegradable).

Similarly, Cox et al. [35] presented characterization of porous bone scaffolds, fabricated using binder jet printing process and composed of hydroxyapatite as well as poly(vinyl)alcohol (PVOH) composite powders (bound with the aid of a water-based binding agent). It was observed that scaffold mechanical stability, microstructure, and porosity would be significantly affected by the flowability of hydroxyapatite:PVOH precursor materials. The fabricated scaffolds were anisotropic, and failure at the boundaries of interlayer bonds was identified as the primary failure mechanism influencing the scaffolds' functional performance. Furthermore, it turned out that *in vivo* osteoconduction and osteointegration would be significantly enhanced by factors, such as scaffold porosity and interconnectivity in addition to powder bed packing and surface roughness.

3. Conclusions

In this chapter, advanced AM processes were analyzed with a focus on PME, FDM, PJP, as well as DLP. In addition, each AM process was characterized in terms of system components, underlying physical phenomena, and influential parameters. The AM processes are inherently complex, despite their benefits and engendered potential applications. The process complexity, to a great extent, stems from complex physical phenomena (such as non-Newtonian material deposition and phase change) as well as nonlinear material-process interactions. Furthermore, the 3D fabrication of bone tissue scaffolds and implants consists of a broad spectrum of design and process parameters, for example, scaffold porosity, material viscosity, and UV light intensity. Hence, investigation of the effects of the process parameters integrated with physics-based process characterization using computational methods, such as CFD and FEA, would be an inevitable need toward obtaining optimal material transport and deposition regimes for the fabrication of bone tissues with tunable medical properties.

Acknowledgements

This work was made possible by the NASA Established Program to Stimulate Competitive Research (EPSCoR), Grant # 80NSSC19M0054. Dr. Salary would like to sincerely thank the NASA West Virginia Space Grant Consortium (Morgantown, WV, USA). In addition, Dr. Salary would like to sincerely acknowledge the Marshall University Research Corporation (MURC) for supporting this work via the John Marshall Scholars Faculty Award. Furthermore, Dr. Salary would like to sincerely thank the West Virginia State's Higher Education Policy Commission and the Community and Technical College System for the Open Education Resources (OER) Grant.

Dr. Salary also would like to thank the College of Engineering & Computer Sciences (CECS) at Marshall University, the Cabell Huntington Hospital and Medical Center (Huntington, WV, USA), as well as FibreTuff Biotechnology Company (Toledo, OH, USA).

Nomenclatures

AJP	aerosol jet printing
AM	additive manufacturing
CAD	computer-aided design
CFD	computational fluid dynamics
DHT	deposition head temperature
DLP	digital light processing
FDM	fused deposition modeling
FEA	finite element analysis
PCL	polycaprolactone
PJP	polymer jet printing
PME	pneumatic micro-extrusion
SLS	selective laser sintering

Author details

Roozbeh (Ross) Salary
Departments of Mechanical and Biomedical Engineering, College of Engineering and Computer Sciences, Marshall University, Huntington, WV, USA

*Address all correspondence to: salary@marshall.edu

References

[1] Chaffins A, Yu M, Claudio PP, Day JB, Salary RR. Investigation of the functional properties of additively-fabricated triply periodic minimal surface-based bone scaffolds for the treatment of osseous fractures. In: Proc. Paper No. 2004, ASME 2021 International Manufacturing Science and Engineering Conference (MSEC 2021), Virtual Conference (Due to the COVID-19 Pandemic); 21-25 June 2021; University of Cincinnati, Cincinnati, OH, USA. American Society of Mechanical Engineers

[2] Zhao D, Hart C, Weese NA, Rankin CM, Kuzma J, Day JB, Salary RR. Experimental and computational analysis of the mechanical properties of biocompatible bone scaffolds, fabricated using fused deposition modeling additive manufacturing process. In: Proc. Paper No. 8511, ASME 2020 International Manufacturing Science and Engineering Conference (MSEC 2020); 22-26 June 2020; Cincinnati, OH, USA. American Society of Mechanical Engineers. DOI: 10.1115/MSEC2020-8511

[3] Klemstine C, Abdelgaber Y, Lawrence L, Day JB, Claudio PP, Salary RR. Characterization of the compressive properties of triply periodic minimal surface (TPMS) polycaprolactone scaffolds for bone tissue engineering. In: Proc. Paper No. IMECE2021-72125, ASME International Mechanical Engineering Congress & Exposition (IMECE 2021), Virtual Conference; 1-5 November 2021. American Society of Mechanical Engineers

[4] Yu M, Yeow YJ, Lawrence L, Claudio PP, Day JB, Salary R. Characterization of the functional properties of PCL bone scaffolds fabricated using pneumatic microextrusion. Journal of Micro and Nano-Manufacturing. 2021;**9**(3):030905 (13 Pages). DOI: 10.1115/1.4051631

[5] Zhao D, Yu M, Lawrence L, Claudio PP, Day JB, Salary RR. Investigation of the influence of consequential design parameters on the mechanical performance of biodegradable bone scaffolds, fabricated using pneumatic micro-extrusion additive manufacturing process. In: Proc. Paper No. 8512, ASME 2020 International Manufacturing Science and Engineering Conference (MSEC 2020); 22-26 June 2020; Cincinnati, OH, USA. American Society of Mechanical Engineers. DOI: 10.1115/MSEC2020-8512

[6] Lawrence L, Day JB, Claudio PP, Salary RR. Investigation of the regenerative potential of human bone marrow stem cell-seeded polycaprolactone bone scaffolds, fabricated using pneumatic microextrusion process. In: Proc. Paper No. 1997, ASME 2021 International Manufacturing Science and Engineering Conference (MSEC 2021); 21-25 June 2021; University of Cincinnati, Cincinnati, OH, USA. American Society of Mechanical Engineers. DOI:10.1115/MSEC2021-63411

[7] Weese NA, Rankin CM, Zhao D, Hart C, Quinlan P, Day JB, Salary RR. Experimental optimization of polymer jetting additive manufacturing process using Taguchi design. In: Proc. Paper No. IMECE2020-24271, V02AT02A046, 7 Pages, ASME 2020 International Mechanical Engineering Congress and Exposition (IMECE 2020); 16-19 November 2020. American Society of Mechanical Engineers. DOI: 10.1115/IMECE2020-24271

[8] Yeow YJ, Yu M, Day JB, Salary RR. A computational fluid dynamics (CFD) study of material flow in pneumatic microextrusion (PME) additive manufacturing process. In: Proc. Paper No. 24325, ASME 2020 International Mechanical Engineering Congress and Exposition (IMECE2020); 16-19

November 2020; Portland, OR, USA. American Society of Mechanical Engineers. DOI: 10.1115/IMECE2020-24325

[9] Yu M, Lawrence L, Claudio PP, Day JB, Salary RR. Pneumatic microextrusion-based additive biofabrication of polycaprolactone bone scaffolds—part II: Investigation of the influence of polymer flow parameters. In: Proc. Paper No. 2002, ASME 2021 International Manufacturing Science and Engineering Conference (MSEC 2021), Virtual Conference (Due to the COVID-19 Pandemic); 21-25 June 2021; University of Cincinnati, Cincinnati, OH, USA. American Society of Mechanical Engineers

[10] Yu M, Yeow YJ, Lawrence L, Claudio PP, Day JB, Salary RR. Investigation of the effects of design and process parameters on the mechanical properties of biodegradable bone scaffolds, fabricated using pneumatic microextrusion process. In: Proc. Paper No. 24252, ASME 2020 International Mechanical Engineering Congress and Exposition (IMECE2020). 16-19 November 2020; Portland, OR, USA. American Society of Mechanical Engineers. DOI: 10.1115/IMECE2020-24252

[11] Gibson I, Rosen DW, Stucker B. Additive Manufacturing Technologies—3D Printing, Rapid Prototyping, and Direct Digital Manufacturing. New York, NY, USA: Springer; 2014. DOI: 10.1007/978-1-4939-2113-3

[12] Spoerk M, Gonzalez-Gutierrez J, Sapkota J, Schuschnigg S, Holzer C. Effect of the printing bed temperature on the adhesion of parts produced by fused filament fabrication. Plastics, Rubber and Composites. 2018;**47**(1):17-24. DOI: 10.1080/14658011.2017.1399531

[13] Jenkins M, Stamboulis A. Durability and Reliability of Medical Polymers. Philadelphia, PA, USA: Woodhead Publishing; 2012

[14] Salary R, Lombardi JP, Tootooni MS, Donovan R, Rao PK, Borgesen P, et al. Computational fluid dynamics modeling and online monitoring of aerosol jet printing process. Journal of Manufacturing Science and Engineering. 2017;**139**(2):021015. DOI: 10.1115/1.4034591

[15] Salary R, Lombardi JP, Tootooni MS, Donovan R, Rao PK, Poliks MD. *In situ* sensor-based monitoring and computational fluid dynamics (CFD) modeling of aerosol jet printing (AJP) process. In: Proc. Paper No. 8535, ASME 2016 11th International Manufacturing Science and Engineering Conference (MSEC 2016); 27 June–1 July 2016; Virginia Tech, Blacksburg, VA, USA. American Society of Mechanical Engineers. p. V002T004A049. DOI: 10.1115/MSEC2016-8535

[16] Salary R, Lombardi JP, Weerawarne DL, Rao PK, Poliks MD. A computational fluid dynamics (CFD) study of material transport and deposition in aerosol jet printing (AJP) process. In: Proc. Paper No. 87647, ASME 2018 International Mechanical Engineering Congress & Exposition (IMECE 2018); 9-15 November 2018; Pittsburgh, PA, USA. American Society of Mechanical Engineers. DOI: 10.1115/IMECE2018-87647

[17] Salary R, Lombardi JP, Weerawarne DL, Rao P, Poliks MD. A computational fluid dynamics investigation of pneumatic atomization, aerosol transport, and deposition in aerosol jet printing process. Journal of Micro and Nano-Manufacturing. 2021;**9**(1):010903. DOI: 10.1115/1.4049958

[18] Du X, Wei D, Huang L, Zhu M, Zhang Y, Zhu Y. 3D printing of mesoporous bioactive glass/silk fibroin composite scaffolds for bone tissue engineering. Materials Science and

Engineering: C. 2019;**103**:109731. DOI: 10.1016/j.msec.2019.05.016

[19] Du X, Yu B, Pei P, Ding H, Yu B, Zhu Y. 3D printing of pearl/$CaSO_4$ composite scaffolds for bone regeneration. Journal of Materials Chemistry B. 2018;**6**(3):499-509. DOI: 10.1039/C7TB02667F

[20] Park JY, Shim J-H, Choi S-A, Jang J, Kim M, Lee SH, et al. 3D printing technology to control BMP-2 and VEGF delivery spatially and temporally to promote large-volume bone regeneration. Journal of Materials Chemistry B. 2015;**3**(27):5415-5425. DOI: 10.1039/C5TB00637F

[21] Cui H, Yu Y, Li X, Sun Z, Ruan J, Wu Z, et al. Direct 3D printing of a tough hydrogel incorporated with carbon nanotubes for bone regeneration. Journal of Materials Chemistry B. 2019;7(45):7207-7217. DOI: 10.1039/C9TB01494B

[22] E3D-v6 Assembly. Open Source Creative Commons (CC BY-NC-SA 3.0) License, UK. Available from: https:// e3d-online.dozuki.com/

[23] Lai Y, Li Y, Cao H, Long J, Wang X, Li L, et al. Osteogenic magnesium incorporated into PLGA/TCP porous scaffold by 3D printing for repairing challenging bone defect. Biomaterials. 2019;**197**:207-219. DOI: 10.1016/j. biomaterials.2019.01.013

[24] Deng C, Yao Q, Feng C, Li J, Wang L, Cheng G, et al. Retracted: 3D printing of bilineage constructive biomaterials for bone and cartilage regeneration. Advanced Functional Materials. 2017;**27**(36):1703117. DOI: 10.1002/adfm.201703117

[25] Hassanajili S, Karami-Pour A, Oryan A, Talaei-Khozani T. Preparation and characterization of PLA/PCL/HA composite scaffolds using indirect 3D printing for bone tissue engineering. Materials Science and Engineering: C. 2019;**104**:109960. DOI: 10.1016/j. msec.2019.109960

[26] Oladapo BI, Zahedi S, Adeoye A. 3D printing of bone scaffolds with hybrid biomaterials. Composites Part B: Engineering. 2019;**158**:428-436. DOI: 10.1016/j.compositesb.2018.09.065

[27] Libonati F, Gu GX, Qin Z, Vergani L, Buehler MJ. Bone-inspired materials by design: Toughness amplification observed using 3D printing and testing. Advanced Engineering Materials. 2016;**18**(8):1354-1363. DOI: 10.1002/adem.201600143

[28] Raines R, Day JB, Salary RR. Experimental characterization of the mechanical properties of medical-grade dental implants, fabricated using vat-photopolymerization additive manufacturing process. In: Proc. Paper No. 85436, ASME 2022 International Manufacturing Science and Engineering Conference (MSEC 2022); 27 June–1 July 2022; Purdue University, West Lafayette, IN, USA. American Society of Mechanical Engineers

[29] Pei X, Zhang B, Fan Y, Zhu X, Sun Y, Wang Q, et al. Bionic mechanical design of titanium bone tissue implants and 3D printing manufacture. Materials Letters. 2017;**208**:133-137. DOI: 10.1016/j. matlet.2017.04.128

[30] Zhang B, Pei X, Zhou C, Fan Y, Jiang Q, Ronca A, et al. The biomimetic design and 3D printing of customized mechanical properties porous Ti6Al4V scaffold for load-bearing bone reconstruction. Materials and Design. 2018;**152**:30-39. DOI: 10.1016/j. matdes.2018.04.065

[31] Zhao L, Pei X, Jiang L, Hu C, Sun J, Xing F, et al. Bionic design and 3D printing of porous titanium alloy scaffolds for bone tissue repair. Composites Part B: Engineering. 2019;**162**:154-161. DOI: 10.1016/j. compositesb.2018.10.094

[32] Inzana JA, Olvera D, Fuller SM, Kelly JP, Graeve OA, Schwarz EM, et al. 3D printing of composite calcium phosphate and collagen scaffolds for bone regeneration. Biomaterials. 2014;**35**(13):4026-4034. DOI: 10.1016/j.biomaterials.2014.01.064

[33] Zhou Z, Buchanan F, Mitchell C, Dunne N. Printability of calcium phosphate: Calcium sulfate powders for the application of tissue engineered bone scaffolds using the 3D printing technique. Materials Science and Engineering: C. 2014;**38**:1-10. DOI: 10.1016/j.msec.2014.01.027

[34] Bergmann C, Lindner M, Zhang W, Koczur K, Kirsten A, Telle R, et al. 3D printing of bone substitute implants using calcium phosphate and bioactive glasses. Journal of the European Ceramic Society. 2010;**30**(12):2563-2567. DOI: 10.1016/j.jeurceramsoc.2010.04.037

[35] Cox SC, Thornby JA, Gibbons GJ, Williams MA, Mallick KK. 3D printing of porous hydroxyapatite scaffolds intended for use in bone tissue engineering applications. Materials Science and Engineering: C. 2015;**47** 237-247. DOI: 10.1016/j.msec.2014.11.024

Section 2

New Technological Solutions

Chapter 4

Perspective Chapter: Composites Manufactured by Stereolithography

Patrick Simpson, Michael Holthaus, Luke Gibbon and Chad Ulven

Abstract

Stereolithography (SLA) is a widely utilized rapid additive manufacturing process for prototypes and proof-of-concept models with high resolution. In order to create structurally sound components using SLA, reinforcement needs to be incorporated in the UV-based resins typically used. However, the introduction of reinforcement into vat-based SLA printers has had limited success due to a host of processing challenges including the creation of a homogeneous resin mixture and UV-inhibiting constituents. The effectiveness of using a dual curing system, consisting of a photo and thermal initiator, for the additive manufacturing of carbon fiber short-fiber composites via vat photopolymerization, was investigated. The necessary processing parameters were developed that resulted in successful printing and curing of composites at a 5% fiber volume. Manufacturing with reinforcements that have different densities from the resin creates separation issues, either suspending to the top or settling to the bottom. Following the approaches discussed in this chapter, an even distribution of short fibers was achieved throughout SLA printed samples using a modified commercial printer. Separation was overcome by inducing a continuous flow of reinforced liquid resin in the printer vat during printing. This flow field adaptation allows commercial SLA printers the ability to produce composite parts with different densities of the constituents utilized.

Keywords: stereolithography, composites, carbon fiber, fiber glass, and dual cure

1. Introduction

A current drawback of additive manufacturing (AM) parts is that the material properties of the parts produced are much lower than that of parts manufactured using traditional manufacturing such as milling, injection molding, and traditional composite manufacturing methods [1–3]. If the material properties of the parts produced using AM could be increased to be similar with that of traditional manufacturing methods, a new era of design could be opened up. This would be accomplished by the ability to use AM to make parts that could not be produced using any other methods due to their complex geometry. With the advent of computer-aided design (CAD) software, the design process was changed, but limitations of what can be made in the virtual world versus what can be manufactured in the real world still exist. With the incorporation of AM into the final production processes, the complexity of the geometries that could be manufactured increases. This increase in complexity of part geometry can allow for a decrease in the over complexity of the part, subassembly, or whole assembly, along with helping to relieve supply chain issues. For instance, by combing multiple parts into one, the need for several fasteners can be eliminated. This decreases the number of parts needed, number of components to analyze, and possible points of failure.

An appealing advantage of vat photopolymerization over material extrusion processes, such as fusion deposition modeling (FDM), is the ability to manufacture parts that have isotropic material properties [4]. For parts produced using FDM, the part has the properties of the material in the plane of printing, but perpendicular to that it becomes dependent on the mechanical adhesion of the polymer layers to each other for the part's mechanical properties [4]. For parts manufactured using vat photopolymerization, it is possible to produce parts that have near isotropic mechanical properties [4]. Although the part is produced in a layer-by-layer process, as like in FDM, the thermoset polymer is not completely cured within each layer before the part is raised and the next layer is printed. This allows for unreacted polymer functional groups in a previous layer to react with the polymer curing in the current layer being printed. Because of the ability of vat photopolymerization to produce parts with near isotropic properties, the orientation of the part while printing does not depend on what direction force will be applied to the finished part, but what orientation of the part will optimize the printing process.

One of the limiting factors of vat photopolymerization is the material properties of the resins used to create parts. The parts produced using these resins are often weak and brittle, limiting their use for many end-use structural applications [5]. One method of improving the properties of a material is the incorporation of a reinforcement material in the creation of a composite. Short-fiber composites have traditionally been produced via injection and compression molding. By using short fibers as a reinforcement, the same manufacturing method used for polymers can be used to form the composites, but with increased material properties [6].

Short-fiber composites usually find their applications in situations where isotropic material properties are desired, typically manufactured via injection or compression molding, but the manufacturing processes can influence the fiber orientation due to the flow of the material during manufacturing final material properties [7]. While the flow-induced alignment can be taken advantage of to some extent, it can be limited due to the requirements of the mold design and can be an undesired effect when isotropic properties are desired [8].

With short fibers already in widespread use as a reinforcement material for traditional manufacturing methods, such as injection and compression molding, they have found their way into use for additive manufacturing methods as well [1, 3, 9–19]. For FDM there are numerous types of reinforcements ranging from nanoparticle to continuous fiber, of both natural and synthetic materials, which are currently being researched and available for sale on a commercial level [1, 3, 16]. Whereas FDM-based methods have a number of publications in the area of short-fiber composite characterization, the area of vat photopolymerization manufactured composites is lacking in published studies and available data using carbon fiber as a reinforcement.

There have been various nanosized reinforcements studied as a method of increasing the material properties of vat photopolymerization manufactured parts such as: cellulose nanocrystals (CNC), multi-wall carbon nanotubes (MWCNTs), and silver nanoparticles (AgNPs) [12–14, 18]. Feng et al. [14] used lignin-coated cellulose nanocrystals (L-CNC) at 0, 0.1, 0.5, and 1 weight percent with an acrylic matrix. The research was carried out using Form+1 (Formlabs, Somerville, MA), which is a bottom-up desktop vat photopolymerization printer. At a loading of 0.5 wt.% L-CNC, there was an increase in the tensile strength and Young's modulus by 3% and 5%, respectively [14]. This was achieved only after a thermal post-cure being carried out on the specimens with the non-post-cured specimens showing unimproved or reduced properties depending on the loading of L-CNC [14]. The decrease in material properties that was seen at higher weight percent was attributed to poor dispersion of the L-CNC and poor adhesion between the L-CNC and matrix [14].

Sandoval et al. [12] investigated a composite made from MWCNTs at 0.025 and 0.1 weight percent with an epoxy-based matrix using the commercial resin, DSM Somos® WaterShed™ 11,120. A top-down printer, the 3D Systems SLA-250/50 machine (3D Systems, Rock Hill, SC) was modified from a 47 liter vat to a 500 ml vat with the sweeping mechanism removed, and a peristaltic pump was used to recirculate the resin mixture [12]. The research looked at the increasing tensile strength and fracture strength of the resin. For 0.025 wt.% of MWCNTs, an increase in tensile strength of 5.7% with an increase in fracture strength of 26% was reported. While at 0.10 wt.%, an increase of 7.5% and 33% in tensile and fracture strength, respectively, was reported, but it was pointed out that at the higher loading of 0.1% MWCNTs, there were issues with agglomeration of the MWCNTs [12]. The elongation at break decreased 28% for the MWCNTs reinforced resin, and the fracture mode was reported as a brittle-type verse as more of a ductile failure mode that was seen in the pure resin [12].

Short glass fibers have been studied more as reinforcement materials for vat photopolymerization in part due to the decrease in opacity when compared with that of carbon fiber [3]. In one study, Cheah, C. et al. [19] looked at using short glass fibers 1.6 mm in length at various fiber volume fractions of 0, 5, 10, 15, and 20% and a urethane acrylic matrix, DeSolite SCR310. The experiment was carried out for comparing molded and 3D printed samples. Although the paper does not state what machine was used to print the samples, it can be inferred that a top-down style was used [19]. Cheah et al. saw improvements for all fiber volumes, with increased mechanical properties achieved by increasing fiber amount, and part shrinkage can be reduced. For a fiber volume of 20%, they were able to achieve an increase in tensile strength of 24% and an 80% increase in the Young's modulus [19]. The top-down vat photopolymerization machine that was employed resulted in the manufacture of composites that were close to unidirectional in fiber orientation due to the scraping step in between layers [19]. While the creation of unidirectional composites is desirable in some applications, the leveling step would prevent the printing of an isotropic part and therefore could limit potential applications and restrict the printing process based on how the part must be printed.

Along with short glass fibers, there have also be studies that have looked into the use of continuous glass fiber as a reinforcement in vat photopolymerization. Karalekas et al. [2] placed a single layer of nonwoven glass fiber mats, of various thicknesses, within specimens produced using vat photopolymerization. This was done by pausing the printer at a predetermined build height, placing the mat of the specimen, and resuming the print [2]. Karalekas et al. were able to show an increase in the Young's modulus, but a decrease in tensile strength for thinner mats. For the thicker mats, the Young's modulus was shown to decrease; this was contributed to the inability for the photopolymer to fully cure with in the thicker mats [2]. While this study was able to show that continuous glass fibers could be placed into the part for reinforcement, the fact that it was added by hand during the build process is inefficient, especially if multiple layers of fiber are to be used in the manufacturing of a part, and would be difficult for the manufacturing of complex parts.

Some research has been carried out using continuous carbon fiber by Gupta et al. [17] where to overcome the issues of fully curing the part, a dual curing system was used. The dual curing system employed a photo initiator for initially curing the

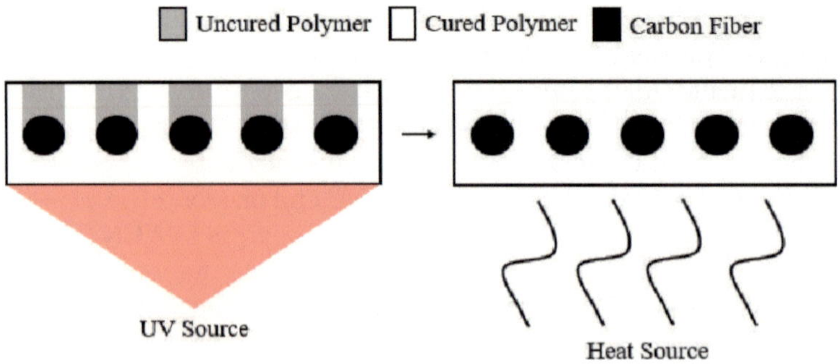

Figure 1.
Dual curing system.

fiber and matrix in the desired geometry and a thermal initiator to cure the remaining resin. While Gupta et al. were able to show that the system was fully cured, they did not report any information on the material properties of the composite produced and appeared to have limited success in incorporating the initiator into the liquid resin [17]. The ability for the thermal initiator to cure the areas that the UV light cannot, due to the opacity of the carbon fiber, could be a promising method of incorporating carbon fiber into vat photopolymerization produced, shown in **Figure 1**.

There are a host of complications from suspending fiber to achieving optimal cure with opaque fibers hindering the UV light. Some success has been achieved by introducing glass fibers into vat printers, while struggling to keep a homogeneous resin mixture as the fibers settle to the bottom of the print bed [20].

SLA composite parts have been approached with different tactics in the past. Some success has been achieved in the smaller particle reinforcement with chemical and rheological approaches. For filler-sized particles, approaches such as suspension with shear thinning polymers have been attempted to produce SLA ceramics [21, 22]. Another tactic used is surface modification of the filler to reduce conglomerations and improve the stability of the dispersion [23]. Other research has shown that certain fibers can be aligned and suspended via shear flow or electric field induction [24–27]. However, using a more common and larger fiber does not allow all these techniques to be applied and requires a different approach.

Fiber suspension is a critical and complicated issue when looking at the disparaging density ranges of fibers to resin combinations. If a homogeneous resin bath cannot be maintained throughout the print, the resulting parts will either have no reinforcement or possibly too much reinforcement to successfully print. An adaptation to a Moai 130 SLA vat printer was designed in this study to provide a homogeneous reinforced resin bath for printing using standard printing parameters. The adaptation ensures the glass fiber is suspended in the build plate region of the vat for the entirety of any print allowing a homogeneous composite component to be manufactured.

2. Equipment and processing

2.1 Opaque reinforcement

A Moai Laser vat photopolymerization printer manufactured by Peopoly, and purchased from MatterHackers, (Lake Forest, CA) as a DIY kit, was assembled and calibrated. The Moai is a bottom-up printer that uses a 405 nm 70-micron spot size laser and is based on an open-sourced design. The photopolymer resin used in this research was Moai Standard Clear resin, by Peopoly purchased from MatterHackers (Foothill Ranch, CA). It is an acrylic-based photopolymer designed to work with the Moai printer being used in this research.

The primary carbon fiber used was Toray T-700. The fiber was purchased from Composite Envisions (Wausau, WI) as a chopped 3 mm fiber and comes sized for epoxies. The fiber was then milled in a Retsch Rotor Beater Mill SR 300 (Retsch GmbH, Haan, Germany) using a 120 μm screen. The milled fiber was then sieved using a stack of screens, with a stacking sequence of 2 mm, 250 μm, 106 μm, and 76 μm. The fibers were collected in between the 106 μm and 76 μm screens. A sample of fibers were observed using an Axovert 40Mat (Carl Zeiss AG, Oberkochen, Germany), with images obtained by a ProgRes C10plus camera (Jenoptik AG, Jena, Germany), to determine the average length along with the length distribution. The processed fibers were then placed in an oven at 80°C and dried for a minimum of 8 hours before use. The distribution of the length of 300 fibers of the milled Toray fibers had an average length 74.1 ± 40.2 μm, and the distribution of the measured lengths is shown in **Figure 2**.

A variety of thermal initiators were investigated in order to identify the optimum thermal initiator to be used in a dual cure system. The thermal initiators were evaluated based off of the thermal initiators solubility with the resin system, the stability of the resin system at room temperature, and the temperatures needed for post-curing of the samples. The thermal initiators investigated were Dilauroyl Peroxide (Luperox LP), Cumene Hydroperoxide, Dicumyl Peroxide, Tert-Butyl Peroxybenzoate (Luperox P), Benzoyl Peroxide (Luperox A98) and were purchased and used as received from Sigma-Aldrich® (St. Louis, MO).

To prepare the resin system, the resin and fiber were first mixed with a high-speed mixer and then sonicated for 5 minutes. After sonication the thermal initiator was added and mixed again with a high-speed mixer. The resin system was then degassed in a vacuum chamber and then immediately used to manufacture samples.

For all samples that are manufactured using the 3D printer for this research, the longitudinal axis of the parts was varied in the x-axis of the printer to examine the effects of part orientation relative to the build surface. For all samples prepared, the finished parts were washed in ethanol after being removed from the build

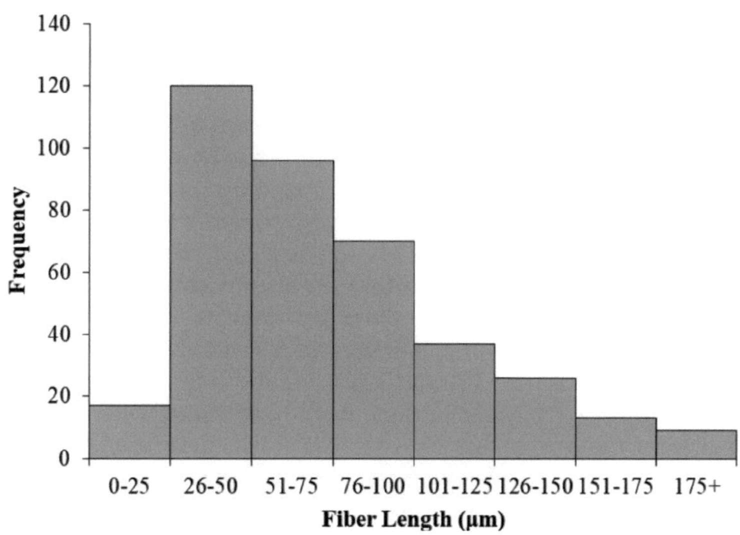

Figure 2.
Milled Toray T-700 fiber length distribution.

platform. This allows for the removal of any uncured resin, and in the case of the fiber-reinforced samples any loose fibers from the surface. The supports that were generated during the printing process are left in place at this time to support the sample while it was being post-cured.

The neat (non-fiber-reinforced) samples were post-cured in an in-house built cure oven, consisting of three 25 Watt LED UV (405 nm) light banks, a heating element, and a rotating platform. The temperature and the time for the post-cure can be adjusted by the use of proportional-integral-derivative (PID) controller. For the neat samples, a temperature of 80°C was used while being exposed to the UV light for 1 hour. All fiber-reinforced samples were post-cured using a VWR Forced Convection Oven (VWR International, Radnor, PA), with the temperate and times determined from the DSC results. After post-curing the supports were removed from the sample, and the area where the supports were attached was sanded with increasingly finer grits of sand paper, ranging from 60 to 600, with care taken to preserve the sample geometry, which varies depending on the desired testing. This was done to diminish the effects of surface defects (due to support material) on the tested samples.

Composite and neat specimens were printed using a Moai 130 inverted SLA printer with the acrylic vat. Printing was performed using factory settings of the Moai 130, with one exemption. The peel step time was extended to allow more time for the higher viscosity resin mixture to flow in the printed area between layers. A standard Peopoly UV acrylic-based SLA resin was utilized as the resin of choice for both the neat specimens and composite specimens.

2.2 High-density reinforcement

A higher-density short-fiber reinforcement, E-glass fiber procured from Fiber Glast Development Corporation (Brookville, OH), was also studied. Fiber glass was selected for its high degree of transparency to UV light to not reduce the effectiveness of the UV cure. Reinforcing the resin while still allowing for good initial cure from the printer and achieving optimal polymerization through post curing was a goal. The particular glass fibers were 0.8 mm fibers denoting that the fibers have a possible length up to 0.8 mm and smaller, resulting in a mean length of 230 μm.

When manufacturing the neat specimens, no unique changes to the printing process occurred. The resin was simply poured into the vat and printing commenced per standard operating procedure for the Moai 130. The composite resin consisted of 85% resin and 15% fiber glass by volume. The composite resin batches were high-speed mixed and degassed prior to introducing the resin to the printer. The viscosity of the neat resin was found to be 517 Pa*s and the composite mixture was 950 Pa*s using a Brookfield DV-II+ Pro Viscometer. This study focused on 15% fiber volume fraction, as higher fiber loading further increased the viscosity, causing premature recirculating pump failures.

With the composite resin system mixed and degassed, the resin was introduced to a modified vat. The resin tank in the Moai 130 SLA printer was adapted to allow for a flow field in the print region. Inlet ports were fixed to the front of the vat, and outlet ports were fixed to the back of the vat. A diagram and picture setup can be seen in **Figure 3**. This setup allows for a constant flow and circulation of resin mixture with a changeover rate faster than the settling of the fibers in the resin.

Pumping the abrasive resin mixture was accomplished using a 12-volt peristaltic pump. The peristaltic pump utilized 6.4 mm inner diameter tube, which was also the same as the vat connections and made for a seamless transition to recirculate the resin mixture. At max output the motor could achieve 400 ml/min with a hose inner diameter (ID) of 6.4 mm.

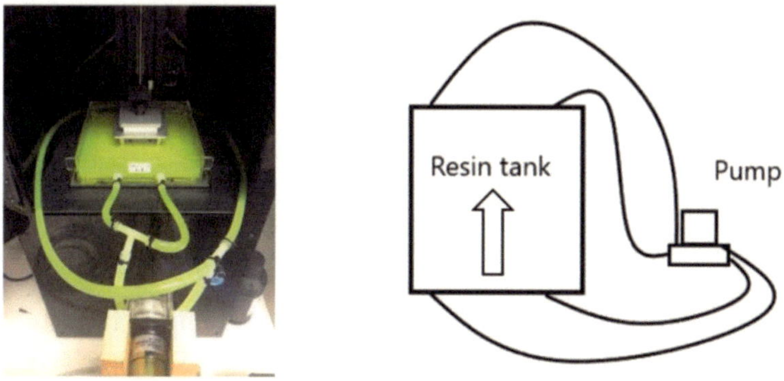

Figure 3.
Diagrams of pumping setup.

After the printing process was complete, the prints where washed with ethanol to remove any UV resin buildup. The print supports were kept in place until after being post cured. To ensure a full cure, all prints were placed in a UV oven for 60 min and 60°C. The oven consisted of a three 25 W LED UV lights and a heating element. The specimens were placed on a rotating platform in the oven. The rotation of the bed allowed for an even distribution of UV Light.

3. Testing and results

3.1 Opaque reinforcement characterization

3.1.1 Testing

The selected thermal initiators were evaluated experimentally via differential scanning calorimetry (DSC) in accordance with ASTM E2160 [28]. The testing was carried out from 25–180°C at a ramp rate of 10°C/min, using a Q20 DSC (TA Instruments, New Castle, DE). For each sample, two runs were carried out. The first was of an uncured sample, and the second was a sample that had been UV cured by placing the DSC pan in the UV cure oven 25.4 mm away from the light source for 1 minute at room temperature. This was done to determine the thermal initiators' onset temperature and to experimentally determine if any reactions are occurring after UV curing. The samples were tested at a fiber volume (V_f) of 5% and a thermal initiator content of 1 wt.% of thermal initiator. The neat Moai resin was also included to determine if there is any activation of the photo initiator at elevated temperatures. Graphs produced from the DSC data were made using TA Universal Analysis (TA Instruments, New Castle, DE).

Tensile testing was carried out using an Instron 5567 load frame (Instron, Norwood, MA), a 25.4 mm extensometer, and a 2 kN load cell. The load frame was controlled and data collected using Bluehill software (Instron, Norwood, MA). While there are no ASTMs directly concerning 3D printed materials, all tensile testing will be carried out referencing ASTM D638 [29] and ASTM 3039 [30]. The specimen geometry was of the type IV according to ASTM D638 [21]. The type IV specimen was chosen due to the limiting size of the build plate (130 × 130 mm) of the Moai printer being used to conduct this research. All samples were tested at a constant cross-head rate of 1 mm/min, so failure of the specimen occurred within 1–10 minutes of testing, per ASTM D3039 [30]. The flexural testing was carried out using an Instron 5567 load frame (Instron, Norwood, MA) with a 2 kN load cell. The load frame was controlled and data collected using Bluehill software (Instron, Norwood, MA). The flexural testing, referencing ASTM D790 [23], was carried out as a three-point bending utilizing center loading, with specimens having a span-to-thickness ratio of 16:1 and a cross-head rate of 1 mm/min. For each sample, five specimens were tested. The maximum tensile strength and Young's modulus were found as specified in ASTM D638 Section 11.2 and 11.4 respectively, and the maximum flexural strength and flexural modulus were found as specified in ASTM D790 Section 12.2 and 12.5, respectively [29, 31].

Scanning electron microscopy (SEM) was utilized to examine the fracture surface of tested samples. This will allow for examination of the fracture surface in determining failure types and to evaluate the dispersion and the orientation of the fibers within the sample [32]. Samples were attached to cylindrical aluminum mounts with colloidal silver paste (SPI Supplies, West Chester, Pennsylvania, USA). The specimens were sputter coated (Cressington 108auto, Ted Pella, Redding, California, USA) with a conductive layer of gold. Images were obtained with a JEOL JSM-6490LV scanning electron microscope (JEOL USA, Inc., Peabody MA, USA); energy-dispersive X-ray information was collected at an accelerating voltage of 15 kV using a Thermo Scientific UltraDry Premium silicon drift detector with NORVAR light element window and Noran System Six imaging system (ThermoFisher Scientific, Madison WI, USA).

3.1.2 Results

In order to fully cure all of the resin within the sample, the use of a dual cure system that uses both a photo initiator and a thermal initiator was investigated. The results of the DSC testing are shown in **Figure 4** (non-UV-cured samples) and **Figure 5** (UV-cured samples) are summarized in **Table 1**.

Based off of the results from the thermal initiator testing, Luperox P was chosen as the thermal initiator to be evaluated for the dual cure resin system and was determined to be effective at a 0.5 wt.%. The 10 hour half-life is the temperature at which half the content of the peroxide is lost after 10 hours. Making all of these peroxides a stable option for the processing will still have peroxide available during the thermal post curing process.

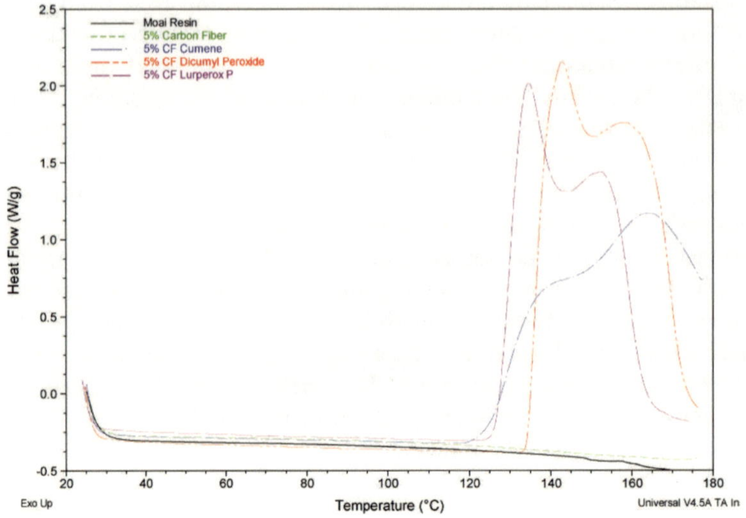

Figure 4.
DSC results for samples without UV curing.

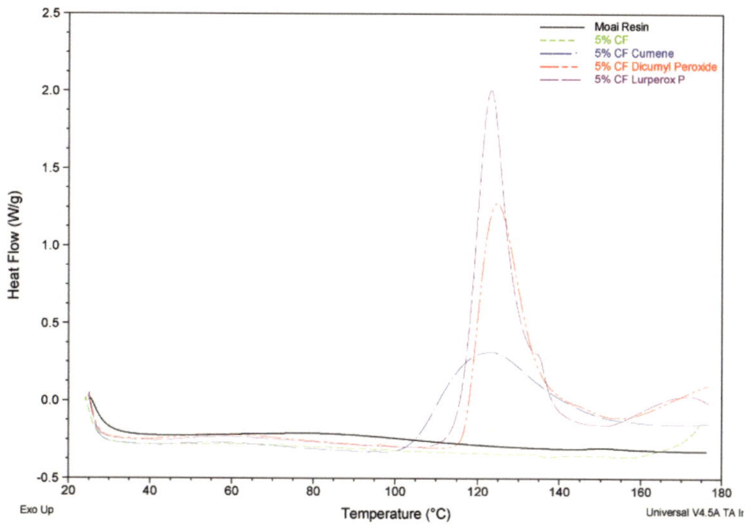

Figure 5.
DSC curves for samples after UV curing.

Thermal Initiator	Onset temperature (°C)	10 Hour half-life (°C) [33]
Luperox P	115.99	104
Dicumyl Peroxide	116.83	114
Cumene Hydroperoxide	104.01	135

Table 1.
Thermal initiator onset temperature from DSC curves.

The fiber volume consistency results are summarized in **Table 2**. The top section of the samples has a higher fiber volume than what the resin mixture was designed to have (5% V_f). This is due to the lowering of the bed plate into the vat trapping more fibers in between the sample and bottom of the vat then desired. The reason believed for the samples to have lower and similar fiber volume gradients despite the different print times is that the area around the sample being printed becomes depleted of fibers as the sample is printed. This slowly lowers the fiber volume of the sample as it is going through the printing process.

Because the flexural specimens have a larger cross-sectional area throughout the entire specimen, the fiber gradient appears more pronounced. Similar burn off testing was carried out with the tensile specimens with the exception that only the two ends of the specimen were tested. The average difference in the fiber volume of the top and bottom portions of the tensile specimens changes by on average of 0.8% for the specimens printed in the 90° orientation. This lower difference could be due to the varying specimen geometry, or that due to their large size, less specimens were printed at the same time to keep the print time short. This allowed for more spacing between the specimens and could have prevented the fiber from becoming depleted from around the specimens being printed.

To aid in sample identification, the follow naming scheme will be implemented for the remainder of this paper. It will consist of the main material parameter being looked at along with the layer height and print orientation. The lettering is in Ref. to the material parameter: M is Moai resin as supplied, LP is Moai resin mixed with thermal initiator (Luperox P), and C consists of carbon fiber, thermal initiator (Luperox P), and Moai resin. The number following the lettering is the layer height

Layer height (μm)	Print time (hours)	Top (% V_f)	Middle (% V_f)	Bottom (% V_f)
100	3.0	5.23 ± 0.02	4.84 ± 0.10	3.57 ± 0.27
50	5.25	5.45 ± 0.15	4.50 ± 0.13	3.88 ± 0.01

Table 2.
Summarized fiber volume consistency results for flexural samples.

(100 μm or 50 μm). The number following the hyphen is the angle (0° or 90°) between the longitudinal axis of the sample and the build platform during the manufacturing process. The summarized results of the tensile and flexural testing are shown in **Table 3**. The samples are identified with a superscript in the first column and identified with the superscripts of other groups in which significant statistical difference between the results was found via an ANOVA analysis and post-hoc Tukey HSD test at a 95% confidence interval. The samples are compared across four different groups: 0° orientation, 90° orientation, 100 μm layer height, and 50 μm layer height.

When comparing the Young's modulus of the samples, there was an increase of 21% between the M100-0 and CF100-0 samples and an increase of 27% for the M50-0 and the CF50-0 with an increase in Young's modulus for the CF50-0 samples when compared with the CF100-0 of 5%. While showing only 4% increase between the LP100-0 and CF100-0 samples and 13% increase for the LP50-0 and CF50-0, evidence of the weak interfacial properties of the short-fiber composite samples can be seen in **Figure 6**.

In **Figure 6**, the smooth channels left behind from the carbon fiber in A, the clean (lack of bonded matrix material) fibers present in B, and the smooth holes in B are all signs of weak interfacial properties due to poor bonding between the matrix and fiber [6, 34]. This limits the ability for the matrix to transfer stress to the fibers and therefore reduce its overall properties [6]. If the composite had better interfacial properties, there would have been evidence of fiber breakage, which was not present in any of the SEM images taken [6, 34].

Sample	Tensile strength (MPa)	Young's modulus (GPa)	Flexural strength (MPa)	Flexural modulus (GPa)
M100-0	58.9 ± 5.0	2.72 ± 0.06	86.0 ± 9.5	2.52 ± 0.07
M100-90	66.1 ± 0.4	2.87 ± 0.05	96.2 ± 1.7	2.42 ± 0.01
M50-0	43.4 ± 7.7	2.81 ± 0.15	82.2 ± 11.2	2.42 ± 0.05
M50-90	64.6 ± 1.6	2.89 ± 0.12	100.8 ± 3.4	2.62 ± 0.06
LP100-0	36.9 ± 7.9	3.17 ± 0.15	27.9 ± 5.4	2.34 ± 0.15
LP100-90	27.6 ± 22.5	3.17 ± 0.26	14.6 ± 3.0	2.47 ± 0.21
LP50-0	50.3 ± 7.7	3.04 ± 0.11	23.9 ± 3.7	2.55 ± 0.23
LP50-90	68.8 ± 5.9	2.83 ± 0.01	13.2 ± 0.6	2.50 ± 0.20
CF100-0	52.4 ± 3.7	3.29 ± 0.29	84.9 ± 4.3	2.59 ± 0.06
CF100-90	41.4 ± 2.6	3.39 ± 0.52	66.1 ± 0.5	2.31 ± 0.11
CF50-0	50.6 ± 6.6	3.46 ± 0.17	43.4 ± 5.5	2.64 ± 0.12
CF50-90	11.69 ± .28	2.92 ± 0.01	19.4 ± 2.5	2.58 ± 0.16

Table 3.
Summarized tensile and flexural testing results.

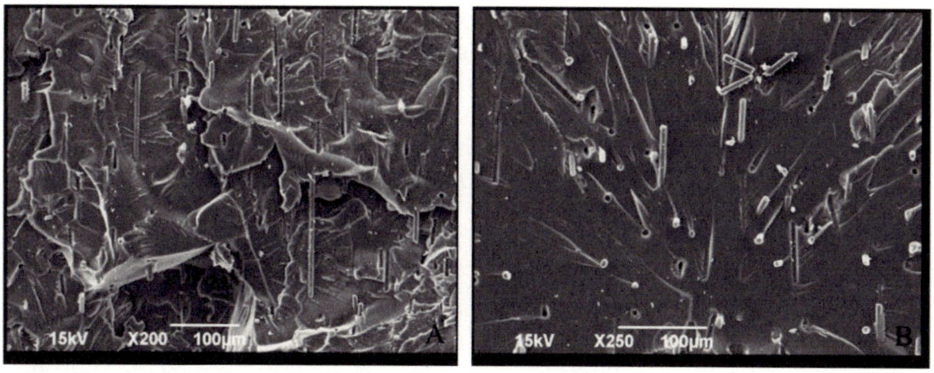

Figure 6.
Fracture surface of CF50-0 (A) X200 and (B) X250 magnification.

When comparing the modulus of the carbon fiber samples printed at different layer heights (100 μm and 50 μm), the increase in modulus of 5% can be attributed to the partial alignment of the fibers in the loading direction. The CF100-90 has a higher modulus than the CF50-90 samples. While this would be expected if the material was going from isotropic material to a special orthotropic material due to fiber alignment, it is not the case for these samples due to cracks being present in the CF50-90 samples [6]. The alignment of the fibers via layer height is demonstrated in **Figure 7** with the yellow arrows highlighting the various fiber orientations.

When comparing the strength of the Moai resin samples, the samples printed in the 90° print orientation for the Moai resin have a higher strength when compared with the same samples printed in the 0° print orientation. This is due to the effects of the support material being removed and leaving behind notches in the specimen. These notches then act as small stress concentrators resulting in lower strength of the sample [35]. Due to these effects, the Moai resin samples tested at a print orientation of 90° are a better representation of the actual ultimate tensile strength of the neat resin material. The samples tested with just the thermal initiator (Luperox P) added also showed flaws the contributed to a lower tensile strength result. These are from cracks that were present on the surface of the samples, which originated during the thermal curing process. These cracks originated from the volumetric shrinkage that occurs during the post-curing process as the degree of conversion within the system increases [36].

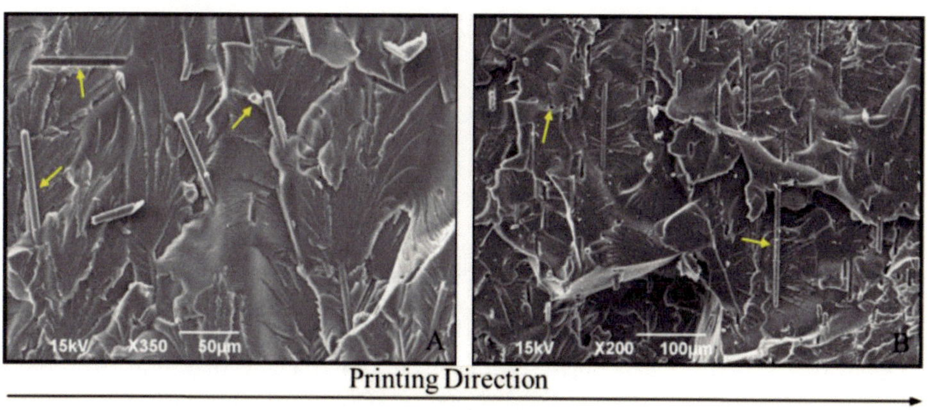

Figure 7.
Fracture surface of (A) CF100-90 and (B) CF50-90 specimens.

The lower strength of the carbon fiber samples is due in part to the end effects of the fiber reinforcement acting as stress concentrators within the composite [6]. This is caused by the large number of very short fibers (much less than the critical length) in the distribution of the lengths, seen in **Figure 2**, being present in the sample [6]. The strength of short-fiber composites is also affected by the length of the carbon fiber being used. To get the maximum amount of reinforcement from the carbon fiber, the fiber needs to be over a critical length to maximize the load transfer between the fiber and matrix [6]. The shorter length of the fibers (76 μm) being used is below the critical length needed for the system (433 μm using a shear lag model) [6]. This results in the load not being fully transferred to the fiber and limiting the composites performance [6].

Overall the flexural modulus of the samples was not improved by the addition of the carbon fiber. This could be primarily due to the lower fiber volume (5%) of the samples, causing them to have results similar to that of the neat resin samples. During the flexural testing, the specimens are subjected to the more complex stress state (compressive, tensile, and shear) when compared with the tensile testing [28, 29].

The effect of the fiber volume gradient that was present in the samples would also have a more pronounced effect of the results when compared with the tensile testing. For the tensile specimens, the gage section had similar fiber volume amount within the measured section. Among the flexural specimens, the whole specimen (within the span length) is tested allowing for failure to occur in the regions of lower fiber volume.

For the flexural testing, the samples are subjected to compressive, tensile, and shear stresses that lead to the multiple failure mechanisms affecting the flexural properties of material [29]. The tensile stresses can lead to fiber breakage and debonding, while compressive stresses can lead to fiber shear and/or buckling and kinking [29]. The fibers exhibited characteristics indicating poor interfacial strength between the fiber and matrix. This along with the low fiber content and weak shear properties of carbon fiber would lead to the failure of the flexural testing to coincide more with that of the compressive type failures, which are governed more by the matrix properties [6, 37, 38].

The samples made with just Luperox P added showed the same type of surface cracking after post-curing as was seen in the tensile samples. This led to them having the weakened flexural strength as seen in the tensile samples made of the same constituents. The carbon fiber samples exhibited poorer flexural strength than the neat Moai resin samples due to the effects of the fiber ends acting as stress concentration, weak interfacial properties, low fiber content, and the presence of voids [6].

3.2 High-density reinforcement characterization

3.2.1 Testing

Printed specimens were characterized for density, 3 pt. flexure, notched fracture toughness, and dynamic mechanical analysis (DMA). The testing conducted is to illustrate the performance enhancements and the quality of cure and consistency of the resulting composite parts.

Density calculations were performed to ensure the fibers were evenly dispersed, following ASTM D792-13 Test Method A [39]. An Ohaus Adventurer (Ohaus Corp., Parsippany, NJ) scale was used to weigh four sections from four different specimens. Using a Mettler Toledo (Mettler Toledo, Columbus, OH) density determination kit, the density comparison was conducted between samples printed using the flow process and one specimen without resin flow. Where A is the weight of the sample in air, B is the weight of the sample in water and the density of water and air with subscripts O and L respectively to calculate the density of the samples.

$$\rho = \frac{A}{A-B}(\rho_O - \rho_L) + \rho_L \tag{1}$$

The flexural testing was carried out using an Instron 5567 load frame (Instron, Norwood, MA) with a 2 kN load cell. The load frame was controlled and data collected using Bluehill software (Instron, Norwood, MA). The flexural testing, referencing ASTM D790 [31], was carried out as a three-point bending utilizing center loading, with specimens having a span-to-thickness ratio of 16:1 and a cross-head rate of 1 mm/min. For each sample, five specimens were tested and the maximum flexural strength and flexural modulus were found as specified in ASTM D790 Section 12.2 and 12.5, respectively [29, 31].

Notch fracture testing was carried out using ASTM D5045 [40] standard with an Instron 5567 load frame and a 2kN load cell. Single-edge-notch bending (SENB) specimens were created as rectangular bars 7 mm × 13 mm x 63 mm. These bars were notched post processing using a vertical mill and a final fine crack initiated with a razor blade. Five neat and five reinforced samples were tested and post-test measured using a Keyence VHX-7000 Digital Microscope for the crack length determination.

Glass fiber is highly transparent to UV light, but DMA was performed to ensure that each part was optimally cured. To perform this, ASTM D5418-15 [41] was followed. Three samples of both neat and glass fiber-reinforced specimens were printed. Duel cantilever DMA was chosen because the Tg can easily be evaluated at the peak of the tan(δ) curve. Each sample was printed at 60 mm x 13 mm × 4 mm to fit in the duel cantilever grips. A TA Instruments DISCOVERY DMA850 was used with a temperature range of 30–150°C, a frequency of 1 Hz, an amplitude of 15 μm, and a heating rate of 1°C/min.

3.2.2 Results

As observed in other research attempts, the high-density fiber glass quickly settles to the bottom of the print. This results in components that have the desired reinforcement at the beginning of the print, but little to no reinforcement after the glass settles to the bottom of the VAT and is unable to flow into the print area. The printer does not have a swipe step or means of resin movement, which allows the fibers to settle to the bottom. With the fiber on the bottom, the fiber free resin is allowed to flow back into the print area with very few fibers present. Therefore, the density of the composite will vary from the top of the print to the bottom of the print without a pumping system. When examining the samples, a gradient of fibers/density could easily be seen without a microscope with the specimens printed without flow.

Comparison of glass content and distribution between prints with no pump and pump can be seen in **Figure 8**. The fiber distribution is consistent from the start to finish for the specimen printed with the pump providing a flowing resin mixture. **Figure 8B** is the end of the print and **Figure 8D** is the beginning of the part but after the scaffolds had already been printed. Similarly, **Figure 8A** is the end of the print and **Figure 8C** is the start of the part but after the scaffolds. It can be seen that even at the start of the part, there is very little glass in the part and diminishes to nearly no fiberglass by the end of the print. No discernable differences were found when

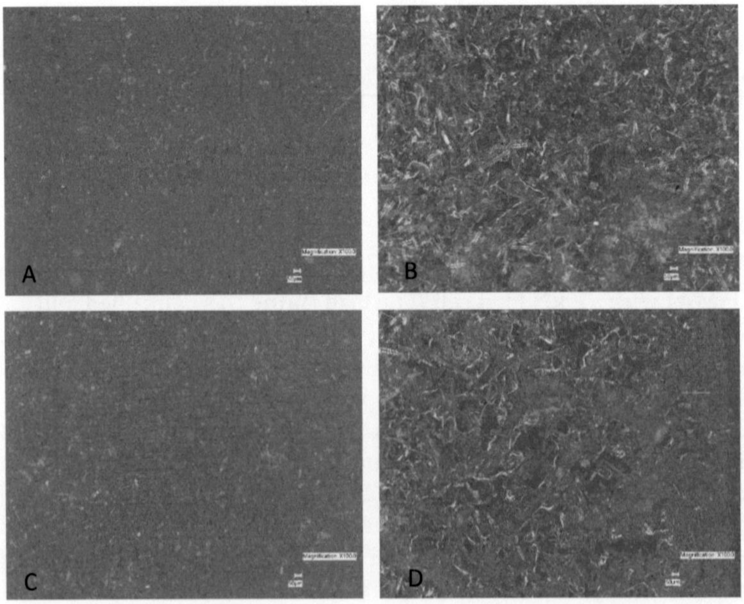

Figure 8.
A, end of print with no pump; B, end of print with pump; C, start of print with no pump; D, start of print with pump.

Process	Density g/cm^3
No pumping	1.26
Sample 1	1.39
Sample 2	1.40
Sample 3	1.39

Table 4.
Densities of specimens examined.

observing parts printed with the circulation pump. **Figure 8** images are taken using a Keyence VHX 7000 digital microscope (Keyence Corp of America, Itasca, IL).

Table 4 shows the density averages of the specimens printed. The theoretical density of the composite specimens at 15% fiber volume fraction, assuming no voids, was 1.42 g/cm^3. The three samples that used the pumping system only had a 1.4% deviation from the target density. As predicted, the density of the no flow specimen was much lower as the majority of the volume did not have any fiberglass present. The composite specimens created with the flow field were consistent and very close to the theoretical density.

For fracture toughness, the Keyence microscope was used to measure the crack length in each specimen. The calibration factor was then calculated for the SENB specimen, and this factor ensured that each specimen was comparable. This calibration factor is denoted as $f(x)$ and takes the geometry of each specimen into account. In this equation K_{IC} is the fracture toughness, P_Q is the max flexure load, *B* is the thickness at the crack and, *W* is the width at the crack.

$$K_{IC} = \left(\frac{P_Q}{BW^{\frac{1}{2}}} \right) f(x) \tag{2}$$

An increase in fracture toughness of 32.5% was observed when adding 15% glass fiber to the resin. These results can be observed below in **Figure 9**. This is a tremendous improvement to one of the historically low properties of a UV curable resin. Also, the standard deviation narrowed in the results when testing reinforced specimens, providing a more consistent performing product.

The increase in fracture toughness by addition of short fibers is not uncommon, and the difference in fracture surface helps discuss the increase. **Figure 10A** is the nearly perfectly smooth brittle failure of the Neat specimen, and **Figure 10B** shows the reinforced fracture surface. The reinforced fracture surface exhibits a multitude of fracture surface redirections because of the obstructing fibers, greatly increasing the fracture surface area. Fiber pullout can also be seen on the surface providing increased toughness and ductility.

The flexural results are a great indicator of a part's quality and encompass the overall performance of the resulting composite. One of the most obvious improvements is the reproducibility of the results that can be seen in **Figure 13**. By introducing fiberglass to the system, there was not only a substantial improvement in modulus but overall a more predictable performance was obtained. During testing, crazing could be observed on the tensile side of the flexural specimen during the plastic deformation portion of the test on just the reinforced specimens. Abrupt brittle failure occurred on all the neat specimens with each specimen having multiple fracture surfaces. The change in the average maximum flexural stress, average flexural modulus, and average max strain is shown in the graphs below (**Figure 11**). The addition of the 15% fiberglass increased the

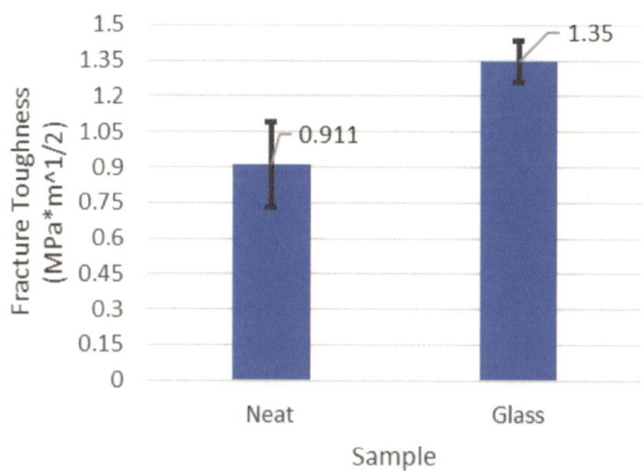

Figure 9.
Average fracture toughness of neat and glass-reinforced samples.

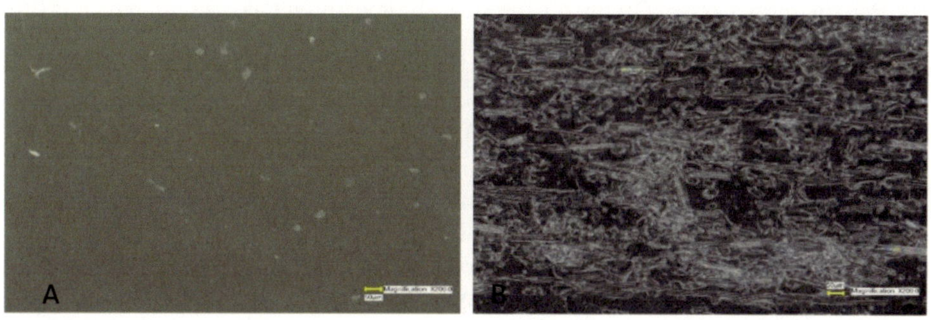

Figure 10.
Fracture toughness surface: A, neat; B, fiberglass.

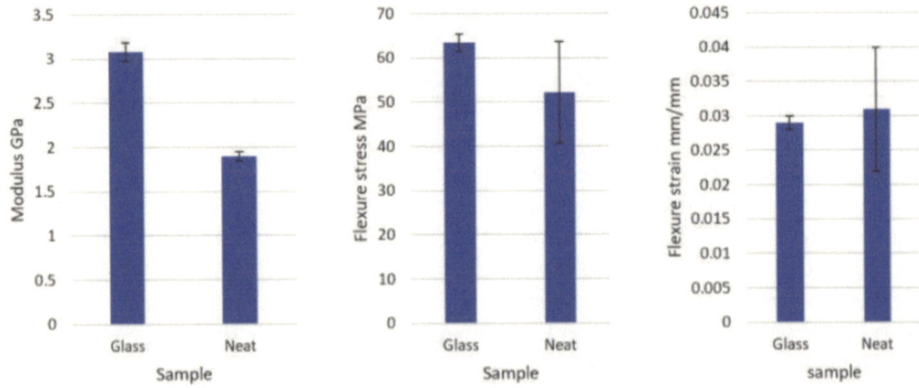

Figure 11.
Comparison of neat and glass fiber-reinforced samples of: Modulus, flexure strength, and flexure strain.

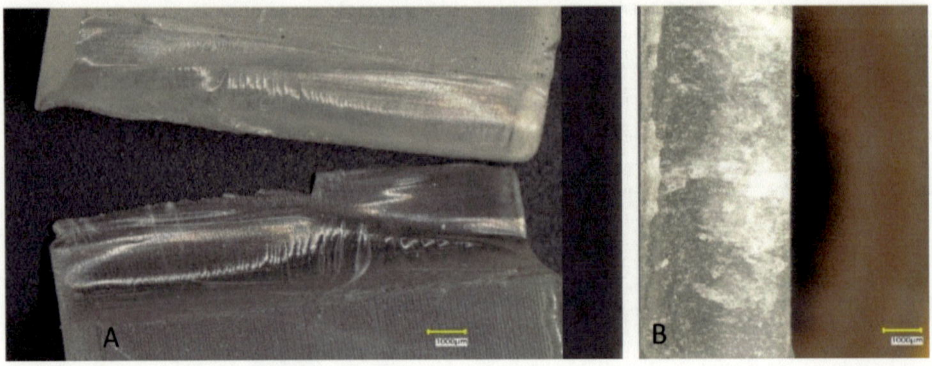

Figure 12.
Flexural specimens: A, neat; B, fiberglass.

average flexure stress by 18%, increased the average modulus by 38%, and effectively did not change the average strain to failure.

The fracture surfaces can be seen in **Figure 12A** for the Neat specimen and **Figure 12B** for the reinforced specimen. The brittle failure on the neat specimens is very clean and consistent. While the fracture surface of the reinforced specimen shows signs of more cumulative damage. In **Figure 12B**, the tensile side of the flexural specimen on the right where the bright white shades on the surface can be seen and at the neutral axis going to the left the compression side shows a clear resin appearance. On the compression side of the specimen, there was little evidence of permanent deformation. On the tensile side, the specimen has crazing, which is the white portion of the specimens on the right side of the neutral axis. This microscopic evidence supports the results of increasing the modulus significantly and improving the toughness of the printed specimen by retaining the strain to failure of the neat polymer. The fibers are providing continued stress transfer and continuity beyond the polymers ability.

Figure 13 is the fracture surface from the **Figure 12B** on the tensile side of the specimen. The image shows clean fiber pullouts and areas of crazing. The fibers are not at a critical length where stress transfer is expected to be significant enough to fracture the fibers in the tensile. Also, seen in this image is the randomness of fiber orientation in the left to right directions, but rarely in the z-direction of the print, which is top to bottom. This orientation is due to the fiber length being greater than the print layer height, allowing for fiber alignment to be reduced to one plane.

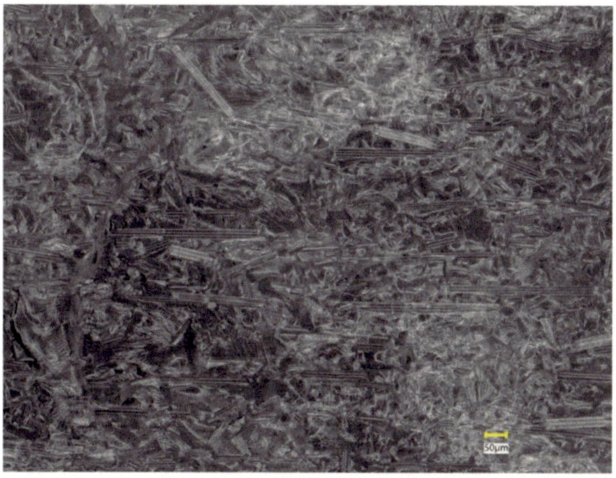

Figure 13.
Flexural fracture surface at 200×.

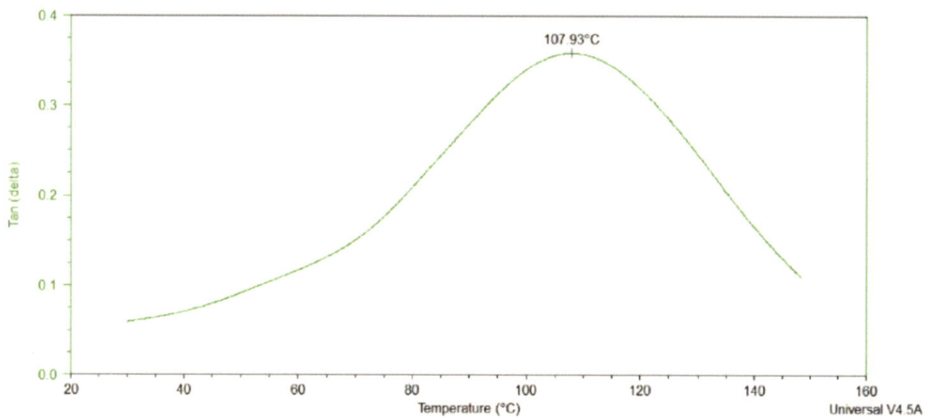

Figure 14.
Neat sample 3 of DMA testing.

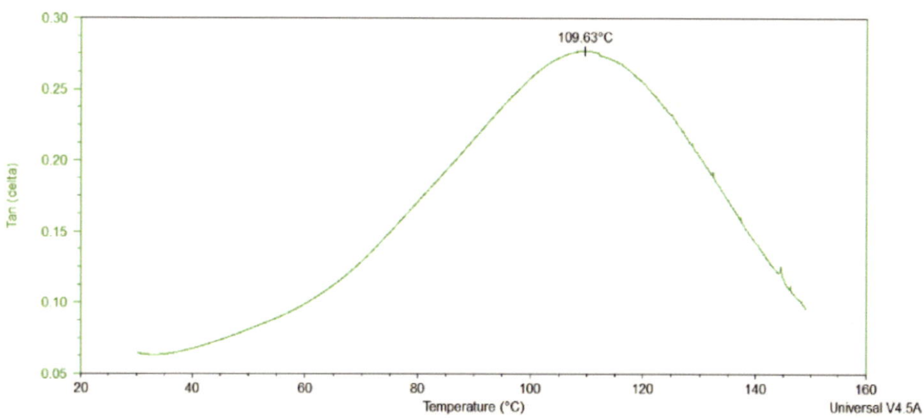

Figure 15.
Glass sample 1 of DMA testing.

To understand if the level of cure of the reinforced specimens was equal to the neat specimens, DMA was performed. The Tg was found using the tan(δ) curve. The neat resin was considered to be optimally cured by the recommended manufacture processing and post curing of the resin, with the reinforced specimens receiving the same processing parameters. Examining the curves of the neat samples, the average Tg was found to 107.59°C. Sample 3 had the closest actual Tg to the average and can be seen below in **Figure 14**.

The three composite specimens yielded an average Tg of 109.19°C. Glass sample 1 had the closest Tg to this value, and the graph can be seen below in **Figure 15**. Adding fiberglass to a neat resin usually results in an improvement in Tg. It was found that by adding the glass fiber to the resin matrix, the Tg increased by 1.6°C. This increase in Tg does confirm that the specimens were comparably cured with and without the fiberglass reinforcement.

4. Conclusions

From the DSC results, it was determine that the neat Moai resin did not exhibit any curing due to thermal processing, but with the incorporation of a thermal initiator, the resin would cure both with and without prior UV curing. Luperox P was chosen as the thermal initiator that yielded the best results from the ones evaluated, from both processing and material properties perspective.

From the tensile testing results, the carbon fiber samples showed an increase in Young's modulus of 21% for a 100 μm layer height, and a 27% for a 50 μm layer height. The change in the modulus due to the lower layer height is due to the fibers being aligned in the loading direction within the sample. This demonstrates that manipulating a part's properties based off of fiber length and layer height could be possible. This increase was less than what was predicted by theoretical models. The lower modulus was thought to be caused by large distribution of fiber lengths and poor interfacial properties between the carbon fiber and matrix. These defects also lead to the carbon fiber samples having a lower tensile strength than the neat Moai resin samples. With improvements in these areas, the tensile properties of the composite could be further improved.

The flexural strength, flexural modulus, and fracture toughness showed no noticeable gains in material properties from the use of the carbon fiber reinforcement. For the flexural testing, this is thought to be because of the low fiber volume (5%) and high fiber volume inconsistency of the samples. While factors such as interfacial strength, fiber length distribution and volume consistency, and low fiber volume content lowered the effectiveness of the carbon fiber as a reinforcement, the main limiting factor is the fiber length itself. The shorter length of the fibers (76 μm) being used is below the critical length needed for the system (433 μm using a shear lag model) [6]. While it would be possible to incorporate fibers of this length into the resin and print parts, they could not have isotropic properties due to forced alignment of the layer height. The effects of fibers folding over within the layers could also affect the ability for the part to print successfully. This could be by limiting the amount of matrix material available to keep the layers attached to one and another. To avoid this, it would require a layer height at least as larger as the fiber length itself. This is outside of the capabilities of even industrial grade vat photopolymerization printers, that typical print at around 200 μm on the top end of layer height [30]. The incorporation of carbon fiber and the subsequent shadowing of the photo initiator from the UV source would even further limit the depth of cure that could be achieved. So while it might not be feasible to use fiber with a length of 433 μm, increasing the fiber length and tightening the fiber length distribution would aid in increasing the material properties for the samples studied in this research.

Additive manufacturing of high-density reinforcement composites using SLA vat printing can be accomplished using the newly developed adaptation of flow induction. Composites with reinforcements having largely disparaging densities can be homogenously manufactured with standard operational settings. This study proved out that the short glass fibers can be incorporated into a 3D printed part resulting in tremendous improvements in mechanical properties while maintaining print cure and consistency. This patent pending design (US Patent Application 63/073,260) can allow for 3D printing to continue to expand its offerings and capabilities.

Acknowledgements

This work was supported by the Army Research Laboratory Award #W911NF-16-0242.

Conflict of interest

The authors declare no conflict of interest.

Author details

Patrick Simpson, Michael Holthaus, Luke Gibbon and Chad Ulven*
North Dakota State University, Fargo, North Dakota, United States of America

*Address all correspondence to: chad.ulven@ndsu.edu

References

[1] Wang X, Jiang M, Zhou Z, Gou J, Hui D. 3D printing of polymer matrix composites: A review and prospective. Composites Part B: Engineering. 2017; **110**:442-458

[2] Karalekas D, Antoniou K. Composite rapid prototyping: Overcoming the drawback of poor mechanical properties. Journal of Materials Processing Technology. 2004; **153-154**:526-530

[3] Parandoush P, Lin D. A review on additive manufacturing of polymer-fiber composites. Composite Structures. 2017;**182**:36-53

[4] Stansbury JW, Idacavage MJ. 3D printing with polymers: Challenges among expanding options and opportunities. Dental Materials. 2016;**32**(1):54-64

[5] Ogale AA, Renault T. 3-D photolithography for composite development: Discontinuous reinforcements. SAMPE Quarterly. 1991;**5**(6):28-38

[6] Agarwal BD, Broutman LJ, Chandrashekhara K. Analysis and Performance of Fiber Composites. New York: John Wiley & Sons; 2017

[7] Mortazavian S, Fatemi A. Effects of fiber orientation and anisotropy on tensile strength and elastic modulus of short fiber reinforced polymer composites. Composites Part B: Engineering. 2015;**72**:116-129

[8] Papthanasiou TD, Guell DC. Flow-Induced Alignment in Composite Materials. Sawton, Cambridge: Elsvier, Woodhead Publishing; 1997

[9] Goh GD, Dikshit V, Nagalingam AP, Goh GL, Agarwala S, Sing SL, et al. Characterization of mechanical properties and fracture mode of additively manufactured carbon fiber and glass fiber reinforced thermoplastics. Materials & Design. 2018;**137**:79-89

[10] Ferreira RTL, Amatte IC, Dutra TA, Bürger D. Experimental characterization and micrography of 3D printed PLA and PLA reinforced with short carbon fibers. Composites Part B: Engineering. 2017;**124**:88-100

[11] Ning F, Cong W, Qiu J, Wei J, Wang S. Additive manufacturing of carbon fiber reinforced thermoplastic composites using fused deposition modeling. Composites Part B: Engineering. 2015;**80**:369-378

[12] Hector Sandoval J, Wicker RB. Functionalizing stereolithography resins: Effects of dispersed multi-walled carbon nanotubes on physical properties. Rapid Prototyping Journal. 2006;**12**(5):292-303

[13] Yang Z, Wu G, Wang S, Xu M, Feng X. Dynamic postpolymerization of 3D-printed photopolymer nanocomposites: Effect of cellulose nanocrystal and postcure temperature. Journal of Polymer Science Part B: Polymer Physics. 2018;**56**(12):935-946

[14] Feng X, Yang Z, Chmely S, Wang Q, Wang S, Xie Y. Lignin-coated cellulose nanocrystal filled methacrylate composites prepared via 3D stereolithography printing: Mechanical reinforcement and thermal stabilization. Carbohydrate Polymers. 2017;**169**:272-281

[15] Tekinalp HL, Kunc V, Velez-Garcia GM, Duty CE, Love LJ, Naskar AK, et al. Highly oriented carbon fiber–polymer composites via additive manufacturing. Composites Science and Technology. 2014;**105**:144-150

[16] Thomas H, David BP, Guido T, Hans NH. State-of-the-art of

fiber-reinforced polymers in additive manufacturing technologies. Journal of Reinforced Plastics and Composites. 2017;**36**(15):1061-1073

[17] Gupta A, Ogale AA. Dual curing of carbon fiber reinforced photoresins for rapid prototyping. Polymer Composites. 2002;**23**(6):1162-1170

[18] Taormina G, Sciancalepore C, Bondioli F, Messori M. Special resins for stereolithography: In situ generation of silver nanoparticles. Polymers. 2018; **10**(2):212

[19] Cheah CM, Nee AYC, Fuh JYH, Lu L. Mechanical characteristics of fiber-filled photo-polymer used in stereolithography. Rapid Prototyping Journal. 1999;**5**(3):112-119

[20] Sano Y, Matsuzaki R, Ueda M, Todoroki A, Hirano Y. 3D printing of discontinuous and continuous fibre composites using stereolithography. Additive Manufacturing. 2018; **24**:521-527

[21] Lasgorceix M, Champion E, Chartier T. Shaping by microstereolithography and sintering of micro-porous silicon substituted hydroxyapatite. Journal of the European Ceramic Society. 2016;**36**:1091-1101

[22] Dufaud O, Marchal P, Corbel S. Rheological properties of PZT suspensions for stereolithography. Journal of the European Ceramic Society. 2002;**22**:2081-2092

[23] Song S, Park M, Lee J, Yun J. A study on the rheological and mechanical properties of photo-curable ceramic/ polymer composites with different silane coupling agents for SLA 3D printing technology. Nanomaterials. 2018;**8**:93

[24] Sakar M, Balakumar S, Saravanan P, Jaisankar S. Electric field induced formation of one-dimensional bismuth ferrite (BiFeO3) nanostructures in electrospinning process. Materials and Design. 2016;**94**:487-495

[25] Liu Y, Chung J-H, Liu W, Ruoff R. Dielectrophoretic assembly of nanowires. The Journal of Physical Chemistry. 2006;**B 110**:14098-14106

[26] Li Y, Wu Y. Coassembly of graphene oxide and nanowires for large-area nanowire alignment. Journal of the American Chemical Society. 2009; **131**:5851-5857

[27] Yunus D, Shi W, Sohrabi S, Liu Y. Shear induced alignment of short nanofibers in 3D printed polymer composites. Nanotechnology. 2016;**27**:49

[28] ASTM. E2160-04 Standard Test Method for Heat of Reaction of Thermally Reactive Materials by Differential Scanning Calorimetry. West Conshohocken, PA: ASTM International; 2018

[29] ASTM. D638-14 Standard Test Method for Tensile Properties of Plastics. West Conshohocken, PA: ASTM International; 2014

[30] ASTM. D3039/D3039M-17 Standard Test Method for Tensile Properties of Polymer Matrix Composite Materials. West Conshohocken, PA: ASTM International; 2017

[31] ASTM. D790-17 Standard Test Methods for Flexural Properties of Unreinforced and Reinforced Plastics and Electrical Insulating Materials. West Conshohocken, PA: ASTM International; 2017

[32] Zhang H, Zhang Z, Breidt C. Comparison of short carbon fibre surface treatments on epoxy composites: I. Enhancement of the mechanical properties. Composites Science and Technology. 2004;**64**(13):2021-2029

[33] Dixon KW. Decomposition rates of organic free radical initiators. In: Polymer Handbook. Vol. 4. New York: John Wiley & Sons, Inc.;1999

[34] Purslow D. Matrix fractography of fibre-reinforced epoxy composites. Composites. 1986;**17**(4):289-303

[35] Anderson TL. Fracture Mechanics: Fundamentals and Applications. Boca Rarton, FL: CRC Press; 2005

[36] Braga RR, Ballester RY, Ferracane JL. Factors involved in the development of polymerization shrinkage stress in resin-composites: A systematic review. Dental Materials. 2005;**21**(10):962-970

[37] Chambers AR, Earl JS, Squires CA, Suhot MA. The effect of voids on the flexural fatigue performance of unidirectional carbon fibre composites developed for wind turbine applications. International Journal of Fatigue. 2006;**28**(10):1389-1398

[38] Parry TV, Wronski AS. Kinking and tensile, compressive and interlaminar shear failure in carbon-fibre-reinforced plastic beams tested in flexure. Journal of Materials Science. 1981;**16**(2): 439-450

[39] "Standard Test Methods for Plane-Strain Fracture Toughness and Strain Energy Release Rate of Plastic Materials", D5045-14. West Conshohocken, PA: ASTM International; 2014

[40] "Standard Test Method for Plastics: Dynamic Mechanical Properties: In Flexure (Dual Cantilever Beam)", D5418-15. West Conshohocken, PA: ASTM International; 2015

[41] ASTM D792-13 Standard Test Methods for Density and Specific Gravity (Relative Density) of Plastics by Displacement. West Conshohocken, PA: ASTM International; 2013

Chapter 5

Perspective Chapter: Breaking the Barriers – Additive Technologies (AX) for Integrated Process Chains and Integrated Devices (IDs) for Hybrid Product Architectures

Matthias Dahlmeyer and Sebastian Noller

Abstract

Additive technology has evolved from rapid prototyping to rapid tooling and manufacturing of load-bearing parts for productive use. Application potential is limited by constituent strengths and weaknesses. To enfold its full potential, research, development, and industrial application have to facilitate combinations of additive and conventional technology. The concept of additive parts manufacturing has to be expanded to a mature technology contributing and facilitating hybrid products and integrated process chains. From a two-dimensional reference model, approaches to integration are derived, and their status is briefly outlined: Efforts to facilitate postprocessing by design for additive manufacturing (DfAM) and hybrid manufacturing have been raised to awareness and are being worked on. Yet, integration of pre-fabricated structures is hardly accounted for, although it bears the potential for a paradigmatic shift in manufacturing: With a wider concept of layer-based processes, Additive Technology could form the core technology for integration of components and functions to Integrated Devices, following the model of the Integrated Circuits and packaging technology in microelectronics and Microelectromechanical Systems. First developments are outlined, but research and development effort has to be dedicated to novel additive processes for this application. Finally, workflows for product developers need to be modified and trained to plan hybrid product architectures already in conceptual phases.

Keywords: additive technologies (AX), integrated process chains, integrated devices (IDs) hybrid product architecture, design for additive manufacturing (DfAM), packaging of integrated structures, additive joining, additive coating, material grading, selective liquid-phase sintering (SLPS)

1. Introduction

Additive processes have developed from rapid prototyping purposes for nonfunctional demonstrators, via rapid manufacturing of specific single parts,

IntechOpen

to a regular manufacturing technology for load-bearing parts on industrial scale. However, additive parts manufacturing (AM) application still focuses mainly on specific parts that are complex, require customization, often with short lead time, and low production volume. Although existing processes are continuously improved and new ones created, it is unlikely that additive processes will overcome their own limitations to actually substitute conventional processes. In order to meet typical requirements of—or even transform—industrial reality from the perspective of a complete product, AM should not be conceived as an alternative to conventional technology, but an interoperable contribution as a specific link in a process chain to realize hybrid products.

This chapter outlines critical factors for process and product integration and expands the current predominant perspective of additive parts manufacturing (AM) to a more comprehensive concept, furtherly referred to as *Additive Technologies (AX)* as a prerequisite to realize an *Integrated Device (ID)*. It identifies factors and approaches to break barriers:

1. between additive and conventional processes, toward integrated process chains,

2. between additive and conventional design, toward integrated product architectures,

3. for a new paradigm in production technology and product architecture that leverages the novel possibilities of additive technology more consequently, and therefore,

4. for the wider industrial adoption and capitalization of AX.

First, the strengths and weaknesses of additive processes and approaches to regard them in product design are recapitulated. Based on a frame of reference, possible dimensions of integrations are introduced to derive approaches for integration. The status for each approach is summarized and illustrated by own examples, and the need for further research on one promising approach is defined.

2. AM processes: light, shadow, and the exploitable potential between

The broad range of established additive processes of today offers significant economic and technologic advantages over traditional formative or subtractive parts manufacturing. Placing material selectively in space allows directly building up complex shapes, with hollow sections, undercuts, delicate, or inner structures—without the costs and building time of part-specific tooling and without the design constraints of conventional processes, e.g., for tool accessibility, material flow, demolding, or machining forces.

On the other hand, additive processes imply constraints of their own, affecting their fitness to fulfill requirements for the growing industrial interest. As additive processes are highly diverse, constraints can vary significantly. For industrial manufacturing of mainly load-bearing industrial metal parts, the main focus here will be processes from the categories powder bed fusion (PBF) and directed energy deposition (DED).

Process constraints include:

- Current process principles require that the building space to be enclosed in the machine.

- The buildup process is sequential and (at least for PBF) unidirectional.
- Overhung geometries require support structures.
- The workpiece, support structures, and the building space continuously alter in geometry and behavior as the buildup proceeds.
- Locally high thermic gradients cause thermal frozen-in tension in typically delicate structures in transient state, influencing the further buildup and reactions to material removal for finishing.
- as well as production, handling, and control of fine-grained powder used as raw material.

These influences affect key target variables, such as

- parts size,
- parts precision and homogeneity—in form, dimensions, surface quality, and material properties,
- manufacturing speed—especially for massive geometries, and
- costs—of fine-grained raw materials and extensive machine time.

Furthermore, PBF processes constitutively require a powder bed of costly raw material to fill the working space—even if only an insignificant fraction of it is actually processed into smaller, delicate, or hollow structures. For single piece or small batch production, the effectively used powder may be a fraction of the excess. If subsequent parts or runs require powder of identical material and grade, the excess powder may be reconditioned (usually by sieving), but efficiency is limited by its coagulation, material alteration, and other contamination (depending on process parameters such as thermal energy applied in the process): Typical is an equal mix of fresh powder, powder from the reservoir, and from the actual bed (distant from the volume affected by the LASER).

DED processes do not require a powder bed, but require a minimum continuous stream of comparably flowable, coarse powder, accelerated by a carrier and shielding gas. This results in comparably high powder delivery rates that have to be matched with a high application heat (LASER) energy. Overall, benefits from higher buildup rates at concession of less resolution and thicker walls result in a "near-net shape" that requires substantial postprocessing. In general, the coarser surface quality may not fulfill functional requirements and may lead to a reduced fatigue life of parts [1].

Overall, AM (as a process group, and each process for itself) has a specific profile of strengths and weaknesses—ideal for some purposes and adverse for others. While the rise of AM may be a disruptive landmark of manufacturing technology, AM does not supersede the range of manufacturing processes but complement them. Many technical applications will require AM to concur rather than to compete with other processes—deploying each process where it is superior and yielding to others where it is inferior. Consequently, a key to capitalize on additive technology will be to expand the focus from process development toward process integration for hybrid architectures of products, assemblies, parts, and features.

3. Integration by design: design for additive manufacturing (DfAM)

The connection between product design and production technology is heavily interdependent: Product design leads its implementation in production—but it has to proactively regard opportunities, limitations, and costs for processes downstream in the value creation chain. Consequently, design rule sets have been developed to design for subsequent processes from early on, ideally already in conceptual phases—allowing for a foresighted, holistic product design. Initial design rules were implemented for manufacturing as *Design for Manufacturing (DfM)*, then for assembly (*DfA*), and then for more aspects (such as transport, testing, service, sustainability), converging to the discipline *Design for X (DfX)*, sometimes paraphrased *Design for eXcellence*. As a new technology, the novel opportunities and limitations as described above require specific design rules and approaches. This is represented in the new research field of *Design for Additive Manufacturing (DfAM)*.

On the one hand, the novel possibilities from selectively placing material spawned visionary approaches and methods, such as topology optimization, lattice or cell-based structures, or multi-material design. While these attracted much attention, there was (and still is) also the need to systemize the specific limitations for reliably designing parts for AM processes and their typical subsequent processes in the more fundamental meaning of DfX. This concerns more the potential and restrictions of specific processes, and how they can be considered to generate parts workably, effectively, and economically. In the past, this has been subject to practical experience and therefore a secret of trade of specialized AM service companies. With regular availability and maturity of processes and machines, publications have increasingly covered systematic Df(A)M guidelines for process limits, achievable feature properties (e.g., dimensions and tolerances of walls, gaps, holes, overlap angle, etc.), and workable and effective buildup strategies (e.g., how to plan part orientation, support structures, etc.). This chapter on integration will focus these more fundamental aspects of DfAM.

Collections of process- and buildup-specific design rules are specified in guidelines such as the German VDI guideline 3405 [2]. A compilation of design rules of various categories, based on the analysis of a complex enclosure for a satellite-based X-ray spectrometer measurement system, was presented in [3].

4. Dimensions of integration: a frame of reference

In order to derive integrative approaches integration, a systematic frame of reference will be used (**Figure 1**).

The *dimension of process sequence* extends the scope along the process chain, opening up the potential for the *integration of precedent and subsequent processes*. There is a growing awareness for *downstream* processes (e.g., finishing, treatment, handling, testing, assembly, and transport).

As the current focus on additive processes is mostly limited to a concept of exclusive generation of parts from scratch (with all downsides), there is no substantial coverage of interoperation with *upstream* conventional processes.

The *dimension of product structure* extends from the AM part up and down the product hierarchy, creating touchpoints for the *integration of subordinate and superordinate product elements*. Due to the drawbacks of AM technology, AM parts had to be considered early on, either as semifinished goods for conventionally finished or treated net shape parts or as components in conventionally joined for *higher-level* assemblies and products (with other conventionally or AM fabricated other parts).

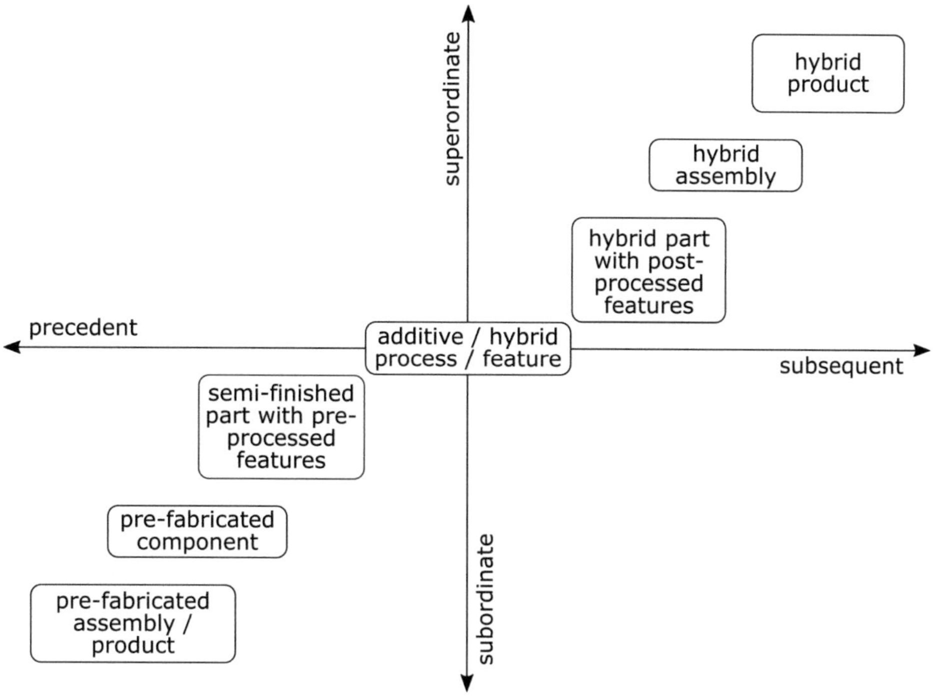

Figure 1.
The origin represents the currently predominant focus on the additive processing and on the additive part. It bears the potential for integration of concurrent hybrid processing (alternating, or maybe simultaneously).

Corresponding with upstream processes, applying AM processes to *lower-level* prefabricated elements (such as semifinished goods, parts, assemblies, or even products) is factually unaccounted for even in DfAM as of today.

Overall, integration of subordinate components, pre-fabricated, precedent to AM processing, would open up potentials currently hardly covered even in research. However, this would require:

- a substantial paradigm shift from the current focus of generative-additive manufacturing of parts (AM) toward a generic concept of *additive technology (AX)* for extended purposes, including, e.g., filling, joining, and finishing with AM processes; and

- the development of technology with new capabilities as a facilitator for this type of application.

From the frame of reference, the following points of process and product integration can be derived:

1. *Facilitate external postprocessing* of additive structures, to chain-link AM processing and structures to conventional (or variant AM) postprocessing for superordinate properties, features, and components.

2. *Facilitate hybrid processing*, to simultaneously create features with combined additive and conventional processes on hybrid machines.

3. *Facilitate externally pre-fabricated elements for additive processing*, to chain-link AM processing and structures to conventional (or variant AM) processed, subordinate properties, features, and components.

4. *Facilitate hybrid product architectures* to methodologically regard the whole process chain and product architecture together from the start.

The following subsections outline the status and potential for each of these aspects.

5. Facilitate external postprocessing of additive structures

For early single-piece rapid prototypes, manual postprocessing such as cutting clean, joining, filling, and painting has been a frequent implicit necessity. Most industrial-scale additive metal parts today still require postprocessing to achieve the required geometric and surface qualities [4]. Therefore, DfAM has increasingly adopted design rules in order to systematically plan for typical postprocesses, such as:

1. cleaning loose and adhering material remains from cavities and surfaces

2. removing the part from the build plate

3. trimming support structures

4. reducing thermal strain by heat treatment

5. finalizing optical or technical quality of surfaces and dimensions, by coating or machining

6. manufacturing conventional parts separately, and preparing AM parts for joining by adding additional, postprocessed features, e.g., for parts handling, for application of mating forces, and for joining features such as threads, contact faces, or sealing faces.

Further successive processes may include testing, transportation, utilization, maintenance, or recycling. However, although these processes need to be regarded

no.	Design rule	Unfit for postprocessing	Fit for postprocessing
1	Provide parallel clamping surfaces for postprocessing of massive parts*.	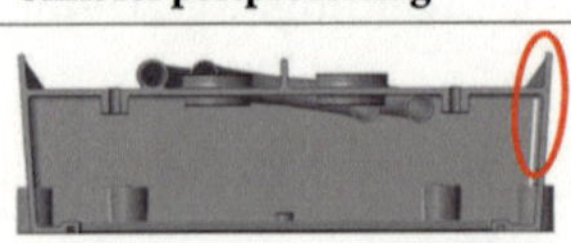	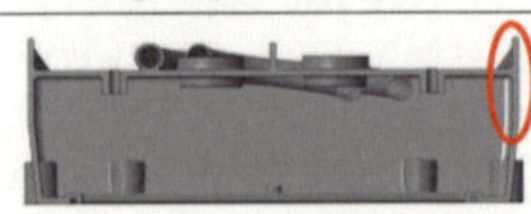
		Parallel clamping surfaces were integrated for postprocessing of the chassis	
2	Provide clamping points for postprocessing of elastic parts.	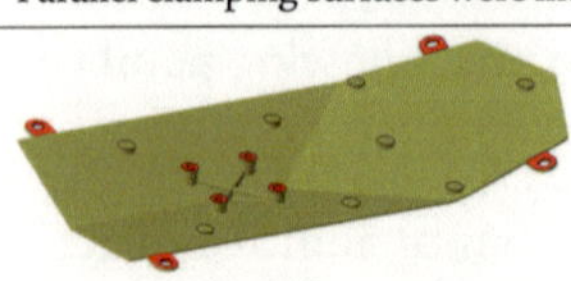	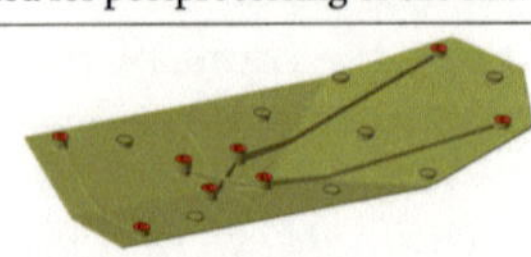
		To avoid removal of clamping points, these are integrated with the part and supported with webs	
3	Provide support points for distinct positioning and stability.	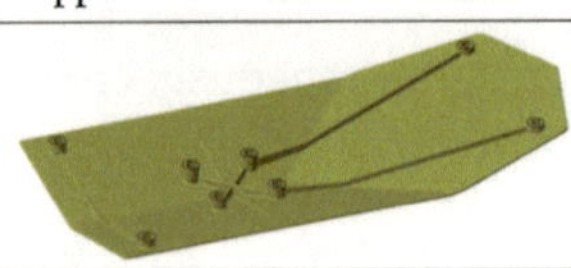	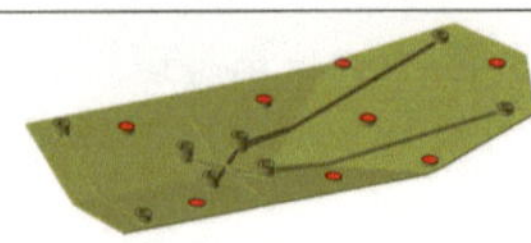
		Multiple support points avoid elastic yielding to machining forces (here of the bottom surface)	

no.	Design rule	Unfit for postprocessing	Fit for postprocessing
4	Stabilize elastic sections for machining forces during postprocessing	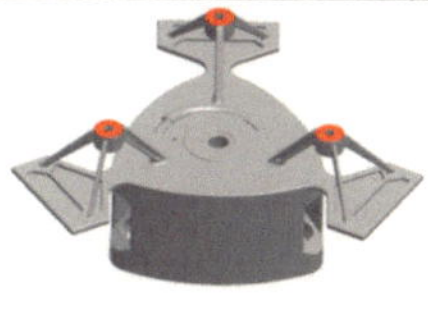	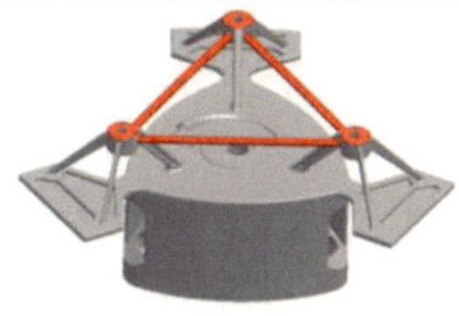
		Avoid elastic yielding when machining the three outer paddle surfaces	
5	Offset fine tolerated surfaces for selective postprocessing*	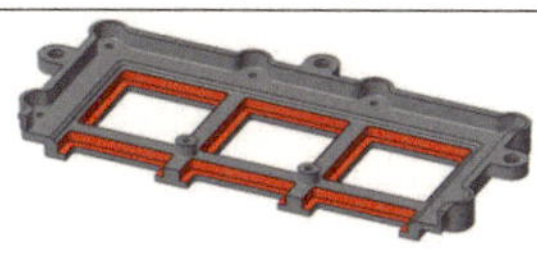	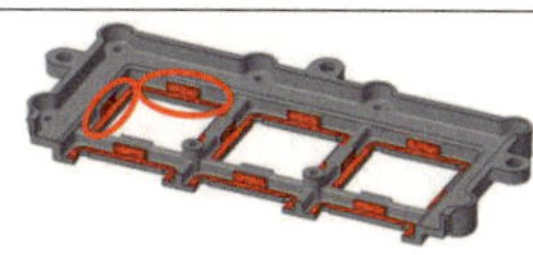
		Reduce contact surfaces for insets to limited contact points	
6	Allow for sufficient tool runout for postprocessing*.		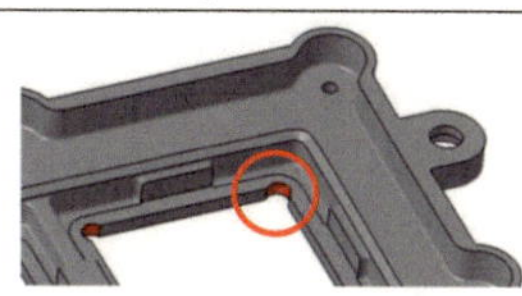
		Recessed radius allows for bigger (more rigid) end mill	
7	Allow access for removal of support structures	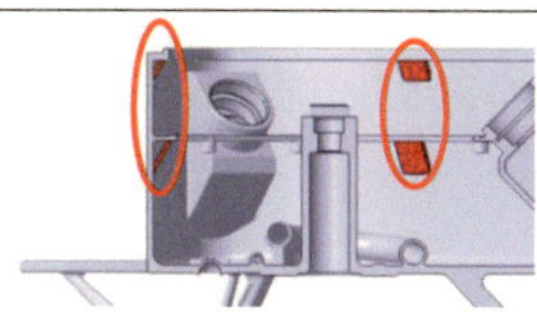	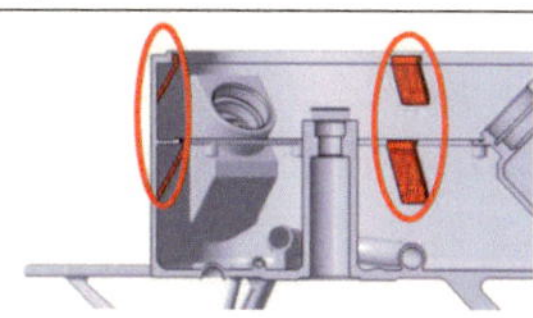
		Support structure inside collar is better accessible than directly below	
8	Substitute geometries prone to warping with massive geometry for later relieve through machining.	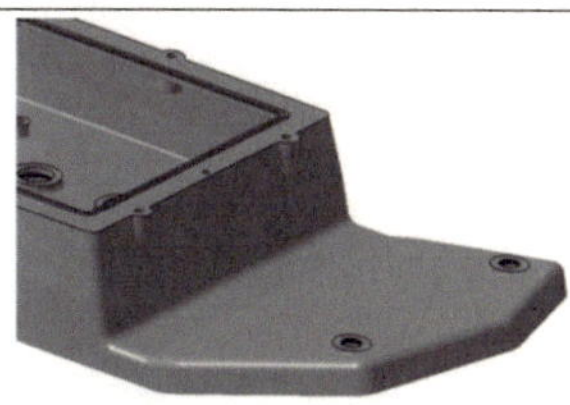	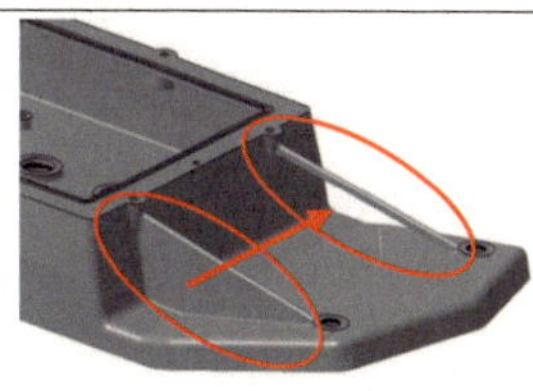
		Long struts are likely to distort during buildup. Instead, a closed surface is built and then cut out without machining forces (wire-cut EDM).	
9	Integrate geometries for in the near-net shape.	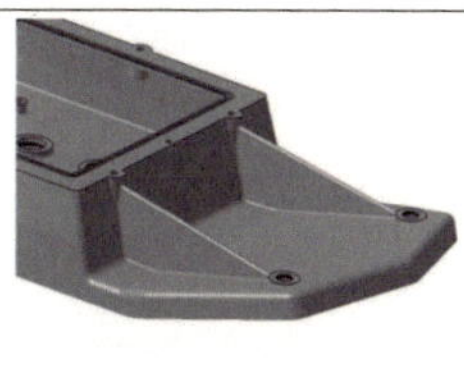	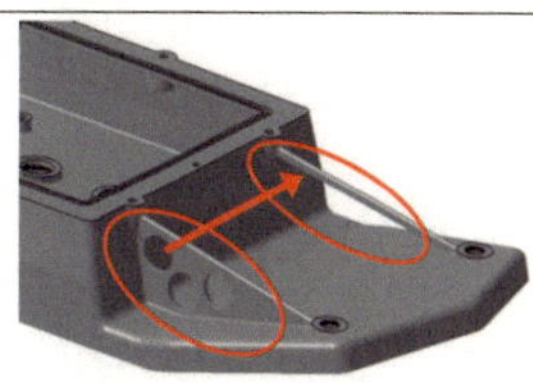
		Apertures for the eroding wire are directly integrated in additive buildup	

**In analogy to typical guidelines for DfM for casting.*

Table 1.
Excerpt of design rules for post-processing of additive parts (based on [3]).

systematically, their systematic consideration is part of the regular established design literature design engineering, specifically *Design for X (DfX)*, and not specifically covered here. An exception may be the testing of AM-specific characteristics such as material properties (e.g., relative density) that require dedicated design considerations.

Lee et al. [1] concludes that based on the existing ISO/ASTM standards, only less than 1% ASTM standards related to the surface finishing of metal AM components and proposes vast R&D opportunity in these areas.

Table 1 shows an excerpt, of design rules specific to postprocessing of additive parts, based on [3].

6. Facilitate hybrid processing

Hybrid AM has been a focus of investigation increasingly over the past decade [4]. Hybrid manufacturing systems compensate the limits of AM processes by alternating or concurrent application of conventional processing to combine advantages of both worlds: high material utilization and complex and internal shapes (AM) combined with high productivity, feature accuracy, and surface quality of conventional machining.

Two approaches can be distinguished:

In a *multi-setup*, additive and nonadditive processes are combined across multiple, separate machines for conventional processing between additive buildup steps. This essentially logistic approach to integration allows to process differently progressed parts in parallel on separate machines (e.g., one part is machined while the next part is still being built up).

However, the main focus of hybrid (additive) manufacturing is a *single setup* in one hybrid machine that combines technologically

- an additive process (usually DED or PBD), reducing excess material for machining, and allowing complex or internal feature or shapes,

- with usually a subtractive machining process, resulting in better "as-built" surface finishing and geometric accuracy, such as milling, turning, or grinding. There are also combinations with heat treatment (conventional or LASER) or surface treatment (e.g., shot peening or LASER shot peening).

Compared with an exclusively additive process or a hybrid multi-setup, advantages of hybrid single setup are as follows:

- Components can be ready-to-use directly after the hybrid manufacturing process instead of being near-net shape.

- This reduces lead time and material (and thus production costs).

- The required surface qualities can be achieved in one clamping (where surfaces are accessible).

- Parts can be designed with less restrictions because there is no need for additional elements, such as clamping surfaces for postprocessing.

- Geometric referencing of the AM component in a downstream machine can be avoided.

- If necessary, accurate interfaces for further positioning and machining can be created.

- Internal geometries (e.g., fluid structures) can be machined during buildup.

- After the additive buildup of each layer, a uniform layer height can be generated by the machining process. Thus, the next layer can be created with reproducible conditions of surface quality and geometry tolerance.

- Since all production steps are automated on one machine via CAD/CAM interface, the probability of human errors is reduced.

On the other hand, limitations and challenges have to be considered, such as:

- In hybrid systems, only additive or conventional processing can be carried out at a time, whereas multi-setups allow parallel processing.
- Chips and coolant can hinder the subsequent additive process or impair the material bond.
- Coolant residues can contaminate the powder.
- Powder residues can hinder the machining process and increase tool wear.
- The clamping of the components must be sufficient, as cutting forces and resulting vibrations occur.
- After the hybrid process, the part's bottom side still has to be machined.

7. Facilitate additive processing of externally pre-fabricated elements

Some problems of AM can be compensated by subsequent or hybrid postprocessing. Other problems remain or originate from adverse effects between additive structures and conventional processing, in the first place. These specific problems can be attributed mainly to the predominant traditional process order "generate or build up material, then subtract from it, and assemble." In consequence, they can be addressed by reversing or stacking this order to "finish material, then add material to it."

This essentially means that otherwise finished parts could be extended by, or be incorporated in, hybrid or monolithic buildups, furtherly referred to as *Integrated Devices (IDs)*.

7.1 Integrated devices (IDs): a new potential for the macromechanical world

In the 1950s, a disruptive invention revolutionized electronic circuitry and their production: Traditionally, discrete electric and electronic components were produced separately, and afterward wired point-to-point, or soldered to a circuit board. Based on preceding developments of a planar process with structured layers on a substrate, the integrated circuit (IC) was developed, as the hybrid IC (incorporating different functional elements into a single buildup, patented by Jack Kilby), and as the monolithic IC (a completely integral buildup of functional elements on one substrate, patented by Robert Noyce). ICs reduced cost, increased processing performance, and paved the road for today's world of ubiquitous microelectronic in industrial and end user products. Especially the hybrid IC with additional packaging processes also facilitated the production of microelectromechanical systems (MEMS).

By applying the same principles to the macromechanical world, the concept of *Integrated devices (IDs)* can be postulated: Rather than fabricating separate components and then assembling them component-to-component, a unit (part, assembly, or product) could be built up by structured layers in an additive process, incorporating premanufactured conventional or additive components, i.e., machined, purchased, or standard parts, complete assemblies up to full products.

IDs could open up completely new application potential for the group of additive processes: Rather than deciding *whether or when* a part should be processed conventionally *or* additively, it could be distinguished *where (which features)* to process conventionally *and* where to build up additively. This would allow to tailor the process chain for a given unit to make use of the advantages of additive technology to where it is superior and to avoid its disadvantages where it is not. Substructures too precise, too massive, or too cost-inefficient for AM could be created externally and then integrated into an additive buildup—even where postprocessing or joining is impossible or uneconomic, due to geometry, accessibility, or structural stability and stiffness to withstand machining or clamping forces with typically complex, delicate parts.

7.2 Additive technology (AX) for IDs

In order to produce IDs that incorporate or package nonadditive features, a processing technology will require not only the primary shaping of structures, but also their extension and alteration (i.e., buildup, joining, and conditioning).

There are examples of packaging-like processes for limited purposes in generative manufacturing, e.g., plastic injection molding around metal inserts. But it is the group of additive processes that bears a striking resemblance to the packaging of ICs and MEMS, by additive buildup of functional units in structured layers.

The current concept of additive manufacturing is mostly focused on (as the designation suggests) parts manufacturing. Additive processes are widely conceived as a subtype or alternative to generative processes. For example, the current (2020) revision draft of the German industry standard DIN 8580 "Manufacturing processes - Terms and definitions, division" [5] attributes group "1.10 Primary shaping by additive manufacturing" clearly to the main group "1. Primary shaping." Consequently, parts integration is implicitly mostly limited to monolithic IDs, where external components and their assembly are substituted by integral AM parts. Also, as-built assemblies (e.g., plastic roller bearings for nonproductive use) from one continuous AM process have demonstrated the fundamental feasibility of monolithic IDs.

However, hybrid approaches to IDs, combining the benefits of additive and conventional processing (as analyzed in Section 2 and 4), are currently not focus of discussion, published research, industrial application, or standardization. This approach would require packaging-like applications of additive processes, such as additive joining and assembly, additive finishing, or additive surface treatment (e.g., melting alloy components into the surface). For distinctiveness of further discussion, the broader concept of *Additive Technology (AX)* shall be used to refer to the technology of applying *additive processes (A)* for *diverse purpose (X)* (not only parts manufacturing).

Table 2 shows a set of examples where design problems can be solved (or solved better) with an AX approach where a pure AM or AM with postprocessing approach is problematic.

Instead of screwing precision-turned elements to complex part with machined threads and joining surfaces, a bolt or nut could be integrated into the additive counterpart for assembly—or the additive section could simply be built up directly around the conventional component without the need for additional joining. Instead of drilling holes and cutting threads, standard purchased nuts could be embedded in a part. Instead of building up massive structures with delicate extensions layer-by-layer, a machined massive core could be extended by additive mountings or surface structures.

Additive manufacturing or postprocessing constraints	Resulting problems	Solution by incorporating pre-fabricated components
additive buildup of locally massive geometries *example: massive core with delicate structures or structured surface*	processing time and costs increased consumption of expensive powder	build up delicate, complex structures on a less costly premachined massive core
complex geometry limits or obstructs access to clamping or working faces, or is incompatible with post-processing. → Split complex part into separate simpler parts for subsequent additive assembly *example: rotational geometry with fine surface requires turning, but global part is not axially symmetric*	impossible or costly to manufacture, or leading to extra work and costs for machining the joining faces and joining.	build up complex, delicate structures on part with machine-finished surface
typical delicate additive structures can deform or warp under clamping and machining forces	making machining unsafe, inaccurate, or cost-inefficient	build up the yielding structure on a finished part
imprinted cooling tension from local thermal gradients during processing can be released when removing surface material	part can warp as a result from postprocessing	pre-fabricate substructure conventionally with finished surface and integrate in subsequent
subsequent postprocessing may require additional features against machining forces, *examples: blind hole threads for joining and clamping, precise contact and seal faces for assembly, supports and fillets*	extra effort and cost for feature creation and eventually removal	buildup additive structure on pre-fabricated parts.
Complex geometry and constraints may prevent standard parts from being joined to buildup	standard features have to be built up as part of an expensive custom part	buildup additive structure on inexpensive standardized parts.

Table 2.
Examples for benefits of the AX approach over AM.

7.3 Developing processes for AX

Established additive processes were developed mainly for parts manufacturing. The extended applications in AX require additive processes that fulfill a different requirements profile, allowing

1. premanufactured parts in the working space,

2. to apply powder from different vectors to inclined surfaces of existing structures,

3. to fuse powder to a wide range of materials,

4. ideally different powder material for grading the added material,

5. to produce precise net shape structures for "as built" use, to avoid the disadvantages of postprocessing, and

6. little impact on material properties of premanufactured parts (unless the purpose of the process is explicitly treatment).

7. Surface quality "as built" has to be sufficiently smooth to avoid stress step-up in the notch grooves, resulting in early fatigue.

Typical PBF or DED processes do not fulfill this profile:

PBF processes can create net shapes of reasonable as-built quality. But they usually do not allow parts to protrude from the powder bed because they would collide with the doctor knife that dresses each layer. The powder bed allows access only from the top, follows a Cartesian working principle with an invariable buildup direction. As the powder is provided as a filled bed, switching to a different material takes a lot of effort. However, the thermal influence on the existing buildup is comparably small.

DED processes, on the other hand, share the working principle with LASER cladding where material is applied to existing parts (e.g., for repair) while it is liquified with a direct energy source (usually, LASER, electron beam, or wire arc). Using a 5-axis-kinematic, material could even be applied from the sides. By using different heads or powder feeds, it is basically possible to alternate between or combine different materials. However, DED processes require a continuous delivery of flowable, coarser powder and carrier (shielding) gas, with adequately high application of energy—resulting in high buildup rates, creating comparatively thick and bulging layers and rough tolerances (depending on the process, up to several millimeters). Results from DED processes are usually designated as "near-net-shape" and require substantial postprocessing.

At the *Hochschule für Technik und Wirtschaft Berlin*—University of Applied Sciences (HTW Berlin), research about AX processes has resulted in two approaches, which are currently in development [6]:

Based on regular PBF processes, workflow and control can be modified to incorporate pre-fabricated inserts with a level top surface. The modified process is described in **Table 3**.

Furtherly, the novel additive process principle "Selective Melt Dispersion (SMD)" was developed in order to fill the niche between PBF and DED. Like DED, it allows parts to protrude out of the buildup and to vary buildup direction, without a voluminous homogeneous powder bed. But the working principle was developed for finer resolutions and tolerances of the as-built net shape: Other than in the regular powder-based DED processes, the powder is not propelled by a shielding gas stream in a mix nozzle and also not heated directly in-delivery. Instead, the directed energy source selectively melts a surface spot of the already built material. Unheated solid metal powder is dispersed onto this melt pool with a separate delivery system, e.g., by a newly developed vibration-operated drop-delivery dispenser. The powder adheres to the liquid surface and is embedded as the working progresses along its continuous track and the pool of metal solidifies as the new layer. This layer is melted again fully to receive the next layer (or for finishing as the final layer).

Separating the powder delivery from the shielding gas stream and the directed energy source allows finer control of each factor, resulting in:

- ability to process fine powder grades, because no mixing nozzle is used that could be clogged with less flowable powder,

	step description	illustration
1	Conventionally machine inserts with a horizontally level top surface at a precisely known height	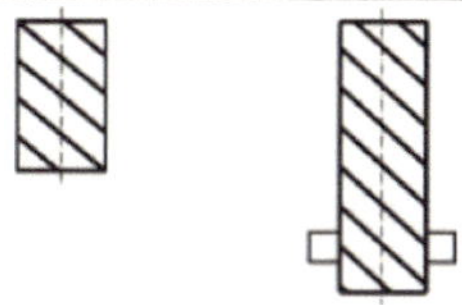
2	Buildup an enclosed cavity in the working space for the insert (in parallel to an additive structure, if required).	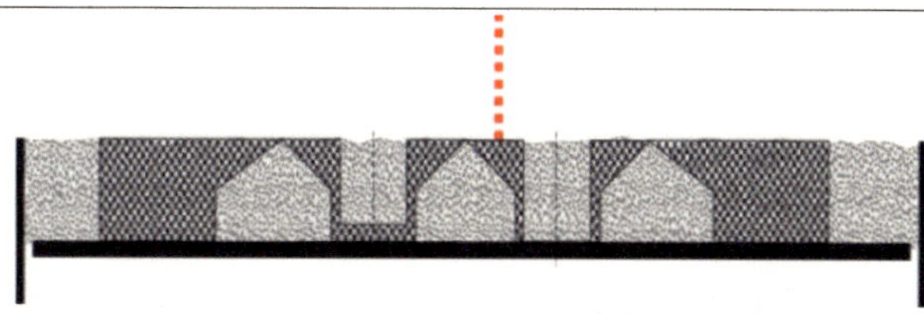
3	Suspend the process when the cavity's height matches exactly the insert's height. Clear the cavity from powder and place the insert with the flat top flush with the current working plane. Fill the remaining gaps around the insert with powder to provide a continuous deck as a support for further layers.	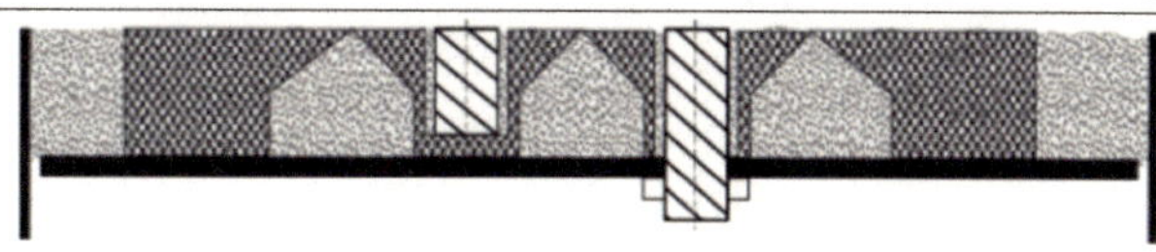
4	Continue the regular process by adding a new layer of powder, fusing it to the insert and all other underlying additive structures, to connecting them with the subsequent buildup and, thus, with, each other.	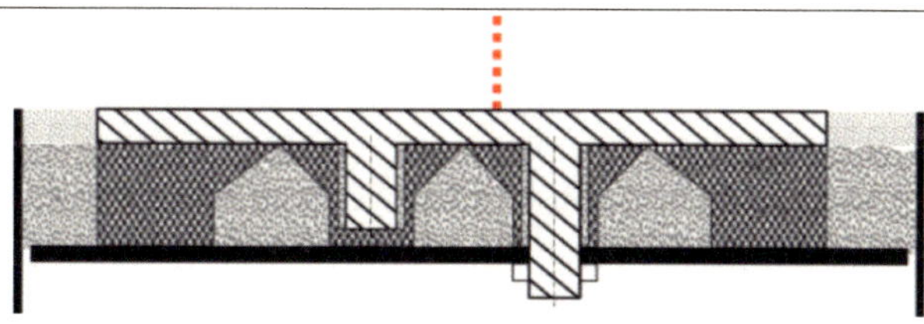

Table 3.
Modified PBF process for pre-fabricated inserts (based on [6]).

- fine control over powder delivery rates, independent of a required minimum gas flow,
- finer resolution and tolerances in the working plane (X and Y axis, e.g., for wall thickness) as directed energy can be applied independently of the amount and focus of a powder gas stream,
- finer resolution and tolerances in buildup direction (Z axis, e.g., for staircase effect) because of fine control of energy application and fine powder grade,
- no track scatter because powder is not liquified in-flight and thus cannot adhere off-track,
- overall less energy induction and thus less thermal gradients in the part, and
- simple reuse of excessive powder without agglomerations or thermal degeneration.

8. Facilitate hybrid product architectures: DfAM methodologies

Introduction of AM to industrial production often fails where regular parts are simply converted to AM parts without a full redesign, starting on product level. This can be attributed to the fact that conventional designs often do not make effective use of the potential of AM, but are substantially affected by its constraints. For workable results, an AM part, its environment, and the complete product should ideally be redesigned or specifically developed.

However, DfAM expertise is still not amply available in many regular product engineering and design departments, and a relatively new part in regular engineering education programs—which is a substantial barrier for a more comprehensive application of AM on industrial scale.

Consequently, the interest of DfAM has expanded in recent years to more integrative approaches, such as the design for postprocessing of AM parts, and product development and design methodologies with specific regard to AM parts as part of the overall product development process [7], more detailed in the thesis [8] (in German).

But overall, AX and DfAM need to be incorporated strategically into regular product and process development departments of branches where AX can make a difference. For regular AM, there are vocational trainings as AM specialist engineers (including modules about design rules and principled for product designers), e.g., since July 2017 by the Association of German Engineers VDI [9]. Also, AM is becoming a regular component of engineering education programs.

9. Conclusions

AM is a mature technology for industrial application. Like all other manufacturing technologies, it has strengths and weaknesses that make it an unlikely "do-all" substitute for conventional processing. Instead, AM is most effectively applied as for specific elements or features, in combination with features from other processes—as one interoperable contribution to an integrated process chain for a complete product. Integration can involve subsequent processes and assemblies, concurrent hybrid processing, and pre-fabricated elements from precedent processes.

Systematic consideration of postprocessing in AM is currently being focused in research, literature, and application. However, it has to be implemented in regular development workflows, requiring systematic competence development, in development staff training and engineering education.

First machines for concurrent hybrid processing are already in the market, research is ongoing. Again, awareness of and competence for the technology and their benefits and constraints have to be systematically implemented in development workflows and the respective staff.

However, the additive working principle bears the potential of a disruptive innovation for micromechanical products: Following the principle of Integrated Circuits (ICs) in microelectronics and MEMS, additive processes can be used to build and package monolithic or hybrid products as Integrated Devices (IDs). In order to package pre-fabricated elements into a hybrid product, the current scope of additive processing would have to be extended from primarily Additive Manufacturing (AM) of parts to novel purposes such as additive joining, additive finishing, and additive treatment. This extended concept of Additive Technology (AX) requires the development of new processes that allow additive processing of pre-fabricated elements.

Research and development will have to acknowledge IDs by AX as a significant future prospect, prioritize efforts to develop processes suited for AX, and systematically implement them in product development workflows. Only then, additive and conventional processes can be combined over the full integrated process chain, and additive and conventional features can be planned on all levels of the integrated product architecture—resulting in a more consequent deployment and capitalization of additive processes.

Acknowledgements

The patent application and content for [6] were developed under participation of David Grüning, M.Sc., before graduating from HTW Berlin, with internal research funding from internal research funding from the HTW Berlin.

Conflict of interest

The authors declare no conflict of interest.

Author details

Matthias Dahlmeyer* and Sebastian Noller
Hochschule für Technik und Wirtschaft (HTW) Berlin, University of Applied Sciences, Berlin, Germany

*Address all correspondence to: matthias.dahlmeyer@HTW-Berlin.de

References

[1] Lee J-Y, Nagalingam A P, Yeo S H. A review on the state-of-the-art of surface finishing processes and related ISO/ ASTM standards for metal additive manufactured components. In: Bártolo P, Chua CK, editors. Virtual and Physical Prototyping. Vol. 16, issue 1. London: Taylor & Francis; 2021. p. 68-96. DOI: 10.1080/ 17452759.2020.1830346

[2] Verein Deutscher Ingenieure e.V. (VDI). Guideline VDI 3405: Additive manufacturing processes, rapid manufacturing - Basics, definitions, processes (original German title: Additive Fertigungsverfahren - Grundlagen, Begriffe, Verfahrensbeschreibungen) [standard]. Berlin: Beuth Verlag GmbH; 2014. Available from: https://www.vdi.de/en/home/vdi-standards/details/vdi-3405-additive-manufacturing-processes-rapid-manufacturing-basics-definitions-processes [Accessed: March 7, 2022]

[3] Dahlmeyer M, Grüning D. Design for Build-Up, Assembly and Function of Productive Components from Additive Manufacturing (Original German Title: Aufbau-, Montage- Und Funktionsgerechte Gestaltung Additiv Gefertigter Produktivbauteile). In: Konstruktion. Vol. 71, issue 5. Düsseldorf: VDI fachmedien; 2019. pp. 93-98. DOI: 10.37544/ 0720-5953-2019-05

[4] Jiménez A, Bidare P, Hassanin H, et al. Powder-based laser hybrid additive manufacturing of metals: A review. In: The International Journal of Advanced Manufacturing Technology. Volume 114, issue 1-2. London: Springer Nature; 2021. pp. 63-96. DOI: 10.1007/ s00170-021-06855-4

[5] Deutsches institut für Normung e.V. DIN 8580:2020-01 – Entwurf: Manufacturing Processes - Terms and Definitions, Division (Original German Title: Fertigungsverfahren - Begriffe, Einteilung) [Standard Revision Proposal]. Berlin: Beuth Verlag GmbH; 2020. DOI: 10.31030/ 3115263

[6] Dahlmeyer M, Grüning D. Patent application WO2019/76910 A3: Method for additive manufacturing of a component and machine for carrying out the method. München: German Patent and Trademark Office; 2019

[7] Kumke M, Watschke H, Vietor T. A new methodological framework for design for additive manufacturing. In: Bártolo P, Chua CK, editors. Virtual and Physical Prototyping. Vol. 11, issue 1. London: Taylor & Francis Online; 2021. pp. 3-19. DOI: 10.1080/17452759.2016. 1139377

[8] Kumke M. Methodisches Konstruieren von additiv gefertigten Bauteilen [AUS, volume 124, at the same time thesis TU Braunschweig]. Wiesbaden: Springer; 2018. p. 248. DOI: 10.1007/978-3-658-22209-3

[9] Verein Deutscher Ingenieure (VDI). Qualifizieren Sie sich zum "Fachingenieur Additive Fertigung VDI". Düsseldorf: Website of the VDI Wissensforum GmbH; 2022. Available from: https://www.vdi-wissensforum.de/lehrgaenge/fachingenieur-additive-fertigung-vdi [Accessed: March 7, 2022]

Chapter 6

Perspective Chapter: Multi-Material in 3D Printing for Engineering Applications

Rajkumar Velu, R. Sathishkumar and A. Saiyathibrahim

Abstract

3D Printing or Additive Manufacturing is one of a novel method in manufacturing of materials with increased accuracy of manufacturing in terms of complexity in parts, design of aerospace and defense parts, light-weighting, etc., This manufacturing method involves layer-by-layer printing or deposition of materials or metals into the perfectly aligned especially in corners, edges and in most complex designs. The design process mostly involved software so that production cost could be estimated in the design stage itself. Additive Manufacturing is one of the most promising approach for small and low-volume productions. The filament used for the process is prominent to the designer, along with the various printing processes. Recent modern printing techniques involve multiple nozzles, whereas designers can use multiple materials on single printing. The use of multi-material in a single part enables the manufacturer to rapidly produce products which have specific applications. This chapter discusses about various multi-material with different mechanical properties that can be used for structural applications through different printing technologies on various precious applications. This technology is quickly adopted by even small-scale industries in recent times.

Keywords: multi-material, 3D Printing, filament, fusion, functionally gradient materials

1. Introduction

Recent advances in 3D printing allow industrialists and researchers to create a whole new foundation for future manufacturing. This new foundation gives multi-material in 3D printing which allows various properties of materials which can be printed that can lead to excellent productivity with very good application like aerospace, bio-applications, spacecraft materials, electronic components, etc., Polymers, ceramics, metals, and biomaterials have all been utilized in several AM techniques to create multi-material structures. In which multi-material within a single component we can achieve properties like hardness, corrosion resistance, tensile and compressive properties needed areas of the product can be achieved and also eliminate the need for complex manufacturing and expensive tooling. These new processes and printing of multi-material can be incorporated in a single part, eliminating the need for multi-part modeling. Multiple material printing can provide increased characteristics, including controlled material anisotropy, that can be essential for functional systems such as those needing surface alignment while still having a high load capacity. Much more progress is required in this sector due to its binding difficulties and usage of printing methods this chapter highlights us to 3D printed metal-metal, metal ceramic, polymer-based, functionally graded material usage, and its application in the structural field is thoroughly discussed.

2. Need for multi-material in structural applications

In recent advancements for specific application-oriented functional needs, there is a growing need for excellent tooling techniques and manufacturing techniques utilized for specific purposes. In order to achieve the particular application, tooling and testing requires more time and multiple tooling which cannot be enabled for the various process while we select new advanced machining techniques. Thus, the new technique 3D printing has given us a wide range of selection of materials or machining techniques that can be resolved. In addition, very specific applications like spacecraft, aerospace, automobile structures, biomechanics, electronic components need a wide range of materials to be machined or incorporated into a single material which eliminates the number of components to be manufactured. Especially in biomechanical application various tissues, muscles, blood vessels, bones are incorporated into a single replacement for various operating and tearing of muscles needed by the surgeon and in electronic components, various chips and motherboards need a separate manufacturing or placing of various sensors to be placed this multi-material printing reach us to a new way of printing single incorporated boards and robotic structures are enabled. For continuous electronics manufacturing, Stereolithography and Direct Ink Write were combined, while for embedded electronics manufacture, Fused filament fabrication and Direct Ink Write were mixed. Products like embedded electronics, sensors, soft robotics, and customized pharmaceutical products might be made more readily for medical or space applications that are not possible with those other traditional production processes. That is why there is a growing need for multi-material 3D printing.

3. Selection of multi-material for 3D printing applications

In the near future, the potential of multi-material 3D printing will be a revolutionary moment in rapid production, customized design, and structural applications. Multi-material 3D printing is compatible with functionally graded materials which has a single structural form, also it has the potential to be used in structural applications to benefit from hybridized/combined material characteristics. Multi-material printing creates quick and durable structures that incorporate the features and functions of all of the hybrid materials which are shown in **Figure 1** [1].

Multi-material 3D printing can be used to build innovative smart 4D structures with specified contour/attributes/operations. Established AM processes, such as FDM, can be enhanced to Hybrid Deposition Manufacturing (HDM) with integrated components to manufacture more sophisticated, interconnected multi-material components than previous approaches [2]. It has been observed that build configuration, manufacturing parameters, and related factors can have a significant impact on the relationship between multilateral interfaces all through 3D printing, thus they should have been adjusted to provide superior thermal, mechanical and surface characteristics [3]. The selection of materials for end-use for performing multi-material 3D printing operation is based on the availability of materials and its machinability, material properties, technique used for printing, processing time and rate of material

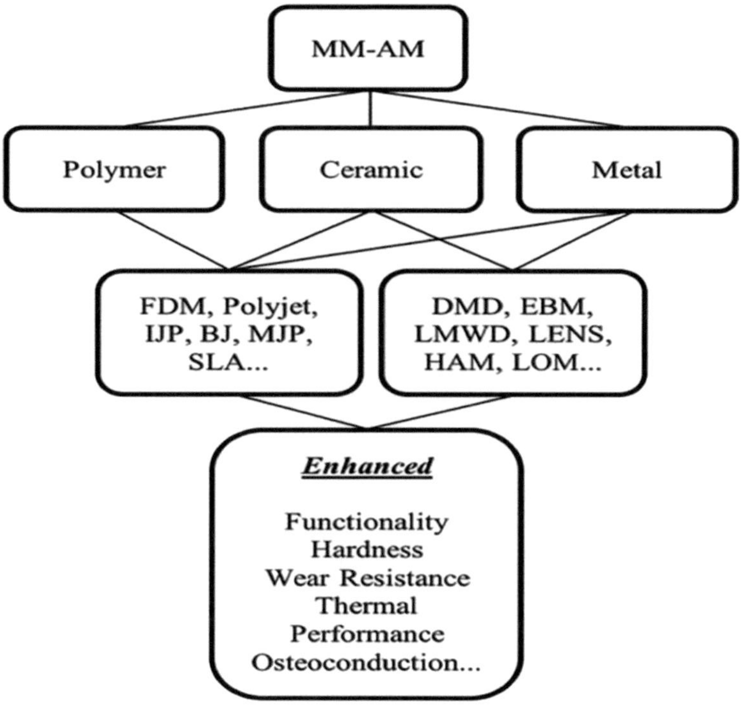

Figure 1.
Overview of some manufacturing and enhancement possibilities for multi-material additive manufacturing.

loss, reengineering and reparability of material, environmental conditions and cost, and the applicational areas such as biomechanics, bioengineering, robotic and space.

3.1 Based on availability of materials and machinability

Availability of the materials for the wide range of applications are the potential thing while selecting a multi material. These are the materials that are easily available and used for general applications that can be printed even on all types of printers like PET – polyethylene terephthalate, PA – polyamide TPU –thermoplastic polyurethane, PC – polycarbonate, HIPS – high impact polystyrene among many others. Investigations on the multi-material parts/components manufactured by 3D printing has been increased significantly in the recent past. Using this multi-material 3D printing, it is possible to generate good quality components through tailoring different layers of different materials up to the desired thickness. This uniqueness made this process stand out when comparing conventional manufacturing methods. Hence, materials used in multi-material 3D printing has to be decided in the design stage itself mainly based on the desired applications. A variety of polymers, ceramics, metals, and biomaterials are available to design and fabricate multi-material components through 3D printing. Some of the investigations conducted are presented here for a better understanding of the availability of materials for multi-material 3D printing [4–6].

Polymer composites filled with carbon or glass fibers wood particles (PLA matrix) sandstone (co-polyester matrix), magnetic (PLA matrix with iron particles), electric conductive (PLA matrix with carbon-based fillers); Photopolymer on the basis of flexible, rigid, and support material widely available one that can be easily machinable in Inkjet printers and Fused Deposition Modeling techniques used for hinge application.

3D printing of silicone elastomers combined with nanosilica (NS) is one type of next-generation advanced structure possessing improved performance and low cost to fabrication. Such multi-material components are made using Direct Ink Writing (DIW) process and are finding their applications in medical devices, flexible electronics, and soft robotics. Silicones are low viscosity materials which are requiring a long curing time during 3D printing processes but their attractive elasticity makes them suitable in the abovementioned areas. To overcome these printability issues, nanosilica is incorporated nowadays with elastomers as a rheology modifier and this addition does not reduce any elasticity of elastomers. The Food and Drug Administration (FDA) is classified this NS is a biocompatible material and hence it is successfully used to manufacture parts for human implants nowadays [7].

Polyethylene glycol (PEG) is an attractive hydrogel with good compatibility to chemical modifications and is used mainly in tissue engineering scaffolds [8]. PEGX – Gelatin multi-material components synthesized by bioink methods are possessing self-supporting layer-by-layer printing quality with robustness (**Figure 2**). PEG can be used with biopolymers to manufacture 3D printed layouts with tailored properties. Also, it is possible to create PEG-polymer and PEG- PEG crosslinked multi-material structures using bioink methods [9].

Recent advancements in human implants especially bone replacement focused on developing a new multi-material which has good compressive strength. Propylene fumarate dimethacrylate (PFDMA) is a macromer that has low viscosity and possesses the required qualities to form High Internal Phase Emulsions (HIPE). This PFDMA polyHIPE has good properties such as biodegradable and cytocompatibility. Lack of compressive properties have been addressed frequently while printing 3D structures during bioink printing processes. To overcome this limitation, a thermoplastic polyester shell is nowadays incorporated to form an outer layer surrounding the trabecular bone scaffold. **Figure 3** shows the PFDMA polyHIPE – polyester shell (PCL) layer by layer printing multi-material 3D printed scaffold for use in human implants [10].

Surprisingly, conductive thermoplastic elastomer filaments have also been developed for multi-material Fused Deposition Modelling (FDM) to produce complicated structures by using more than one polymeric filament. Lately, multi-material 3D functional prototypes were produced with recycled thermoplastics such as High Impact Polystyrene (HIPS), Acrylonitrile Butadiene Styrene (ABS), and Polylactic Acid (PLA) using FDM. HIPS has high impact resistance and good machinability and hence can be useful for structural parts. A very important property of PLA is its excellent biodegradability which promotes them to biomedical applications. As a familiar thermoplastic, ABS contains good thermal and heat resistance hence can be suitable for civil structures. Combining all these properties in different ratios, it is possible to form 3D multi-material structures in FDM. Multi-material layered

Figure 2.
3D bioprinting PEGX-Gelatin (a) extrusion through 200 μm tip, (b) four printed layers, 15 × 15 mm and (c) 20 layers porous hexagon shape ≈5 mm thick [9].

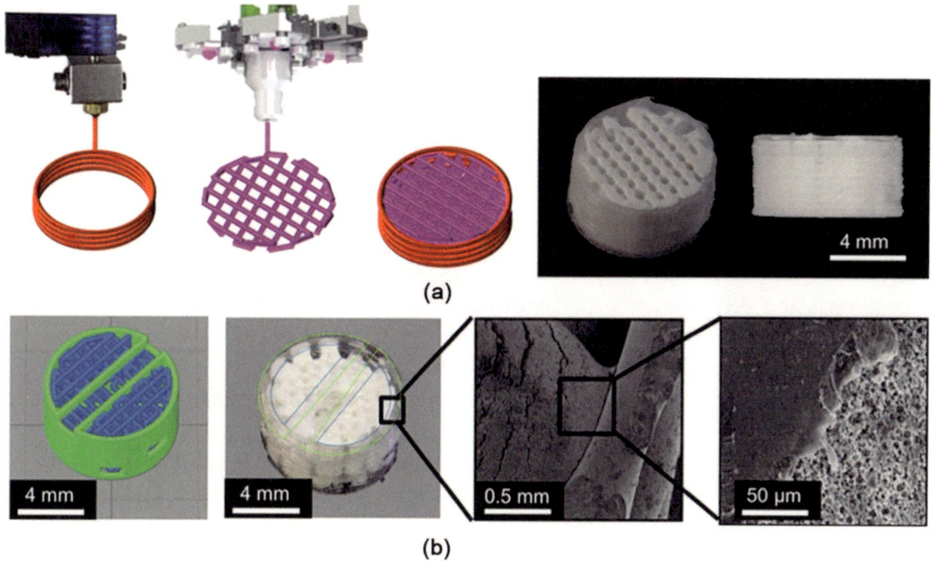

Figure 3.
3D printing process of (a) thermoplastic polyester outer shells and HIPE emulsion ink inner material, and (b) integration between the emulsion ink and thermoplastic (PCL) shell.

ABS/PLA/HIPS flexible functional prototypes synthesized using FDM are shown in **Figure 4** which are finding their desired interests in soft electronics and robotics [1] .

It may be noted that multi-material 3D printing of multiple thermoplastic polymers is possible for operational devices and can enhance their mechanical characteristics. In the recent past, many investigations are going on the development of more unique combinations of materials for structural entities in 3D printing processes.

3.2 Based on material properties

Mechanical performance of multi-material additively produced components is often superior to that of single-material printing. The development of voids across succeeding layers of printed components might impact their mechanical characteristics due to a reduction in interfacial adhesion within printed layers. Another typical issue of multi-material AM is the difference in mechanical behavior between horizontal tension or compression and vertical tension or compression. Robust 3D printing methods, such as micro-additive stacking, are required to ensure layer consistency and enhance surface quality to the specifications required for their respective purposes.

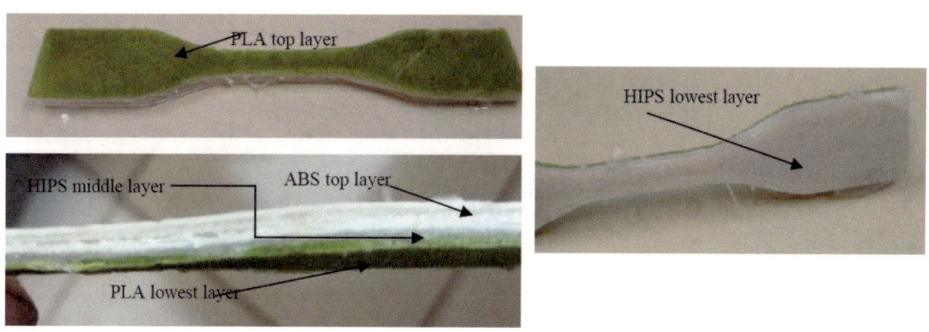

Figure 4.
Multi-material layered ABS/PLA/HIPS flexible functional prototypes.

3D printed polyamide-based composites filled and laminated with continuous and short carbon fibers have very good synergistic reinforcement on the mechanical properties. Thermal, morphological, and mechanical tests for the printing drives were performed initially in order to determine the characteristics of the printing elements that provide good continuous carbon fiber strength. Carbon Fiber bundles soaked with polycarbonate solution have been created during the embedding procedure. Inseminated carbon bundles were inserted in 3D printed Polycarbonate to boost mechanical strength that may be utilized in building health monitoring constituents with markedly better material characteristics. ABS – acrylonitrile butadiene styrene, PLA polylactic acid, and PVA – Polyvinyl alcohol, each one with different purposes and performance characteristics, excellent mechanical properties, less smell while printing, used for varying support structures, and can be easily removed without damaging the structure [11].

Furthermore, by leveraging AM's design flexibility, the computational optimization-based design technique may be extended to multiple design scales for diverse characteristic composites, potentially improving the functionality of developed composite materials structures.

Novel multi-material bilayer absorbing composites comprised of graphene, Li0.35Zn0.3Fe2.35O4, polymethyl methacrylate (GFP) as the corresponding layer and graphene, carbonyl iron powder, polymethyl methacrylate (GIP) as the absorption layer (**Figures 5** and **6**) were created using multi-material digital light processing (DLP) 3D printing method, wherein graphene has been used as a conductive, highly resistant, transparent material and hardness [12].

The successive coating of all-inorganic viscoelastic TE inks comprising $Bi_xSb_{2-x}Te_3$ particles and customized utilizing $Sb2Te_4^{2-}$ chalcogenidometallate binders resulted in the multi-material 3D printing of composition-segmented BiSbTe materials.

By modifying the composition of $Bi_xSb_{2-x}Te_3$ particles, the peak ZTs of the 3D-printed substances were controlled to switch from ambient temperature to 250°C. Fabricated ideally constructed Thermo Electric Generators (TEG) comprised of the 3D-printed, composition-segmented tri-block $Bi0.55Sb1.45Te_3/Bi0.5Sb1.5Te_3/Bi0.35Sb1.65Te_3$ TE leg, as shown in **Figure 7**, which then expands the peak ZTs and fulfills complete compliance throughout the whole temperature scale, achieving record-high efficiency in multi-material 3D printing of composition-segmented BiSbTe materials by sequential deposition of all-inorganic viscoelastic TE inks comprising $Bi_xSb_{2-x}Te_3$ particles customized with $Sb2Te_4^{2-}$ chalcogenidometallate binders. By modifying the composition of $Bi_xSb_{2-x}Te_3$ particles, the apex ZT_s of the 3D-printed materials were controlled to switch from ambient temperature to 250°C.

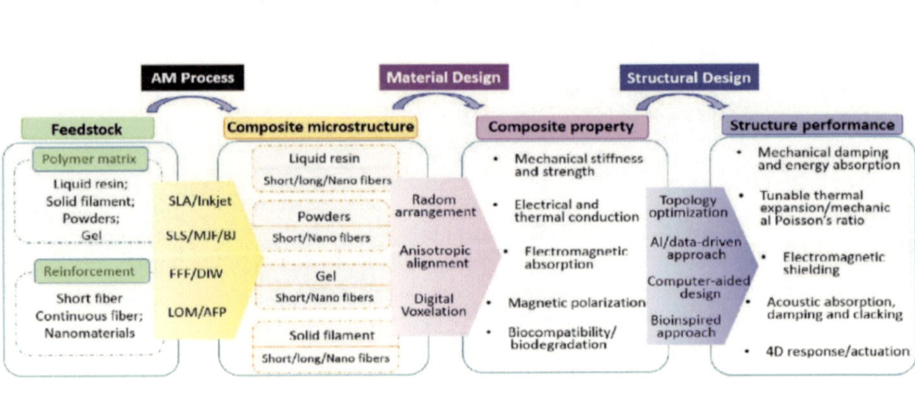

Figure 5.
A framework of process structure properties performance in polymeric composites via AM.

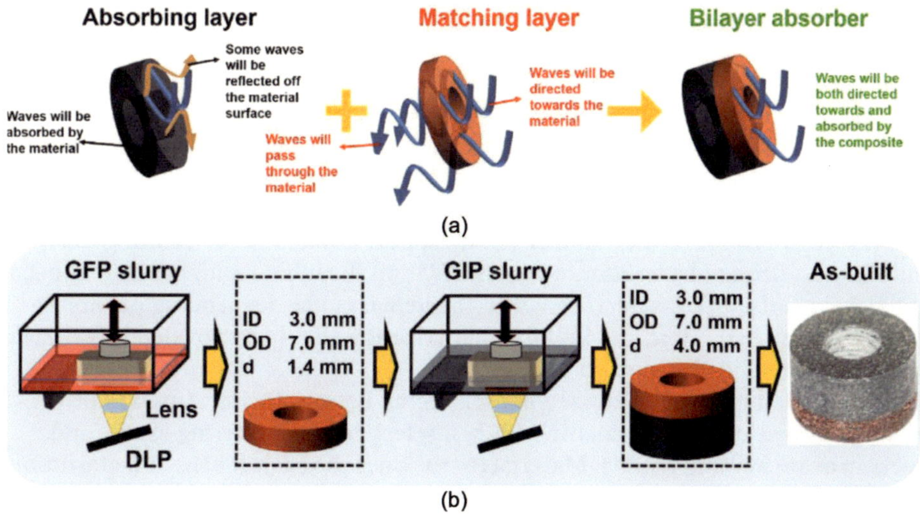

Figure 6.
(a) Schematic diagram of the design principle of the bilayer absorber; (b) schematic diagram of the DLP 3D printing process of the GFP þ GIP bilayer absorber and a photograph of the GFP þ GIP bilayer absorber. (A color version of this figure can be viewed online.) [12].

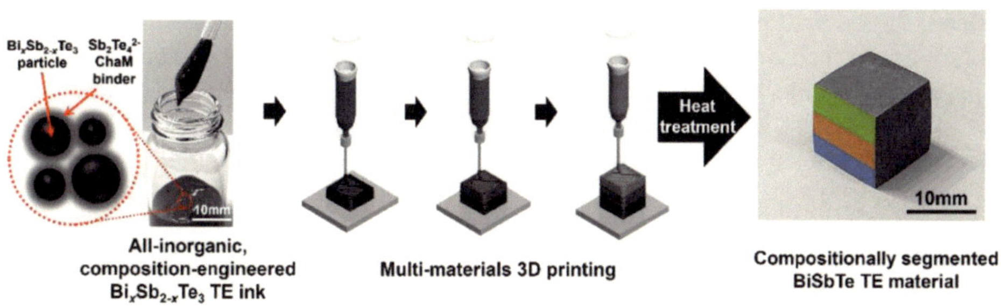

Figure 7.
Scheme of a sequential 3D printing of multi-segmented TE materials. Green (Top), orange (Middle), and blue (Bottom) colored patterns correspond to each block in the compositionally segmented TE legs by multi-material 3D printing using TE inks containing Bi0.55Sb1.45Te3, Bi0.5Sb1.5Te3, and Bi0.35Sb1.65Te3 particles, respectively.

The optimized TEG with the 3D-printed, composition segment enhances the apex ZT_s and fulfills complete compatibility over the whole temperature spectrum, attaining a record-high efficiency for electrical and conductive multi-material systems.

Functional gradient digital composites may also be created and manufactured using 3D printing multi-material, which can increase the effectiveness of the manufactured structural design [13], for instance, have envisioned a digital laminate composite optimization method. The suggested digital design flow can keep updating both the macroscale shape of the constructed laminate configuration and the alignment of short reinforced fibers within the topography at the same time. Aside from short-fiber reinforced composites, AM can also be used to digitally design and construct functional gradient continuous fiber-reinforced laminates. Numerous design strategies have been introduced to regulate the alignment and volume fraction of AM manufactured continuous fiber composites. Their findings demonstrate that the optimal functional gradient laminate outperforms standard CFRP laminates with homogeneous orientation angles and fiber densities in terms of stiffness and strength [14].

3.3 Based on various techniques of printing

In the initial stages printing of material is limited to the extrusion properties and growing printing techniques thus it has been extended to multi-material with very good machinability structures. PVA – polyvinyl alcohol, ABS – acrylonitrile butadiene styrene, and PLA – polylactic acid was the primary alternatives in the early phases, each having its own set of applications and functional properties. Whilst ABS is recognized for its mechanical qualities, PLA is noted for its printing ease and lack of a strong plastic odor or fumes. PVA, on the other hand, has mostly been utilized to print support structures when the item includes suspending portions. Powder fusion, extrusion, and liquid polymerization all rely on powder, liquid-solid transitions, or solid-liquid-solid transitions [15].

Various methods can be utilized in each sector. Powder fusion is used in processes such as selective laser melting (SLM), selective laser sintering (SLS), and electron beam melting (EBM). Material extrusion is the basis for fused deposition modeling (FDM) (fusion then solidification of the material). **Figures 8** and **9** depict the different 3D printing processes that are now available. Finally, stereolithography is based on a liquid's polymerization (liquid-solid transition).

3.3.1 Vat Photopolymerization

The three primary vat polymerization techniques are digital light synthesis (DLS), digital light processing (DLP), and stereolithography (SL). Although the vat photopolymerization method does not always enable multi-material applications, due to its numerous benefits such as surface quality, dimension precision, and the ability to print on a wide range of materials, vat photopolymerization has indeed been modified to enable multi-material printing [16].

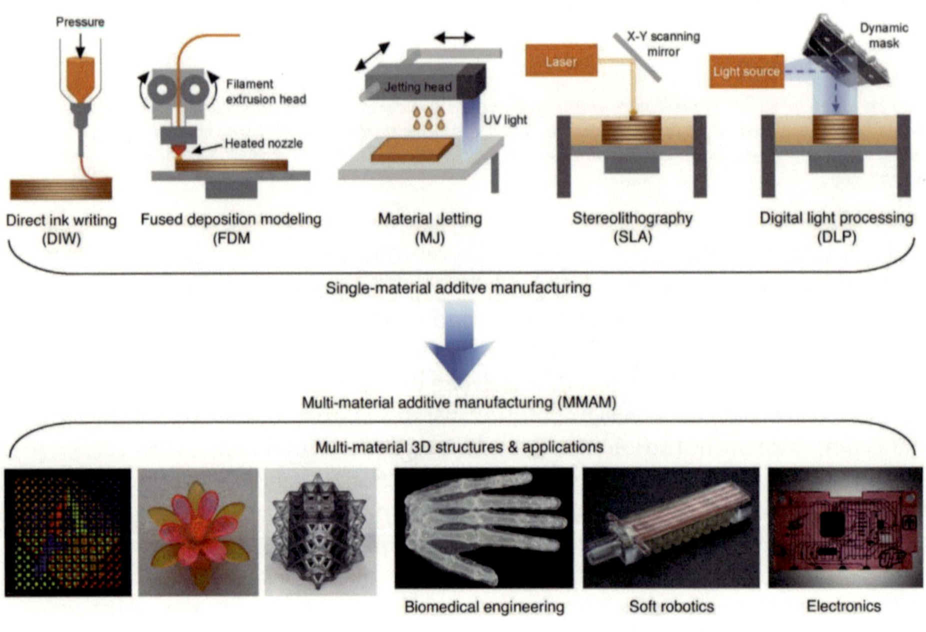

Figure 8.
Multi-material 3D printing of structural from various powerful engineering techniques.

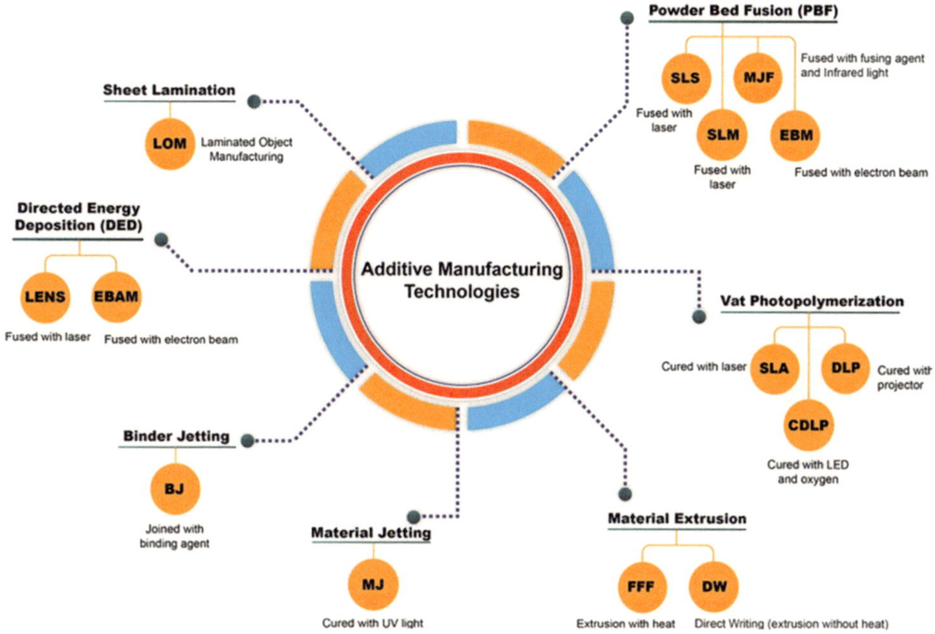

Figure 9.
The illusion of various additive manufacturing techniques.

3.3.2 Stereolithography

In a vat photopolymerization, SL employs a photopolymer liquid as the source material. By lowering the construction platform into the vat and curing it with a UV laser, liquid plastic is converted layer by layer into a 3D item [13, 17, 18]. Advantage of this stereolithography can build large parts with very good accuracy and surface finish but Works with photopolymers which are not stable over time and do not have well defined mechanical properties which has casting patterns,

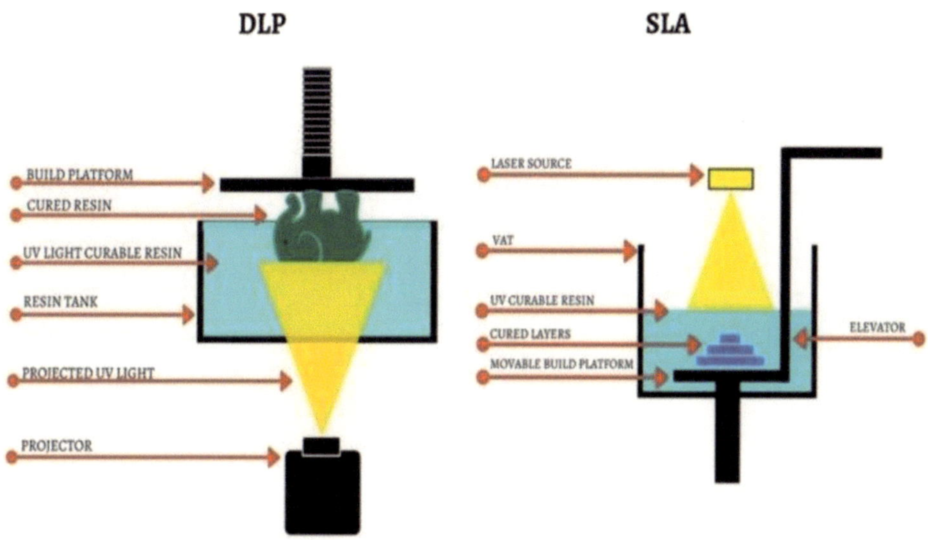

Figure 10.
Differences of DLP and SLA schematic.

Prototypes, jewelry, and medical applications. **Figure 10** depicts the stereolithography process with DLP.

3.3.3 Digital light processing (DLP)

DLP technology is very similar to SL but uses a different light source and makes use of a liquid crystal display panel. In this two or more digital micromirroring devices are used for delivering multi ink materials for multi-material applications, thus **Figure 11** depicts the Digital Light Processing technique [20, 21]. A tiny microfluidic chip had also been employed for multiple material supply and interchange, allowing for direct printing of multicellular cell-laden microstructures. Both blue light (470 nm) and green light (530 nm) are used for radical curing, but only blue light is used for cationic curing. Printing of multi-material 3D structures with specified and spatially defined mechanical and chemical characteristics may be efficiently accomplished by selectively projecting pictures with two wavelengths of light [19]. This has the benefit of higher print performance relative to SLA, outstanding laying precision, low-cost printing but the insecurity of the consumable material, high cost of materials with applications in prototypes, casting, patterns, jewelry, dentistry, and medical area.

3.3.4 Extrusion

It may be divided into two major subgroups based on the temperature necessary or appropriate for extrusion: fused filament fabrication (FFF) or fused deposition modeling (FDM) for extrusion of melted thermoplastic polymers and direct ink writing (DIW) for extrusion without melting.

3.3.5 Fused Filament Fabrication

A nozzle is used to extrude a plastic filament that has been melted. Layer-by-layer, objects are constructed and it can construct completely functioning components out of common polymers [16, 22–24], however anisotropy in the z-direction (vertical direction) and a step-structure on the surface are characteristics of printed components, as seen in **Figure 12**. The applications that can be printed

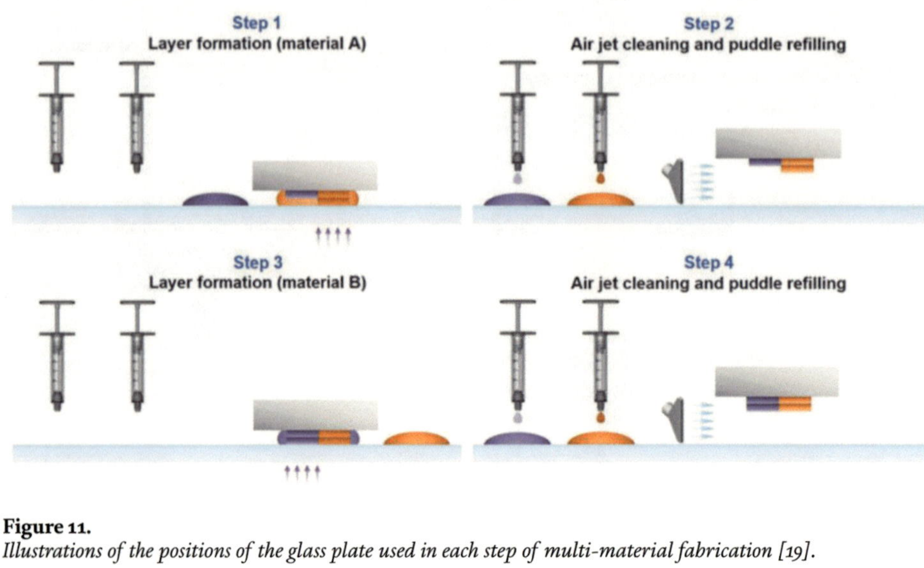

Figure 11.
Illustrations of the positions of the glass plate used in each step of multi-material fabrication [19].

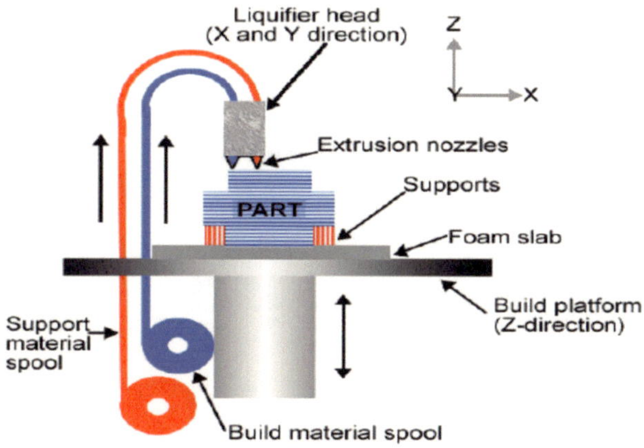

Figure 12.
Fused Filament Fabrication schematic.

include prototypes, support components (jigs, fixtures), and small series items. Furthermore, multi-material printing capabilities for FDM is required to build sacrificial support structures for printing components with overhang characteristics. Multi-material FDM, like multi-material DIW technology, is accomplished by utilizing several extrusion heads with independently controllable nozzle temperature, printing speed, and resolution. This bi-extruder features an intermixing with thick blades that create passive mixing of two melted filaments. The mixing also allowed for the printing of functionally graded materials as well as the improvement of interfacial bonding strength between two distinct materials by increasing the mechanical interlocking of the materials at the interface. Printing with a carbon nanotube (CNT)-coated filament and microwave heating after printing has been demonstrated to improve interfacial bonding, but they have not been used to connect diverse materials. BS + carbon fiber, ABS + Fe/Cu, PCL + TCP, PLA + carbon fiber are some of the multi-material that can be printed using FFF or Fusion Deposition Modeling.

3.3.6 Direct ink writing (DIW)

Material in the form of a semi-liquid or paste can be extruded via a nozzle and utilized to print the cross-sections of a sliced 3D model in this method. The material mixer also enables the printing of a mixture of different materials, as seen in **Figure 13**. Furthermore, by precisely controlling material flow via specific pressure lines, a mixing ratio may be adjusted to the required concentration, enabling 3D printing of functionally graded materials with customized properties (e.g., mechanical, chemical). This has the benefit of highest resolution for an extrusion system Ideal for research settings and medical (bone) applications, but with limited part geometry, the system is expensive. Small build volume applications include solid monolithic components, scaffolds, physiologically compatible tissue implants, customized composite materials and ceramics [20, 25–27].

3.3.7 Binder jetting

Inkjet printing heads spray a liquid-like bonding agent onto the powder's surface. The item is built up layer by layer by connecting the particles together, as seen in **Figure 14**. Binder jetting is best suited for big printing and the production of low-cost metal components. Binder jetting takes place at ambient temperature,

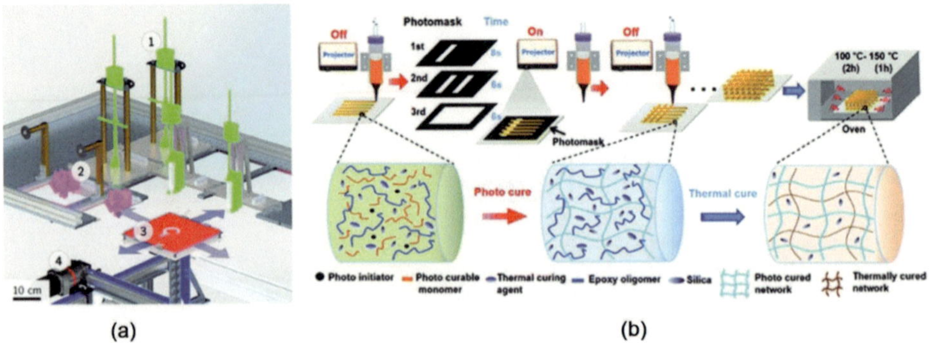

Figure 13.
(a) The scheme shows a multi-phase additive manufacturing system that combines DIW components (1) with FDM modules (2) heated platform (3) and with gear-boxes (b) Schematic showing the processes involved in a dynamic photomask-assisted DIW multi-material system. The DIW-printed structure is placed under the dynamic photomask which consists of a projector with a set of light patterns [25].

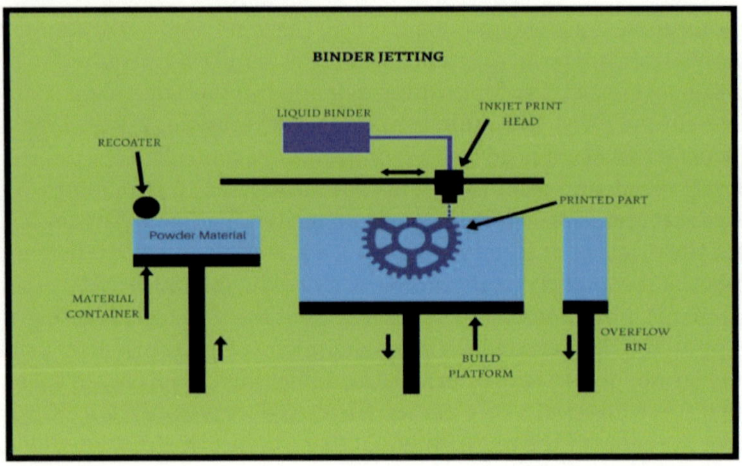

Figure 14.
Schematic of Binder Jet Printing.

which avoids warping and curling issues. Powder size, layer thickness during binding, part orientation in bed, heater power, roller speed, curing temperature, curing duration, sintering time, sintering temperature, and sintering environment are examples of these. To investigate the impact of each of these factors and their interactions, a massive experimental design with hundreds of samples and testing would be required [28–30]. BJ enjoys the benefits of A relatively quick and inexpensive technique, with a wide range of material types and the ability to produce full-color parts, however parts produced straight from the machine have restricted mechanical characteristics and applications such as prototypes, casting patterns, molds, and cores.

3.3.8 Material jetting—Multijet modeling (MJ)

Inkjet printing head jets molten wax onto a printing bed. Once the material is cooled and solidified, it allows to fabricate layers on top of each other shown in **Figure 15**. This method has the benefit of obtaining very high precision and surface qualities, but it only works with wax-like materials for prototypes, casting, and patterns.

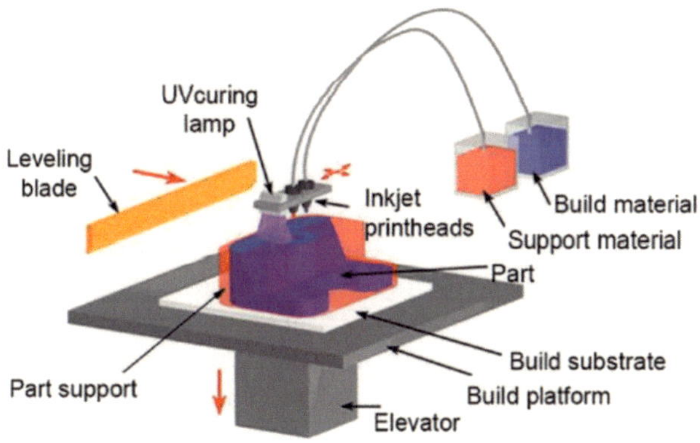

Figure 15.
Schematic of Material jetting.

3.3.9 Polyjet modeling

Unlike multijet, the printing head jets liquid photopolymers onto a printing bed. The material is quickly cured and cemented by UV light, allowing layers to be built on top of one other. Different materials may be jetted together to create multi-material and multi-color things, but it works with UV-active photopolymers, that are not long-lasting.

3.3.10 Powder bed fusion

Powder bed fusion (PBF) [31–33] is an AM technique that operates on the same fundamental concept as milling in that components are produced by adding material rather than removing it through traditional forming processes. The PBF method starts with the development of a 3D CAD model that is numerically sliced into many distinct layers. Because the heat source is generally an energy beam, a heat source scan path is generated for each layer, which specifies both the boundary contour and some sort of fill sequence, commonly a raster pattern. The benefits of power bed fusion include reduced material waste and expense (superior buy-to-fly ratio), Improved manufacturing development timelines, fast prototyping and low-volume production Capable of producing functionally graded components, Parts that are fully customized on a batch-by-batch basis, removing set designs, when compared to other additive manufacturing methods, it has a high resolution. The unmelted powder may be recycled effectively.

3.3.11 Laser sintering (LS)

SLS and LS have certain commonalities. A laser is used to selectively melt a tiny layer of plastic powder. The components are built up layer by layer in the powder bed. This may make parts in standard plastics with good mechanical characteristics, a continuously expanding list of materials available, but products do not have the same qualities as their injection-molded counterparts possessing application of prototypes, support parts, small series parts.

3.3.12 Selective laser sintering (SLS)

Selective laser sintering (SLS) is an additive manufacturing (AM) technique that utilizes a laser as the power source to sinter powdered material (typically nylon or polyamide), intending the laser automatically at points in space described by a 3D model, binding the material together so that creates the solid structure as seen in **Figure 16**. It is comparable to selective laser melting; the two are instantiations of the same principle but differ in technical specifics [34–46]. SLS is a newer technology that has mostly been utilized for fast prototyping and low-volume manufacturing of component parts. As the commercialization of additive manufacturing technology progresses, the number of production positions grows. Polymers such as polyamides (PA), polystyrenes (PS), thermoplastic elastomers (TPE), and polyaryletherketones (PAEK). Polyamides have been the most widely utilized SLS materials because of their excellent sintering tendency as a semi-crystalline thermoplastic, leading to components with desired mechanical characteristics. Polycarbonate (PC) is a material of great interest for SLS owing to its excellent toughness, thermal stability, and flame resistance; nevertheless, such amorphous polymers are difficult to produce.

3.3.13 Electron beam melting

The EBM method makes use of a high-power electron beam to produce the energy required for high melting capacity and productivity. The hot process produces components with minimal residual stress, while the vacuum guarantees a clean and regulated atmosphere. By scanning the concentrated electron beam to selectively melt certain powder regions, the EBM system constructs structures from the bottom up. It takes data from a 3D CAD model and applies successive layers of powdered material. The technique is repeated until the final layer of the component is constructed. It takes place in a vacuum, making it an excellent method for fabricating structures made of reactive materials which cannot be released into the atmosphere.

In material extrusion systems, dual or multi-extruder printing heads are commonly utilized to print multi-material components at the same time. It is standard practice, for instance, to utilize one extruder to print dissolvable endorses that may be readily detached from the main printed structure, or to print in two colors or two materials that will be included in the finished print. Conversely, dual and multi-extrusion printers generally have a few constraints: the existence of an extra extruder (second or more) reduces the printing area that would have been accessible with a single extruder; the possibilities of oozing and stringing increase. A dual-extruder had been utilized to print components out of PLA, ABS, and high

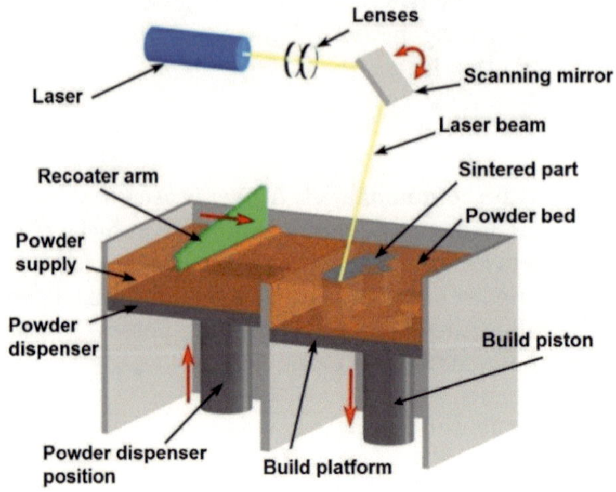

Figure 16.
Schematic of Selective Laser Sintering.

impact polystyrene and the findings demonstrate that mechanically interlocked extrudates significantly minimize adhesion failures inside as well as between filaments, as seen in **Table 1**.

3.4 Based on less processing time and material losses

The 3D printing technology was used to create a huge nozzle 40 inches in diameter and 38 inches tall, with integrated cooling channels for space applications. This nozzle had been created in approximately 30 days, as opposed to nearly a year using standard welding processes. The next image, **Figure 17**, clearly indicates the reduced processing time of additive manufacturing has indeed been given with less machinability.

When compared to other 3D printing technologies, digital light processing (DLP)-based 3D printing is a low-cost, high-speed, and high-resolution 3D printing technology that is derived on a localized photopolymerization process stimulated by the projection of digitally masked UV patterns on to the liquid surface, allowing us to achieve near-zero material loss. Because the printing process begins in a

Technique	Principle	Material	Advantages	Limitations
Fused deposition modeling (FDM)	Extrusion-based	Thermoplastics (ABS, PLA, PC, PA, etc.); glass (new); eutectic metal; ceramics; edible material, etc.	Simple using and maintaining; easily accessible; multi-material structures; low cost	Rough Surface; low resolution; high cost (for glass and metal)
Directly ink writing (DIW)	Extrusion-based	Plastics, ceramic, food, living cells, composites	Versatile	Low resolution; requires post-processing
Stereolithography apparatus (SLA) & (Digital light procession)DLP	Photocuring	Photopolymers	High accuracy; simple	Single material; unbiocompatible
Laminated object manufacturing (LOM)	Lamination	Sheet material (paper, plastic film, metal sheets, cellulose, etc.)	Versatile; low cost; easy to fabricate large parts	Time-consuming; limited mechanical properties; low material utilization; design limitations
Selective Laser Sintering (SLS) & Selective Laser Melting(SLM)	Powder-based laser curing	Powdered plastic, metal, ceramic, PC, PVC, ABS wax, acrylic styrene, etc.	High accuracy; wide adaptation of materials; high strength	Limited mechanical properties; high cost
Photopolymer jetting(Ployjet)	Inkjet-based	Liquid photopolymers	High accuracy	High cost
3D Powder Binder Jetting (3DP)	Inkjet-based	Any material in particulate form, plaster, ceramics, sugar, etc.	No need for support material; versatile; lower cost; colorful printing	Low strength; post surface treatment; limited mechanical properties

Table 1.
Summary of each printing method.

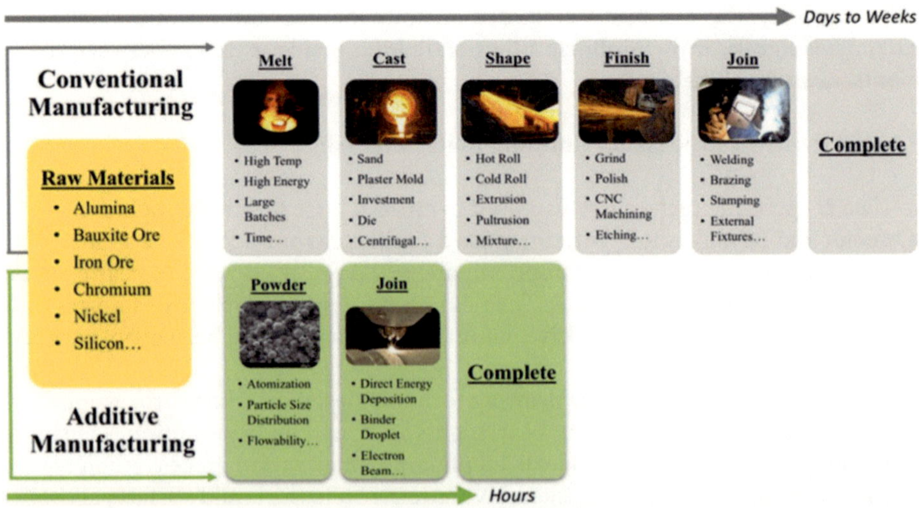

Figure 17.
Process comparison of conventional manufacturing processes versus AM for creating multi-layered structures.

liquid environment, this approach avoids the need for any support elements in the production of porous or hollow structures, pneumatically actuated soft robots, and other structures and devices having trusses or cavities. Projection microstereolithography, which generated micron-scale printing resolution, continuous liquid interface production, which enabled 100 times faster printing, and large area projection microstereolithography, which manufactured 3D features with feature sizes ranging from nanometers to centimeters, are among the remarkable innovations in this technology. In an effort to minimize fabrication time, topdown exposing DLP with several resin containers were used, however, the use of cleaning solutions to eliminate uncured resin proved detrimental to features finer than 300 m. It also was discovered that regulating the liquid levels in the numerous containers was challenging, and the process was still rather sluggish in the manufacture of complicated multi-material components.

3.5 Based on biomechanics and bioengineering products

Alginate-gelatin, collagen, chitosan, cellulose, titanium alloys, synthetic polymers (e.g., polycaprolactone [PCL], ABS, PLA, PA, polydimethylsiloxane [PDMS], polyether ether ketone [PEEK]) are the most commonly utilized alloys in 3D printing techniques. Composite materials can be readily photographed and are less corrosive than metal alloys. The use of laser cladding in additive manufacturing to combine Ti and Mo (15%) powder. The microstructure and hardness for changing Mo content were studied by constructing compositionally graded structures, and it has been found that the greatest hardness of 450 HV [47–49].

The HA/Ti6Al4V (Hydroxyapatite/Titanium alloy) composite powder, as well as bind, has high biocompatibility, allowing it to merge with bone and enhance osteointegration and bonding strength throughout time just after the initial stage of implantation [50–56]. To make the braided carbon/PEEK composite compressive bone plates, a micro-braiding manufacturing process was used to achieve high and uniform implantation of the matrix into reinforcing fibers. Three different braiding angles were used to examine the four-point bending characteristics of composite bone plates [57–61]. Despite the fact that the bending characteristics of braided carbon/PEEK composites indicated good potential for bone plate usage. For the treatment of fractured bones, a composite braided cast with a Kevlar/Cold cure composite had also proven to be both practical and promising (**Figure 18**).

Braided casts have indeed been studied as a possible alternative for compression bone plates. ABS plastic, which has great impact resistance and hardness, has been selected for the polygonal tablet of nacre-like composite, whereas softer plastic Poly Lactic Acid (PLA) and Thermoplastic Polyurethane (TPU) rubber-like materials are chosen for the adhesive and cohesive layer.

Defense graded Aluminum (AA5083-H116) platelets are glued together via thin vinylester adhesive layers. Cohesive and sticky layers assist to attenuate and absorb the energy given by the shock wave, reducing plastic degradation to the composite tablets. Bioactive composites composed of 85% poly(-caprolactone) (PCL) and 15% nanometric hydroxyapatite (HA) derived from biogenic sources have been 3D printed using an extrusion-based technique to create porous scaffolds appropriate for bone regeneration.

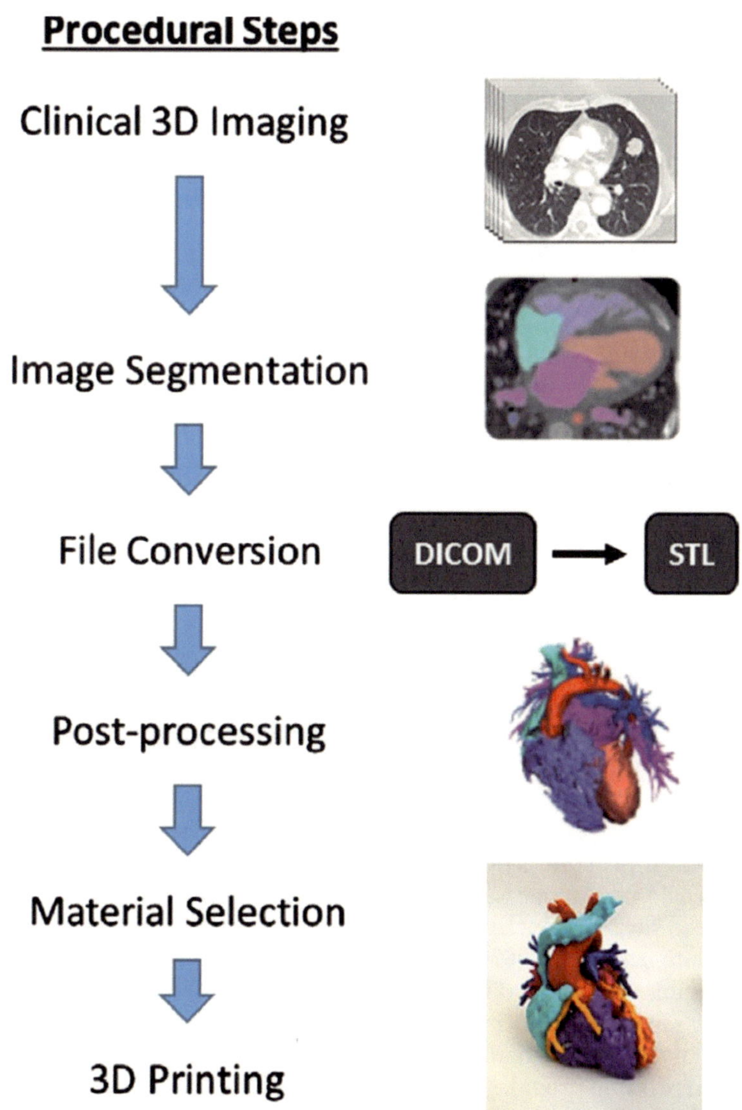

Figure 18.
Procedural steps for Biological 3D Printing from STL files.

PCL/bio-HA scaffolds have better mechanical characteristics and bioactivity. Biocompatible composites of 80 wt% polyvinyl alcohol (PVA)-(20 wt%) polyvinylpyrrolidone (PVP) blend with various concentrations of bioactive nanohydroxyapatite, $Ca_{10}(PO_4)_6(HO)_2$ (HAP) [80PVA-20PVP-xHAP] offered to continue to support cues for the increase in cell viability and biocompatibility with antibacterial.

PA12/ceramic (PA12/ZrO2/HAP) composites have been discovered to function well with the FDM method, allowing the fabrication of medical implants with satisfactory mechanical capabilities for non-load bearing purposes [62]. The hybrid bio-ceramic filled PA12 composite feedstock filament for craniofacial reconstruction via FDM framework, in which the fixed 15 % loading of zirconia (ZrO2) and also various weight fractions (30, 35, 40 wt%, etc.) of beta-tricalcium phosphate (β-TCP) have always been compounded with PA12 to custom tailor the printability of the PA12 composites.

A multi-material 3D printer is used to create an in vitro drug-testing device for heart tissue. Six distinct materials were utilized to create this complicated gadget, each with its own function: (1) dilate dextran as a release layer for the cantilever, (2) dilute thermoplastic polyurethane (TPU) as the cantilever, (3) carbon black and TPU mixture as a strain gauge wire, (4) shear-thinning soft PDMS (polydimethylsiloxane) as a feature to guide tissue orientation, (5) shear-thinning mixture of gold and polyamide as electrical contact pads, (6) shear-thinning PDMS and insulation. This work exemplifies how 3D printing various materials can provide multifunctional devices that can inspire the construction of complex tissues that will improve the area of tissue engineering.

PDMS (Polydimethylsiloxane) lines with 4 wt% CF reinforced PDMS composite provide high biocompatibility strength [63]. Furthermore, due to the remarkable strength and biological properties of PEEK and its CHAp composite, 3D-printed PEEK offers numerous biomedical uses, and its biological macromolecular behavior leads to health sustainability. Organic solvents and extremely high temperatures are not always used in multi-material bioprinting.

3.6 Based on robotic and space applications

Rapid manufacturing of soft robotic systems, such as combustion-powered jumpers, multilegged robots, and stiffness-tunable actuators, is a significant benefit of 3D printing for robotics. Recently, shape-memory alloy and motor tendons actuated 3D printed soft crawling robots have also been created and demonstrated a substantial advantage in the easy and rapid manufacturing that significantly requires multi-material application. In which sensors, tubes, pipelines, metals, caps, motility models, and so on are combined into a single multi-material part, where 3D printing offers significant advantages.

Fabrication of a low-profile antenna for spatial applications with an integrating artificial magnetic conducting (AMC) ground plane. This system has two deposition heads, with one print head dispensed Polycarbonate (PC) that used a filament extrusion technique to print all dielectric components while the second head dispensed silver conducting parts using a micro-dispensing technology. A silver palladium paste (ESL 9694-SA) has been used as a conductive material in soft robotics wearables because it displays good adhesion, great solderability, and high conductivity. There is a necessity for viscosity reduction so that multi-material resin or ink-like structure 20% ethanol has been added as a solvent, and mixing well for 20 minutes in printing provides flexible and stretchable enough even for wearables.

To create an innovative caterpillar-inspired pneumatically-driven soft crawling robot that can then be directly 3D printed without the need for a complex assembly process. This allows us to lessen the number of parts from a pneumatic bellow-type body, 12 anisotropic frictional feet, as well as two end caps into a single part while also putting additional feature synergy locomotion. Translucent photopolymer (Agilus 30 Clr) is utilized as the principal material, and to increase the viscosity of the substance, low-yield polymer Journal Pre-proof 7 (SUP706B) is employed as that of the backup material with good mobility.

Figure 19 illustrates the functionally gradient material which can be utilized for smooth hand grippers, roller wheels, and chairs that are used for robotic applications. To construct multi-material components and FGMs with complete 3D flexibility, we require voxel-level spatial control over component placement. Several existing AM methods on the market seem to be capable of this level of compositional control, although they are largely targeted toward prototype or one-off component production.

Rapid Analysis and Manufacturing Propulsion Technology (RAMPT) will assist in the development of methods to 3D print metal rocket engine components utilizing blasted powder-directed energy deposition. The upgraded system provides academics, researchers, and manufacturers, ranging from machine and metal injection molding (MIM) shops to high-volume manufacturers. Binder jetting uses a digital file to inkjet a binder into a bed of powder particles, resulting in a variety of binding solid parts being created one layer at a time. When printing metals, the finished item must be sintered to fuse the particles all together in a solid object with varying material properties.

3.7 Based on recycling and remanufacturing

Very few studies on the recycling and remanufacturing of multi-material have been undertaken. Fiber-reinforced thermoplastic composites are a popular multi-material that may be recycled and remanufactured completely. A closed-loop material recycling scheme for carbon fiber and matrix had been employed

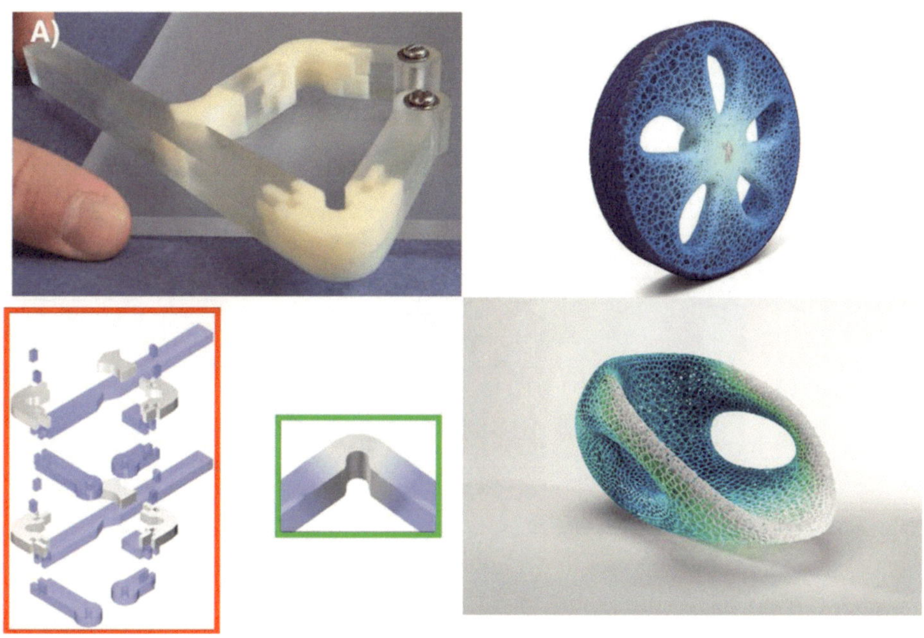

Figure 19.
(A) Polymer-polymer FGMs could simplify the construction of flexible joints connecting rigid elements for soft robotics, combine supple viscoelastic materials with stiff engineering polymers for applications like airless tires, and bring furniture design using multi-material.

for continuous FRTPs, long FRTPs, short FRTPs, and ultimately powder reinforced plastics has been presented and experimentally shown in **Figures 20** and **21**. A notion of direct structural recycling of thermoplastic composites was also proposed, in which big composite goods may be broken into small-size structural parts that can be directly utilized to make smaller composite products. The recycling of FRTP is well illustrated in the figure below. Nonetheless, completely recovering continuous fiber and matrix remains a problem for the development and use of fiber-reinforced thermoplastic composites. Development of novel and improved recyclable composite materials, as well as manufacturing processes, is required for future development in order to fulfill end-use characteristics and recyclability at the same time. Due to the ordered distribution of continuous carbon fiber tows, 3D printing of continuous carbon fiber reinforced thermoplastic composites (CFRTPCs) provides both high mechanical performance and the potential of entirely recycling and even remanufacturing. Given that the recycled carbon fiber filaments already included a few thermoplastics, the filament feed rate was decreased significantly from 100 mm/min to 80 mm/min to maintain a fiber composition comparable to the initially printed samples. Several measures were used to test the performance of 3D printed CFR PLA composites. The universal testing equipment was used to evaluate flexural and interlaminar shear strength. Furthermore, after the carbon fiber had been recycled into an impregnated filament and used again in the remanufacturing process, there has been no discernible improvement in tensile strength and modulus, most likely due to the unidirectional nature of the produced composites, wherein the tensile force was sustained primarily by the carbon fiber rather than interfaces. The modulus of

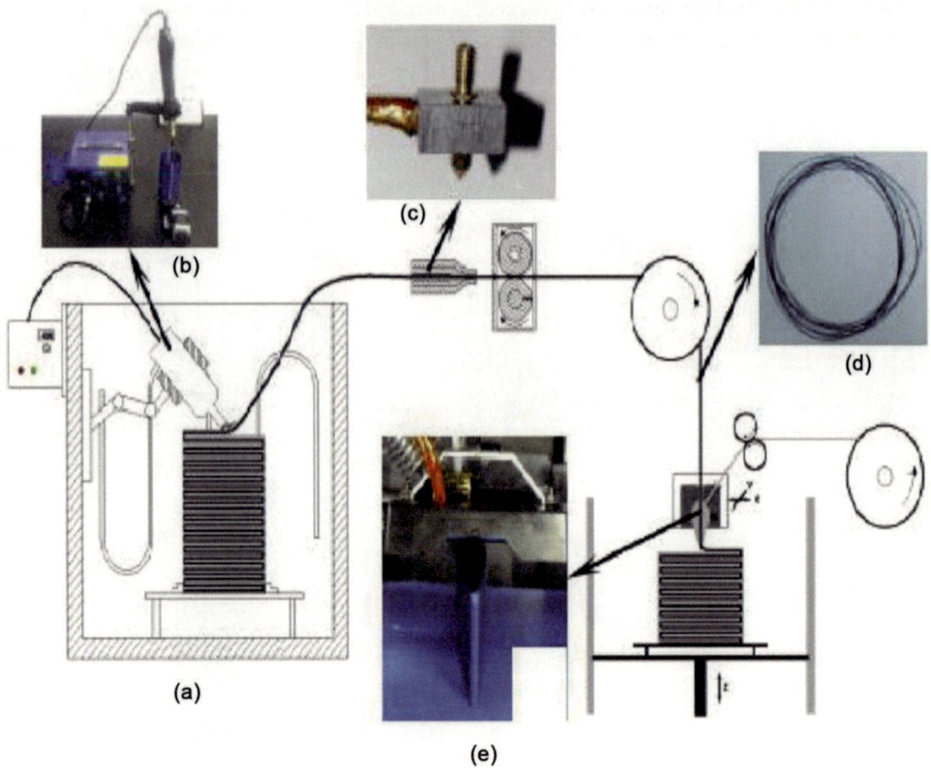

Figure 20.
Scheme of recycling and remanufacturing of 3D printed CFRTPCs (a), and key elements for each step: (b) hot air gun system, (c) remolding nozzle, (d) recycled impregnated filament, and (e) remanufacturing process.

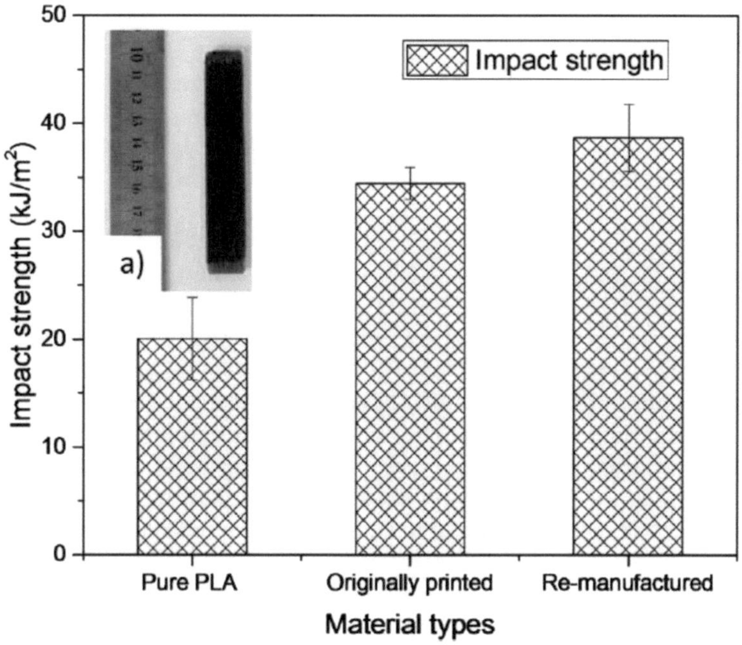

Figure 21.
Impact strength of pure PLA, originally printed and re-manufactured composites specimens, inserted pictures are standard specimens (a), broken specimen from the original printing.

remanufactured composite specimens was somewhat reduced. This opens up the prospect of creating a "green" composite for the future via 3D printing without the need for carbon fibers. Aside from mechanical performance, economic and environmental concerns might be addressed. Moreover, whenever high-performance natural fibers like flax are utilized in the 3D printing process, true green composites will be generated in the future. This unique idea addresses all elements of minimizing the environmental effect of composites, including lowering the cost of the expensive raw material-carbon fiber, employing renewable resources, fully recycling composites to a higher level, and remanufacturing with greater performance [19].

3.8 Significance of multi-material over other the recently available material

Multi-material printing is becoming more popular in the medical field for a variety of applications. Wearables with multi-material capabilities are being developed to detect cancer at an early stage. It will be useful in identifying malignancies that are typically not discovered at an early stage, such as mesothelioma. This cancer is caused mostly by asbestos inhalation, and the early symptoms are mild enough to have been disregarded. Early diagnosis of cancer using multi-material enabled wearables combined with Artificial Information that can help patients live longer lives. Their lower life span, which might offer some risks, can be readily addressed by multi-material devices such as sensors and transmitters integrated with multi-material medical implants. Numerous design ideas of multi-material printing may be answered in the software itself through decreasing according to the demands of the patients particular and applications. Furthermore, this application has the benefit of creating two or more material advantages in a single part above typical high strength or accessible materials. This technique lowers the pain associated with any existing traditional methods or materials.

4. Conclusion and future research

Multi-material Additive Manufacturing provides a variety of material properties in a single component while also allowing us to tool several components into a single component. Multi-functional 3D components have the potential to revolutionize several industries, including biomedical engineering, soft robotics, electronics, spatial, and aerospace applications. Despite significant advances in MMAM over the last few years, there are many other outstanding problems to be encountered, such as low production throughput, poor scalability and surface finish, limited material selection, high cross-contamination, and low interfacial bonding among different materials. In addition to this as well, a fundamental scientific knowledge of materials science, particularly MMAM, FGM, kinetics, and mechanics, is required to promote and improve MMAM research. As the processing complexities of MMAM rise rapidly because of the diversity of materials involved, the creation of new manufacturing products and new advancements in advanced manufacturing must be addressed. Advances in MMAM and its numerous new searches in the advanced study will lead us to varied combinations of new materials over existing material models. We can affect the manufacturing sector with new breakthroughs in the developing 3D printing technology by gaining a better knowledge of diverse material understanding and binding forces, as well as conquering rigorous printing time reduction with very excellent processing time.

Author details

Rajkumar Velu[1*], R. Sathishkumar[2] and A. Saiyathibrahim[3]

1 Additive Manufacturing Research Laboratory, Indian Institute of Technology Jammu, Jammu and Kashmir, India

2 C.A.R.E Engineering College, Tiruchirappalli, India

3 Karpagam Institute of Technology, Coimbatore, India

*Address all correspondence to: rajkumar7.v@gmail.com; rajkumar.v@iitjammu.ac.in

References

[1] Singh R, Kumar R, Farina I, Colangelo F, Feo L, Fraternali F. Multi-material additive manufacturing of sustainable innovative materials and structures. Polymers. 2019;**11**(1):62. DOI: 10.3390/polym11010062

[2] Bittner SM, Guo JL, Melchiorri A, Mikos AG. Three-dimensional printing of multilayered tissue engineering scaffolds. Materials Today. 2018;**21**(8): 861-874

[3] Vu IQ, Bass LB, Williams CB, Dillard DA. Characterizing the effect of print orientation on interface integrity of multi-material jetting additive manufacturing. Additive Manufacturing. 2018;**22**:447-461

[4] Lopez B, Moises D, Ahmad R. Tensile mechanical behaviour of multi-polymer sandwich structures via fused deposition modelling. Polymers. 2020;**12**(3):651

[5] Alghamdy M, Ahmad R, Alsayyed B. Material selection methodology for additive manufacturing applications. Procedia CIRP. 2019;**84**:486-490

[6] Rafiee M, Farahani RD, Therriault D. Multi-material 3D and 4D printing: A survey. Advanced Science. 2020;7(12): 1902307

[7] Zhou L-y, Gao Q, Jian-zhong F, Chen Q-y, Zhu J-p, Sun Y, et al. Multimaterial 3D printing of highly stretchable silicone elastomers. ACS Applied Materials & Interfaces. 2019;**11**(26):23573-23583

[8] Joas S, Tovar GEM, Celik O, Bonten C, Southan A. Extrusion-based 3D printing of poly (ethylene glycol) diacrylate hydrogels containing positively and negatively charged groups. Gels. 2018;**4**(3):69

[9] Rutz AL, Hyland KE, Jakus AE, Burghardt WR, Shah RN. A multimaterial bioink method for 3D printing tunable, cell-compatible hydrogels. Advanced Materials. 2015;**27**(9):1607-1614

[10] Sears N, Dhavalikar P, Whitely M, Cosgriff-Hernandez E. Fabrication of biomimetic bone grafts with multi-material 3D printing. Biofabrication. 2017;**9**(2):025020

[11] Lopes LR, Silva AF, Carneiro OS. Multi-material 3D printing: The relevance of materials affinity on the boundary interface performance. Additive Manufacturing. 2018;**23**:45-52. DOI: 10.1016/j.addma.2018.06.027

[12] Zuo Y, Su X, Li X, Yao Z, Yu T. graphene / carbonyl iron composites with superior microwave absorption properties and adjustable bandwidth. Carbon N. Y. 2020;**167**:62-74. DOI: 10.1016/j.carbon.2020.05.071

[13] Bandyopadhyay A, Heer B. Materials Science & Engineering R Additive manufacturing of multi-material structures. Materials Science & Engineering R: Reports. 2018; **129**(March):1-16. DOI: 10.1016/j.mser.2018.04.001

[14] Yuan S, Li S, Zhu J, Tang Y. Additive manufacturing of polymeric composites from material processing to structural design. Composites: Part B. 2020, 2021;**219**(November):108903. DOI: 10.1016/j.compositesb.2021.108903

[15] Eun S et al. Nano Energy Composition-segmented BiSbTe thermoelectric generator fabricated by multimaterial 3D printing. Nano Energy. 2021;**81**(November):105638. DOI: 10.1016/j.nanoen.2020.105638

[16] Flores I, Flores I. Design and additive manufacture of functionally graded structures based on digital materials. Journal of Additive

Manufacturing. 2019;**30**:100839. DOI: 10.1016/j.addma.2019.100839

[17] Xu Y, Wu X, Guo X, Kong B, Zhang M, Qian X. The boom in 3D-printed sensor technology. Sensors. 2017;**17**(5):1166. DOI: 10.3390/s17051166

[18] Ngo TD, Kashani A, Imbalzano G, Nguyen KTQ, Hui D. Additive manufacturing (3D printing): A review of materials, methods, applications and challenges. Composites Part B Engineering. 2018;**143**(December): 172-196. DOI: 10.1016/j.compositesb.2018.02.012

[19] Tian X, Liu T, Wang Q, Dilmurat A, Li D, Ziegmann G. Recycling and remanufacturing of 3D printed continuous carbon fi ber reinforced PLA composites. Journal of Cleaner Production. 2017;**142**:1609-1618. DOI: 10.1016/j.jclepro.2016.11.139

[20] Zorzetto L et al. Properties and role of interfaces in multimaterial 3D printed composites. Scientific Reports. 2020;**10**:22285. DOI: 10.1038/s41598-020-79230-0

[21] Wagner MA, Huang J, Okle P, Paik J, Spolenak R. Hinges for origami-inspired structures by multimaterial additive manufacturing. Materials and Design. 2020;**191**:108643. DOI: 10.1016/j.matdes.2020.108643

[22] Dunn C et al. Digital light processing 3D printing of conductive complex structures Digital light processing 3D printing of conductive complex structures. Additive Manufacturing. 2017;**18**(September): 74-83. DOI: 10.1016/j.addma.2017.08.011

[23] Schwartz JJ, Boydston AJ. Printing. Nature Communications. 2019;**10**(791): 1-10. DOI: 10.1038/s41467-019-08639-7

[24] Rocha VG, Saiz E, Tirichenko IS, García-Tuñón E. Direct ink writing advances in multi-material structures for a sustainable future. Journal of Materials Chemistry A. Chicago. 2020;**8**(31):15646-15657

[25] Doyle M, Agarwal K, Sealy W, Schull K. Effect of layer thickness and orientation on mechanical behavior of binder jet stainless steel 420 + bronze parts. Procedia Manuf. 2015;**1**:251-262. DOI: 10.1016/j.promfg.2015.09.016

[26] Han D, Lee H. ScienceDirect Recent advances in multi-material additive manufacturing: Methods and applications. Current Opinion in Chemical Engineering. 2020;**28**:158-166. DOI: 10.1016/j.coche.2020.03.004

[27] Karakurt I, Lin L. ScienceDirect 3D printing technologies: Techniques, materials, and and post-processing. Current Opinion in Chemical Engineering. 2020;**28**:134-143. DOI: 10.1016/j.coche.2020.04.001

[28] Wang D. High-efficiency high-resolution multimaterial fabrication for digital light processing-based three-dimensional printing. April. 2018;**2020**. DOI: 10.1089/3dp.2018.0004

[29] Roach DJ, Hamel CM, Dunn CK, Johnson MV, Kuang X, Qi HJ. The m 4 3D printer: A multi-material multi-method additive manufacturing platform for future 3D printed structures. Additive Manufacturing. 2019;**29**(February):100819. DOI: 10.1016/j.addma.2019.100819

[30] Gómez A. ScienceDirect ScienceDirect Development of a multi-material additive manufacturing process for Development of a multi-material additive manufacturing process for electronic devices Costing models for capacity optimization in Industry. Procedia Manuf. 2017;**13**:746-753. DOI: 10.1016/j.promfg.2017.09.180

[31] Wen X, Gao S, Feng J, Li S, Gao R, Zhang G. Chest-wall reconstruction

with a customized titanium-alloy prosthesis fabricated by 3D printing and rapid prototyping. Journal of Cardiothoracic Surgery. 2018;**13**(1):1-7. DOI: 10.1186/s13019-017-0692-3

[32] King WE, Anderson AT, Ferencz RM, Hodge NE, Kamath C, Khairallah SA. et al. Laser powder bed fusion additive manufacturing of metals; physics, computational, and materials challenges. Applied Physics Reviews. 2015;**2**(4):041304

[33] Horng MH, Kuok CP, Fu MJ, Lin CJ, Sun YN. Cobb angle measurement of spine from x-ray images using convolutional neural network. Computational and Mathematical Methods in Medicine. 2019;**2019**:1-18. DOI: 10.1155/2019/6357171

[34] Huang H, Liu J, Wang L, Fan Y. A critical review on the biomechanical study of cervical interbody fusion cage. Med. Nov. Technol. Devices. 2021;**11**(February):100070. DOI: 10.1016/j.medntd.2021.100070

[35] Velu R, Calais T, Jayakumar A, Raspall F. A comprehensive review on bio-nanomaterials for medical implants and feasibility studies on fabrication of such implants by additive manufacturing technique. Materials (Basel). 2020;**13**(1):92. DOI: 10.3390/ma13010092

[36] Velu R, Vaheed N, Ramachandran MK, Raspall F. Correction to experimental investigation of robotic 3D printing of high-performance thermoplastics (PEEK): A critical perspective to support automated fibre placement process. International Journal of Advanced Manufacturing Technology. 2020;**108**(4):1027. DOI: 10.1007/s00170-019-04763-2

[37] Whenish R, Antony MM, Balaji T, Selvam A, Ramprasath LS, Velu R. Design and performance of additively manufactured lightweight bionic hand. AIP Conference Proceedings. February, 2021;**2317**. DOI: 10.1063/5.0036119

[38] Selvam A, Mayilswamy S, Whenish R, Velu R, Subramanian B. Preparation and evaluation of the tensile characteristics of carbon fiber rod reinforced 3d printed thermoplastic composites. J. Compos. Sci. 2021;**5**(1):1-15. DOI: 10.3390/jcs5010008

[39] Velu R, Vaheed N, Raspall F. Design and robotic fabrication of 3D printed moulds for composites. Solid Free. Fabr. 2018 Proc. 29th Annu. Int. Solid Free. Fabr. Symp. – An Addit. Manuf. Conf. SFF 2018. 2020:1036-1046

[40] Velu R, Vaheed NM, Venkatesan C, Raspall F, Krishnan M. Experimental investigation on fabrication of thermoset prepreg composites using automated fibre placement process and 3D printed substrate. Procedia CIRP. 2020;**85**:293-298. DOI: 10.1016/j.procir.2019.09.049

[41] Venkatasudhahar M, Velu R, Logesh K. Investigation on the effect of flyash on tensile, flexural and impact strength of hybrid composite material. International Journal of Mechanical and Production Engineering Research and Development. 2018;**8**(4):117-122. DOI: 10.24247/ijmperdaug201813

[42] Velu R, Fernyhough A, Smith DA, Joo Le Guen M, Singamneni S. Selective laser sintering of biocomposite materials. Lasers Eng. 2016;**35**(1-4):173-186

[43] Velu R, Kamarajan BP, Ananthasubramanian M, Ngo T, Singamneni S. Post-process composition and biological responses of laser sintered PMMA and β-TCP composites. Journal of Materials Research. 2018;**33**(14):1987-1998. DOI: 10.1557/jmr.2018.76

[44] Velu R, Raspall F, Singamneni S. 3D Printing Technologies and Composite

Materials for Structural Applications. Green Composites for Automotive Applications. Elsevier Ltd; Jan 2019:171-196

[45] Venkatesan C, Velu R, Vaheed N, Raspall F, Tay TE, Silva A. Effect of process parameters on polyamide-6 carbon fibre prepreg laminated by IR-assisted automated fibre placement. International Journal of Advanced Manufacturing Technology. 2020;**108**(4):1275-1284. DOI: 10.1007/s00170-020-05230-z

[46] Yagi M, Akilah KB, Boachie-Adjei O. Incidence, risk factors and classification of proximal junctional kyphosis: Surgical outcomes review of adult idiopathic scoliosis. Spine (Phila Pa 1976). 2011;**36**(1). DOI: 10.1097/BRS.0b013e3181eeaee2

[47] Yagi M, Nakahira Y, Watanabe K, Nakamura M, Matsumoto M, Iwamoto M. The effect of posterior tethers on the biomechanics of proximal junctional kyphosis: The whole human finite element model analysis. Scientific Reports. 2020;**10**(1):1-8. DOI: 10.1038/s41598-020-59179-w

[48] Ledet EH, Liddle B, Kradinova K, Harper S. Outcomes: A review. Innov Entrep Heal. 2018:41-51. DOI: 10.2147/IEH.S133518.Smart

[49] Yu J et al. Current advances in 3D bioprinting technology and its applications for tissue engineering. Polymers. 2020;**12**(12):1-30. DOI: 10.3390/polym12122958

[50] Bogu VP, Ravi Kumar Y, Khanara AK. Modelling and structural analysis of skull/cranial implant: Beyond mid-line deformities. Acta of Bioengineering and Biomechanics. 2017;**19**(1):125-131. DOI: 10.5277/ABB-00547-2016-04

[51] Ćuković S, Devedžić G, Luković V, Anwer N, Zečević-Luković T, Subburaj K. 3D modeling of spinal deformities shapes using 5th degree b-splines. 12th Int. Sci. Conf. MMA 2015 – Flex. Technol. 2015;**18**:2-5

[52] Anitha D et al. Effect of statistically iterative image reconstruction on vertebral bone strength prediction using bone mineral density and finite element modeling: A preliminary study. Journal of Computer Assisted Tomography. 2019;**43**(1):61-65. DOI: 10.1097/RCT.0000000000000788

[53] Sollmann N et al. Multi-detector computed tomography (MDCT) imaging: Association of bone texture parameters with finite element analysis (FEA)-based failure load of single vertebrae and functional spinal units. Quantitative Imaging in Medicine and Surgery. 2021;**11**(7):2955-2967. DOI: 10.21037/qims-20-1156

[54] Sollmann N et al. MDCT-based finite element analyses: Are measurements at the lumbar spine associated with the biomechanical strength of functional spinal units of incidental osteoporotic fractures along the thoracolumbar spine? Diagnostics. 2021;**11**(3):455. DOI: 10.3390/diagnostics11030455

[55] Más Y, Gracia L, Ibarz E, Gabarre S, Peña D, Herrera A. Finite element simulation and clinical followup of lumbar spine biomechanics with dynamic fixations. PLoS One. 2017;**12**(11):e0188328. DOI: 10.1371/journal.pone.0188328

[56] Rohlmann A, Pohl D, Bender A, Graichen F, Dymke J. Activities of everyday life with high spinal loads. PLoS One. 2014;**9**(5):98510. DOI: 10.1371/journal.pone.0098510

[57] Jones AC, Wilcox RK. Finite element analysis of the spine: Towards a framework of verification, validation and sensitivity analysis. Medical Engineering & Physics. 2008;**30**(10):1287-1304. DOI: 10.1016/j.medengphy.2008.09.006

[58] Niinomi M. Titanium Spinal-Fixation Implants. Woodhead Publishing Series in Biomaterials, Elsevier Inc.; 2018:347-369. DOI: 10.1016/B978-0-12-812456-7.00016-0

[59] Allain J, Dufour T. Anterior lumbar fusion techniques: ALIF, OLIF, DLIF, LLIF, IXLIF. Orthopaedics & Traumatology, Surgery & Research. 2020;**106**(1):S149-S157. DOI: 10.1016/j.otsr.2019.05.024

[60] Cooper L et al. Experimental platform to facilitate novel back brace development for the improvement of spine stability. Computer Methods in Biomechanics and Biomedical Engineering. 2019;**22**(15):1163-1173. DOI: 10.1080/10255842.2019.1645837

[61] Kurtz SM, Lanman T. Dynamic Stabilization and Semirigid PEEK Rods for Spinal Fusion. 2nd ed. In: Kurtz SM, editor. PEEK biomaterials handbook. UK: Oxford, William Andrew Publishing; 2019:281-289

[62] Más Y, Gracia L, Ibarz E, Gabarre S, Peña D, Herrera A. Finite element simulation and clinical follow-up of lumbar spine biomechanics with dynamic fixations. PloS one. 2017;**12**(11):e0188328

[63] Velu R, Raspall F, Singamneni S. Selective laser sintering of polymer biocomposites based on polymethyl methacrylate. Journal of Materials Research. 2014;**29**(17):1883-1892. DOI: 10.1557/jmr.2014.211

Chapter 7

Functionally Modified Composites for FDM 3D Printing

Smith Woosley and Shyam Aravamudhan

Abstract

Fused Deposition Modeling (FDM) 3D printing is an additive manufacturing technique used to fabricate solid thermoplastic polymer objects directly from computer-modeled designs. The current uses for this technology are restricted due to a limited choice of materials, which offer minimal functionality to the printed 3D parts. To expand the application space for FDM-based 3D printing, this chapter is aimed to add functional attributes to printable polymers through the creation of thermoplastic composites. The work focuses on a simple fabrication method to create composite for FDM printing and analytical techniques to characterize dispersion, thermal, and mechanical properties of the nanocomposite. Lastly, the functional characteristics of the FDM printed nanocomposite including their conductivity, ferromagnetism, and radiation shielding properties were studied.

Keywords: functional additive manufacturing, nanocomposites, 3D printing, thermoplastics, printable polymers

1. Introduction

Fused deposition modeling (FDM) 3D printing is an additive manufacturing technology used to construct solid objects directly from computer-modeled designs. In an FDM process, a solid thermoplastic polymer (as the feedstock) is extruded from a nozzle to build the 3D object. When the thermoplastic polymer is heated above its melting temperature, it flows as a viscous liquid, and can be patterned into thin layers. The layers cool quickly after leaving the nozzle, enabling successive layers to be deposited on top of each other, and thus forming a final 3D object. This technology offers an inexpensive and efficient method to produce customized parts with intricate geometries using a simple printing process. Despite the wide range of potential applications from consumer products and medical to industrial, automotive, and aerospace, FDM has been mostly limited to fabrication of prototype items with no inherent functionality. This limitation has restricted FDM technology within the 3D printing and manufacturing industries, as it prevents fabrication of end products and functional systems [1, 2]. The goal of this chapter is to describe a simple and optimized method for improving the functionality of FDM printing through creation of printable composite materials and to demonstrate a broad application space for functional FDM 3D printing.

IntechOpen

The principal reason for the constraint in FDM technology is the limited choice of materials available for FDM printing. Thermoplastic polymers, namely polylactic acid (PLA), acrylonitrile butadiene styrene (ABS), polyethylene terephthalate (PET), Nylon, thermoplastic polyurethane (TPU), and polycarbonate (PC), are primarily used for their thermal processability, which is necessitated by the FDM extrusion process [3]. Very few commercial thermoplastic polymers provide functional characteristics beyond just printability and prototypes. For FDM 3D printing to progress to the next level and compete with traditional manufacturing, direct production of useful and functional products is necessary. Therefore, future research must focus on imparting functionality into materials used in the FDM 3D printing process.

One method that can be used to achieve this goal is the creation of thermoplastic composites, where thermoplastic polymers are mixed with additives with one or more functionality such as improved electrical conductivity, enhanced ferromagnetism, electrochemical or radiation shielding capabilities. The resulting combination will continue to work as a printable material while simultaneously providing the desired functional attribute. This technique has gained popularity recently as researchers work toward developing new materials to broaden the application space for FDM 3D printing. Functional additives studied in the past include fibers of carbon [4–6] or glass [7], and particles, such as iron [8], graphene [9], or silicon carbide [10]. Most of the above studies used compound mixing to distribute the additive within the thermoplastic matrix, while a few have used a fiber encapsulation technique. Several factors must be considered and evaluated to achieve true thermoplastic polymer composites, including composite processability, printability with optimal thermal and mechanical characteristics, dispersion, and uniformity, and finally, improvement in the functional properties. Functional properties include electrical conductivity [11–14], magnetism [15, 16], and bacterial/biofilm resistance [17–19] and have been demonstrated in FDM composite materials. One of the least studied functional attributes in relation to FDM printing is the radiation shielding capabilities [20]. Radiation shielding is important attribute for component manufacturers with stringent radiation protection requirements such as aerospace, unmanned aerial vehicle (UAV), and satellite manufacturers. Among the many ionizing radiations, neutrons are considered to be extremely severe as they can degrade not only material and electrical components but can also cause damage to biological tissues [21, 22]. By infusing functional attributes such as radiation shielding in FDM composites, the radiation-sensitive component manufacturers can directly create complete functional FDM components, rather than just form prototypes.

Here in this chapter, we report on (a) novel protocol for fabricating FDM composite filaments using a thermoplastic polymer as the matrix material and functional additives as the filler; (b) analytical characterization of resulting functional FDM composite to assess particle dispersion, material uniformity, thermal behavior, and mechanical stability; and (c) assessment of functional properties of fabricated filaments and printed 3D parts. In this work, different functional additives, namely carbon black, nickel, boron nitride, and gadolinium particles, were added to an acrylonitrile butadiene styrene (ABS) polymer matrix to form functional composites that could be directly FDM 3D printed. In the rest of the chapter, these functional additives will be independently evaluated (using analytical and functional techniques) to assess improvements in their functional properties. The significance of this work is that by using simple solution mixing and desktop filament extruder approaches, functional characteristics including electrical conductivity, ferromagnetism, and unique radiation shielding properties were tested in FDM printed ABS composite samples.

2. Methodology

2.1 Materials

Thermoplastic Polymer Matrix: Acrylonitrile butadiene styrene (ABS) is one of the widely used thermoplastic polymers in FDM 3D printing [23]. The reasons being its low cost, printability, and reasonable material strength. Therefore, in this work, ABS was chosen as composite matrix for FDM printing. ABS was purchased in the pellet form from IC3D Printers.

Functional Additive: As the functional additive, various micro/nanoparticles and powders were selected based on the intended functional property such as electrical conductivity, ferromagnetism, and radiation shielding capability. For electrical conductivity functionality, Timical Super C65 (EQ-Lib-Super65) carbon black in fine powder form was purchased. For ferromagnetic functionality, nickel nanopowder (of size <100 nm and > 99% purity) was purchased from Sigma-Aldrich (#577995). For radiation shielding capability, boron nitride nanoparticles (purity >99.8% in hexagonal structure) and gadolinium particles (of size <149 μm and purity >99.9%) were purchased from US Research Nanomaterials, Inc. (#US2019) and Sky Spring Nanomaterials, Inc. (#3580DX), respectively. Boron nitride and gadolinium particles have been demonstrated to be effective in blocking neutron radiation [24, 25]. This is a result of the high neutron capture cross section, an indicator of the ability of an element to block and absorb incident neutrons. For example, boron has a value of approximately 760 barn, the 10th highest of all elements, and gadolinium at 490,000 barn is the highest of all elements [26].

2.2 Fabrication of functional nanocomposite filament for FDM printing

Nanocomposite mixing: After several unsuccessful attempts involving solid mixing of ABS (in either powder or pellet form) and functional additives, followed by thermal compounding, a solution processing method was finally optimized for the fabrication of functional composite filaments.

In this method, first ABS pellets (**Figure 1**) were dissolved in acetone. Hundred grams of ABS pellets were added to 400 mL of acetone, left to react for 24 hours, and then mixed mechanically (using paddle mixer and ultrasonication) to form a homogeneous polymer solution. One of the functional additives, namely carbon

Figure 1.
Optical image of as obtained ABS pellets.

black, nickel, boron nitride and gadolinium particles, was added to the solution, mixed, and sonicated to evenly disperse the particles. It was determined that the composite material properties (including mechanical and thermal) degrade significantly for greater than 20% by weight of the functional additive (irrespective of the additive). Therefore, three percentages (corresponding to their equivalent masses), namely 5%, 10%, and 20%, by weight were used in this study. For example, in the case of boron nitride (BN) nanoparticles, 5%, 10%, and 20% by weight correspond to 5.27, 11.11, and 25.0 grams of BN, respectively. First, the solution mixture was solvent evaporated on a glass plate. Next, the dried mixture was sliced into approximately 5 mm x 5 mm pieces and dried at 80°C. Next, to obtain in the composite powder, the pieces were grinded in a hammer mill. Lastly, the composite powder was stored at 60°C with periodic agitation daily for 7 days to ensure no residual solvent was left in the composite powder.

Nanocomposite filament extraction: Next, for use in standard FDM printer, the composite powder was extruded into a filament form. To accomplish this task, a Filastruder filament extruder was used. First, 1.75 mm nozzle and 190°C were set in the extrude. This was followed by feeding of the powder into the hopper as the screw pushed it toward the nozzle. In this process, the powder was heated to its molten state, pressurized, and filament was extruded at a slow extrusion rate of around 22–25 cm/min. Next, the extruded filament was wound onto a spool using an automated guide after it was air cooled. Finally, 1.75 mm pure ABS and composite filaments (ABS with functional additives) that can be used in any standard FDM 3D printer were obtained. The accuracy of the extruded filament (cross-sectional diameter) was measured (discussed in Section 3). In general, the different composite filaments were found to be 1.75 ± 0.13 mm.

2.3 Analytical, thermal, mechanical, and functional characterization

Analytical characterization: Pure ABS and functional composite filaments were set in EpoThin2 epoxy resin and polished for cross-sectional imaging. EpoThin2 epoxy is a low-viscosity, low-shrinkage resin that is ideal for mounting samples for optical and electron microscopy imaging. Scanning electron microscopy (SEM) and energy dispersive X-ray spectroscopy (EDS) measurements were performed on a Zeiss Auriga Field Emission Scanning Electron Microscope (FESEM) (Carl Zeiss AG, Germany) to analyze the dispersion or aggregation of particles and its elemental composition. Horiba XploRA Raman system was used to obtain Raman spectrum of selected nanocomposite filaments (particularly boron nitride filaments). The collected Raman spectra were plotted using Horiba LabSpec 6 Spectroscopy suite. Pure boron nitride particles were also characterized, and the obtained spectra were compared with literature values.

Thermal characterization: Two techniques were used to assess thermal properties: thermogravimetric analysis (TGA) and differential scanning calorimetry (DSC). Thermal degradation onset points and residual mass percentages were calculated on a TA Q500 Thermogravimetric Analyzer. Titanium hang-down pans were loaded with 10 mg of the samples. Heat rate was ramped at 10°C/min from 30–600°C under a nitrogen environment at a 60 mL/min flow rate. This analysis is typically done under nitrogen (or inert) environment to prevent other gases from interfering with the sample during the thermal treatment (i.e., oxidation). Using differential scanning calorimetry (DSC) measurements, the glass transition temperatures (T_g) were analyzed on a TA Q200. Here, 5 mg of each composite was ramped from 35–180°C at 10°C/min. Then, using half heat flow extrapolation, T_g was calculated from the temperature versus heat flow curves.

Mechanical characterization: Mechanical properties, namely strength and modulus of elasticity of both pure ABS and functional composites, were measured. An Instron 5900R material test frame with 5 kN load cell was used to perform mechanical testing. Four-inch filament sections were cut during the extrusion process. These sections were epoxied to plastic tabs as attachment points for the instrument grips, leaving a 2-inch gauge length for tensile testing. Tensile stress was measured using a load cell and plotted as a function of percent strain. Slope of the stress–strain curve in the elastic zone was calculated to determine modulus of elasticity.

Next, ASTM D638 test specimens were cut using a water jet from the printed 3D sample for further mechanical testing. The specimen dimensions adhered to guidelines for type V samples. Dynamic mechanical analysis (DMA) was performed on a Perkin Elmer DMA 8000. This analysis was used to study composite's mechanical properties at different temperatures. Here, the 3D printed samples were cut into rectangular sections of dimensions 50.5 x 12.6 x 3.6 mm. Dual cantilever testing was performed at a frequency of 0.05 Hz with a temperature ramp from 30 to 150 C. The storage modulus values were studied.

Functional characterization: Lastly, functional properties of different nanocomposites and different functional 3D printed samples were measured. 3D printing of both pure ABS filaments and the functional composite filaments was done on a Fusion3 F306 3D printer. The sample dimensions were 65.3 x 65.3 x 3.4 mm at 100% infill and + 45°/−45° raster pattern. The print head temperature was maintained at 245°C printing on a heated bed set at 110°C.

To test the electrical functionality, electrochemical impedance spectroscopy (EIS) was used to assess electrical impedance of the carbon black infused filaments (**Figure 2**). EIS was performed on a VersaSTAT 4 potentiostat in a frequency range of 0.1–10 KHz with an amplitude of 5 mV RMS. Next, to test ferromagnetic functionality of ABS/nickel nanocomposites, a Quantum Design physical property measurement system (PPMS) was used to study magnetization of the nanocomposite 3D prints. The magnetic moment was plotted as a function of applied magnetic field. Finally, in order to evaluate neutron attenuation characteristic, the 3D printed composite samples were exposed to neutron particle beam at the Breazeale

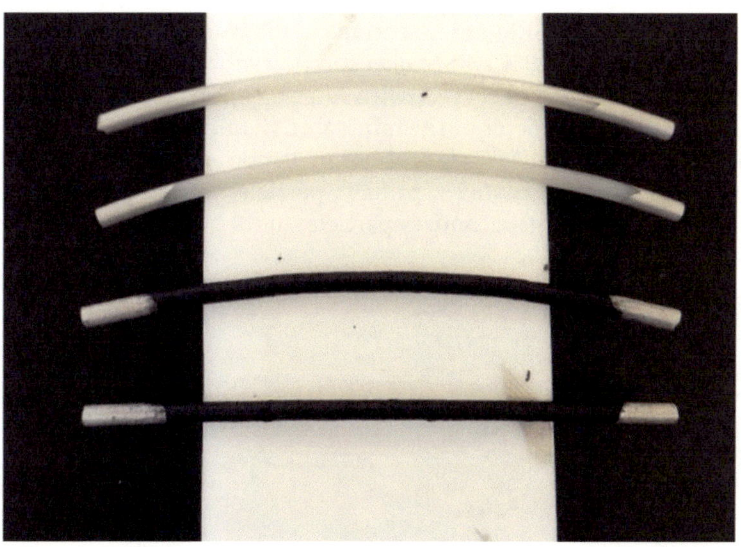

Figure 2.
Functionally modified composite filaments (ABS/carbon black) for electrical impedance measurements.

Nuclear Reactor, part of the Radiation Science & Engineering Center at Penn State University. A detector was then used to measure incident neutrons before and after exposure of the 3D printed composite samples. The test methodology is as follows. First, using a carrier system, the 3D printed samples were moved into the collimated neutron beam consisting of thermal neutrons with average energy of 0.025 eV. This beam was in between BF_3 detectors and the reactor. Next, a minimum of 15,000 counts of neutrons (counting time of at least 30 seconds) were allowed to penetrate the 3D printed sample at multiple spots. For control and for data normalization, both "blank" (with 0% shielding capability) and "black" (with 100% effective shielding) were measured.

3. Results and discussion

Analytical characteristics of different nanocomposite filaments: First, pure ABS extruded filament was studied. **Figure 3a** and **b** show optical image of the pure ABS filament (after the extrusion process) and its corresponding Raman spectrum. Caliper measurements confirmed the uniformity of filament diameter during extrusion, with readings of 1.75 ± 0.05 mm. Raman spectra confirmed pure ABS with definitive peaks at 1629, 1696, and 2262 cm^{-1}. The 1629 cm^{-1} peak corresponds to C – C and C=C vibrations in the styrene aromatic ring, the 1696 cm^{-1} peak demonstrates C=C bonds in the butadiene monomer, and the 2262 cm^{-1} peak shows the presence of C ≡ N bond in acrylonitrile. The 654 and 1031 cm^{-1} peaks are related to aromatic C-H bending.

Next, the different extruded composite filaments were studied. As discussed in Section 2.2, all cross-sectional SEM images were obtained after setting the nanocomposite filaments (about 3–5 cm sections) in EpoThin2 epoxy resin and surface polished to expose the "first" layer below the surface. **Figure 4a** and **b** shows SEM cross-sectional and optical images of the carbon black in composite filament after 20% ABS/carbon black solution processing. Carbon black is in fine powder form, with submicron primary and agglomerates averaging around 20 μm. The measured diameter of the fabricated filament was 1.60 ± 0.15 mm. Arrows in all the optical images indicate a general area (around the middle of the nanocomposite filament) where the SEM cross-sectional images were taken.

Figure 5a and **b** show SEM cross-sectional and optical images of ABS/nickel composite filament after 20% ABS/nickel solution processing. A broad nickel particle dispersion profile was observed, with some agglomerates larger than 100 μm in size. The measured diameter of the fabricated filament was 1.68 ± 0.12 mm.

Figure 6a and **b** show SEM images of boron nitride and gadolinium particles. In the case of boron nitride particles, primary particles were observed in the nano range (70–80 nm), and large secondary particles in the range of 10–50 μm were

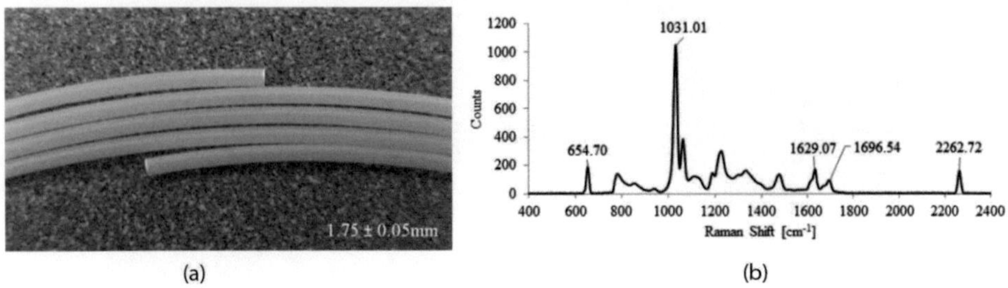

Figure 3.
a) Optical image of the extruded ABS filament and b) its Raman spectra.

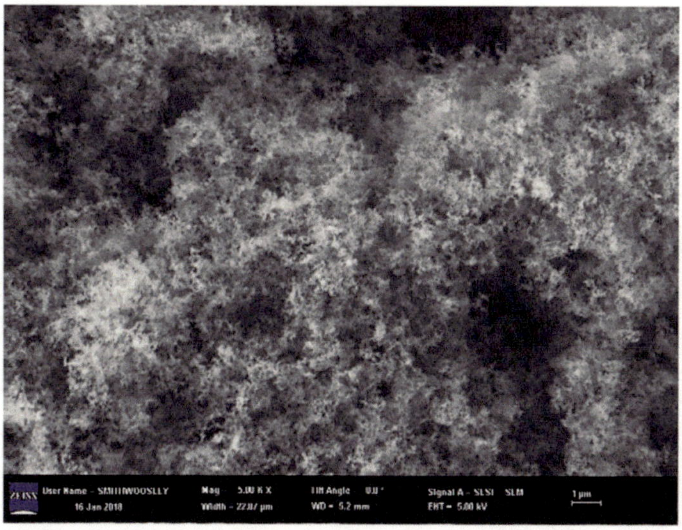

(a)

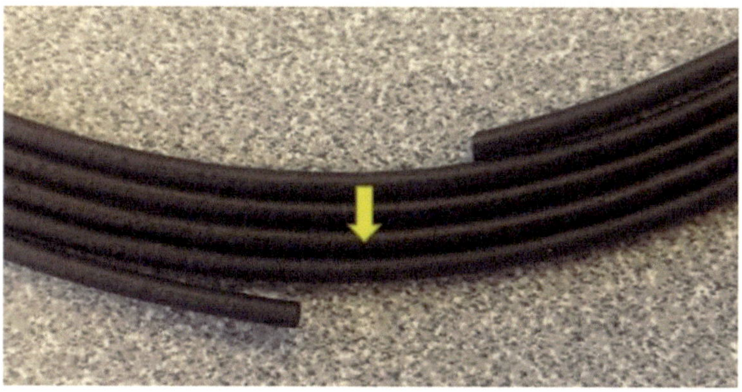

(b)

Figure 4.
a) SEM cross-sectional image of 20% ABS/carbon black composite filament and b) optical image of the 20% ABS/carbon black composite filament.

apparent. In the case of gadolinium particles, a broad size range was apparent with large particles in the range of 10–100 µm.

Figure 7a, b, and **c** show the ABS/boron nitride composite filaments of 5%, 10%, and 20% by weight, respectively. The particles in the composite appeared to be evenly distributed with 5% and 10% concentrations exhibiting smooth surface finish (Ra ~20–30 µm), while surface of the 20% concentration appeared to be rougher (Ra ~80–100 µm). With the respect to the composite filament diameters 5%, 10% and 20% composite filaments were in the range of 1.79 ± 0.06 mm, 1.77 ± 0.04 mm, and 1.76 ± 0.08 mm, respectively. These results were consistent and were within necessary tolerance for FDM printing. **Figure 7d, e**, and **f** show the ABS/gadolinium composite filaments of 5%, 10%, and 20% weight, respectively. The material appeared to be consistent and showed a smooth surface finish at 5% concentration (Ra ~40–60 µm), slight surface roughness at 10% (Ra ~80–120 µm), and significant irregularity at 20% (Ra >250 µm). The diameter measurements for the 5% composite were in range of 1.83 ± 0.08 mm, 1.79 ± 0.12 mm for the 10% composite, and 1.69 ± 0.21 mm for the 20% composite. The severe roughness of the 20% sample resulted in a very inconsistent diameter. The combination of severe surface roughness and inconsistency resulted in this composite being unsuitable for F306 FDM printing.

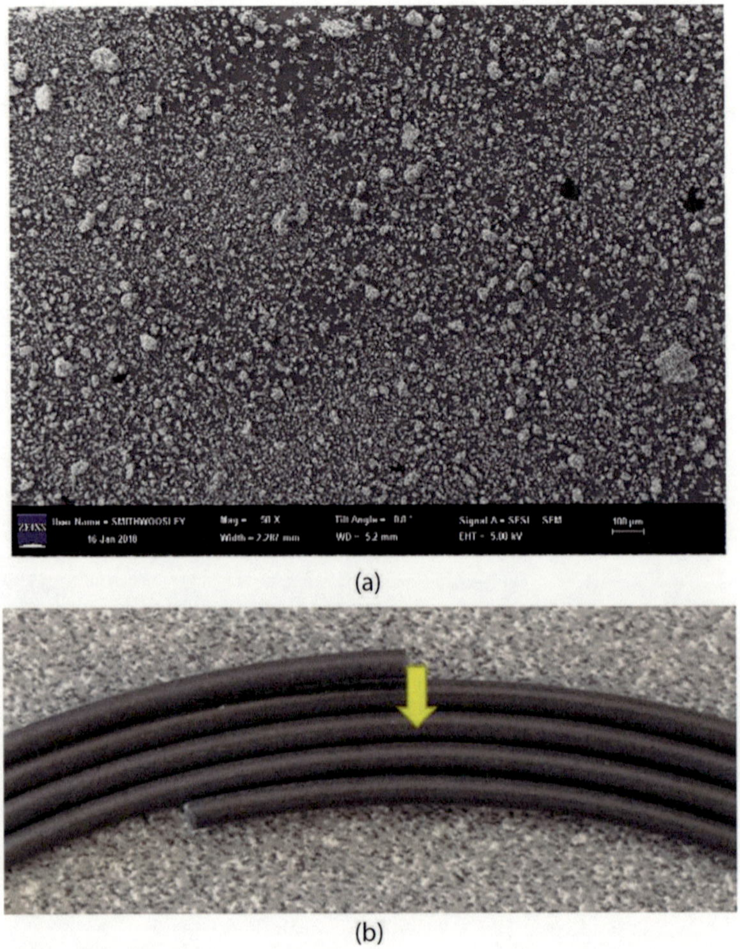

Figure 5.
a) SEM cross-sectional image of 20% ABS/nickel composite filament and b) optical image of 20% ABS/nickel composite filament.

Figure 6.
SEM images of a) boron nitride particles and b) gadolinium particles.

Raman spectroscopy was used to verify the presence of boron nitride within the nanocomposite filament. **Figure 8** shows the Raman spectra of pure BN particles, pure ABS, and 5%, 10%, and 20% BN composites, respectively. In the case of pure

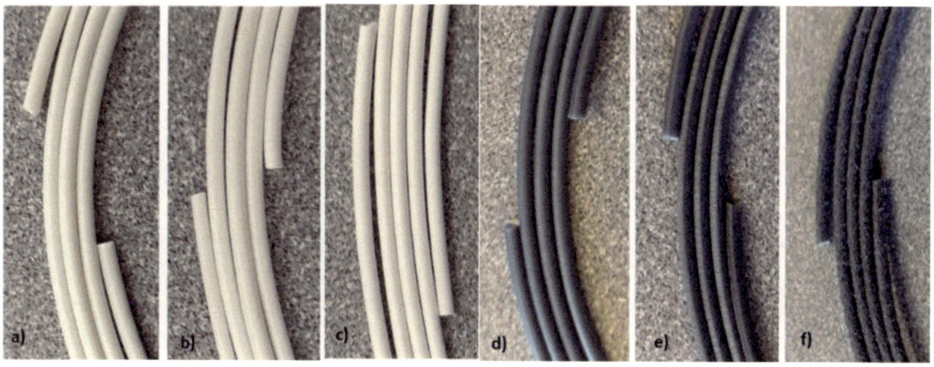

Figure 7.
ABS/boron nitride composite filaments, a) 5%, b) 10%, c) 20% and ABS/gadolinium composite filament, a) 5%, b) 10%, c) 20%.

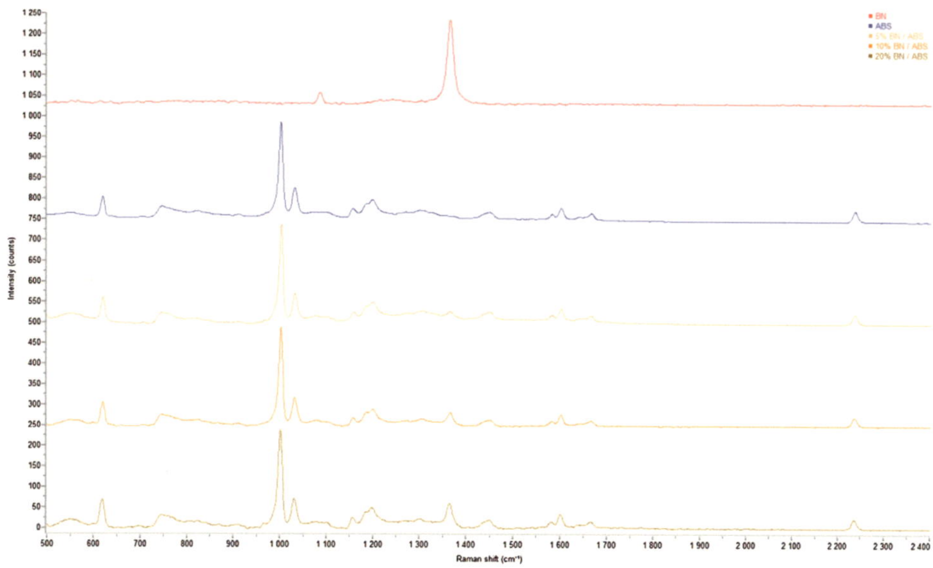

Figure 8.
Raman spectra of boron nitride (BN), pure ABS, and 5%, 10%, and 20% ABS/BN nanocomposite filaments.

boron nitride (BN), a distinct peak at 1364 cm^{-1}, as reported in literature [27], is clearly visible. 1364 cm^{-1} Raman peak is the result of vibrational bond between boron and nitride in their native hexagonal lattice. The Raman spectra of pure ABS spectra also agree with the reported values in literature [28]. This indicates that boron nitride as an additive is present in the composite and does not interfere with any bonding between monomers, as all vibrational modes are also present in the spectra. As expected, at higher additive percentages (10% and 20%), the Raman peak becomes more pronounced because of higher particle concentration.

Thermal characteristics of ABS/BN nanocomposite filaments: Using TGA plots, the dispersion uniformity was studied for pure ABS and the different composite filaments. ABS achieves an average residual mass of approximately 1% at 533°C, while the 5%, 10%, and 20% boron nitride composite samples reach an average residual mass percentage of 6.02 ± 0.07%, 10.23 ± 0.05%, and 20.12 ± 0.13% residual mass, respectively. It is clearly evident that the residual mass is primarily attributed to the boron nitride additive (remaining) added to the ABS matrix, as it is known that ABS will thermally degrade before boron nitride. In addition, the measurements also confirm that the residual mass percentage closely matches the intended 5%, 10%, or 20% by weight of boron nitride. Lastly, when the above measurements were conducted at different time points during the filament extrusion process, it resulted in small standard error (<1%). These results suggest that the boron nitride as an additive in ABS is evenly dispersed in the filament. Similarly, even dispersion results were obtained for other functional additives, namely carbon black, nickel, and gadolinium particles.

From the TGA plots, the thermal onset points were also determined for the different composites. The onset point, which is the temperature at which thermal degradation begins, is also used to assess a material's thermal stability at high temperatures. It is calculated by extrapolation of the slopes of predegradation and post-degradation points. The onset point for ABS was measured to be 384.66 ± 1.13°C. The addition of boron nitride did not significantly change the degradation point of ABS/BN composites compared with pure ABS. ABS/BN composite onset point for 5%, 10, and 20% of BN were 385.55 ± 0.48°C, 384.25 ± 1.15°C, and 390.90 ± 2.01°C, respectively (**Figure 9**). Onset point for 5% and 10% ABS/BN composites closely matched that of pure ABS, while 20% ABS/BN composite showed even a slightly larger thermal onset point. This may indicate that higher concentration of additives may be preventing early thermal degradation. Next, to further understand the thermal properties of ABS/BN composites, differential scanning calorimetry (DSC) measurements were performed to study their glass transition temperature (T_g). This value related to the temperature at which a polymer begins to experience chain movement due to an increase in thermal energy and is an important consideration for 3D printability. The detailed DSC results are reported elsewhere [20].

Mechanical characteristics of ABS/BN nanocomposite filaments: The mechanical characteristics of ABS and ABS/BN composite samples were studied. This was done to determine the effect of BN additive on ABS polymer strength. Following ASTM D638 tensile tests, the stress–strain curve of ABS and ABS/BN composites is shown in **Figure 10**. The variations in stress–strain curves for different additive concentrations were clearly evident. Pure ABS shows the highest tensile stress of 34.01 ± 0.55 MPa, while 5%, 10%, and 20% ABS/BN composite samples have an average tensile stress of 23.85 ± 0.62 MPa, 18.86 ± 0.91 MPa, and 23.16 ± 1.16 MPa, respectively. It is evident from these results that BN addition does degrade the ultimate tensile stress of the ABS polymer. Furthermore, pure ABS showed a superior modulus of elasticity at 1.365 ± 0.05 GPa, while 5%, 10%, and

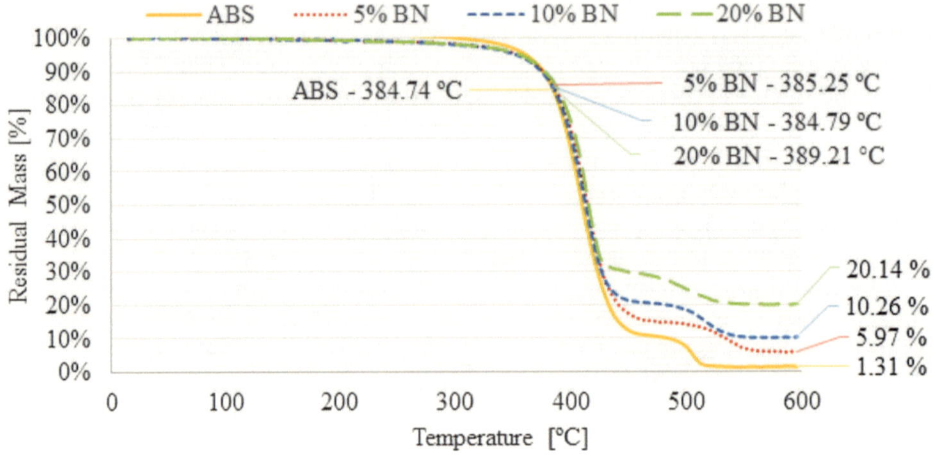

Figure 9.
TGA plot of ABS and ABS/boron nitride nanocomposites.

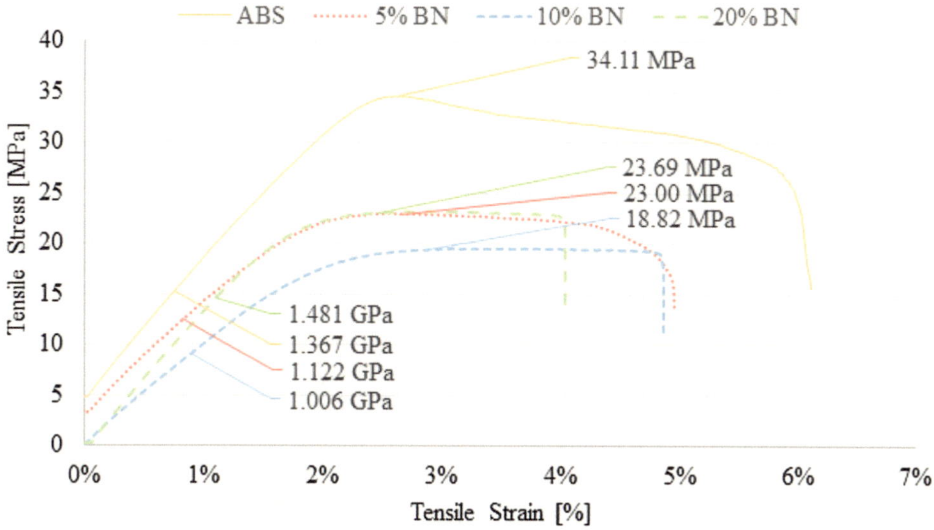

Figure 10.
Stress–strain curves of pure ABS and ABS/boron nitride nanocomposite filaments.

20% ABS/BN composite samples showed 1.179 ± 0.07 GPa, 0.979 ± 0.06 GPa, and 1.420 ± 0.06 GPa, respectively. It is also evident from these results that a smaller percentage of BN as additive influences the composite's elasticity as a softer material while at higher concentrations, it may stiffen the 3D printed composite. The samples were tested to evaluate the storage modulus of the 3D composite as a function of temperature using dynamic mechanical analysis. Detailed results are reported elsewhere [20]. In summary, pure ABS and ABS/BN composites show nearly similar mechanical characteristics. Pure ABS is seen to maintain a glassy region modulus of 2.67 GPa from room temperature to around 105°C, which is the glass transition temperature (Tg) for ABS. After Tg and around 140°C, the modulus falls to rubbery region modulus of 2.15 MPa. Similarly, ABS/BN composites' storage moduli up to 105°C ranges between 2 GPa and 2.54 GPa and at 140°C, it reaches a minimum of 1.31 MPa. It may be concluded that this degradation of ultimate tensile stress may be due to interactions between the additive (BN) and the polymer (ABS) chains, whereby the BN as an additive may be interfering with ABS polymer bonding. Despite the degraded mechanical properties in ABS/BN composites, all composite filaments were printed on a standard FDM printer without any problem. Slight variations in the material strength did not affect the printability of any of the composite filaments, carbon black, nickel, boron nitride, and gadolinium particles.

Functional characteristics of different nanocomposite filaments: Functional assessment was performed to assess conductivity (impedance), magnetism, and radiation shielding characteristics of pure 3D printed ABS and 3D printed functional composites.

First, electrical conductivity of ABS/carbon black functional composites was studied. In this case, the pure ABS and 20% carbon black composite was compared with copper wire. **Figure 11** shows the plot of impedance as a function of frequency for samples of each material. The results show high impedance values for pure ABS, as expected with no electrically conductive components. By adding 20% weight of carbon black, the impedance was lowered by 4–7 orders of magnitude, depending on the frequency, to a value of 2.8 k Ω. This indicates that the composite material is far more conductive than pure ABS. Even though, the carbon black/ABS composites were not as conductive as copper, which has an additional 5 orders of magnitude

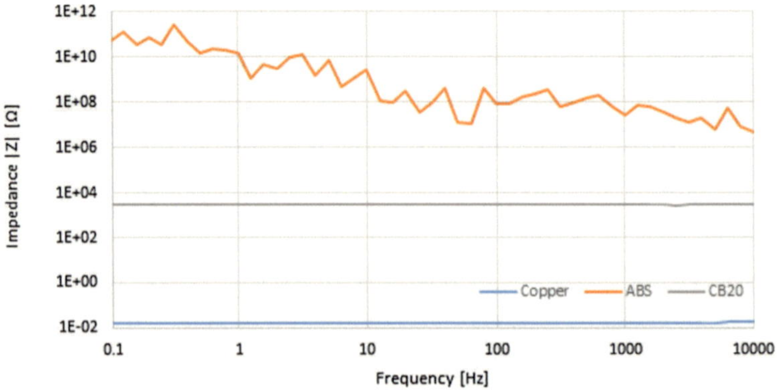

Figure 11.
EIS plot of ABS, 20% carbon black, and copper.

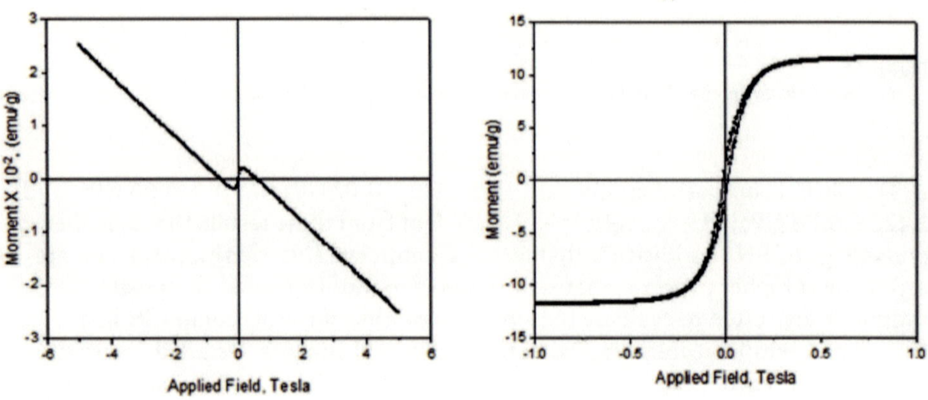

Figure 12.
Magnetic moment plots of (left) pure ABS and (right) ABS/nickel composites.

lower impedance value, close to 15.8 mΩ. Nonetheless, the fabricated 3D functional composite showed marked improvement in electrical conductance.

Next, magnetic characterization of pure ABS and ABS/nickel composites was performed on a physical property measurement system (PPMS). A magnetic field was applied to pure ABS and 5%, 10%, and 20% ABS/nickel composites to measure the resulting magnetic moment. The plots of moment as a function of applied field at room temperature are shown in **Figure 12**, where (a) is pure ABS, and (b) is 20% nickel/ABS composite. Pure ABS showed an inverse relationship between field strength and moment, indicating diamagnetic properties. This is typical of polymeric materials and is a result of absence of any paramagnetic or ferromagnetic response. With increasing nickel additive concentration, hysteresis responses were observed, as well as a magnetic saturation point, indicating the presence of ferromagnetic function. Magnetic saturation point for 20% nickel/ABS composite was 11.5 emu/g.

Finally, radiation shielding properties of pure 3D printed ABS and 3D printed composites of boron nitride and gadolinium particles were studied. **Figure 13** shows the attenuation percentage data for pure ABS, boron nitride/ABS, and gadolinium/ABS composites. Pure ABS showed 50.37% neutron attenuation capability, which means that around half of the neutrons passed through the shielding panel. This was not unexpected. Pure ABS, which contains a large

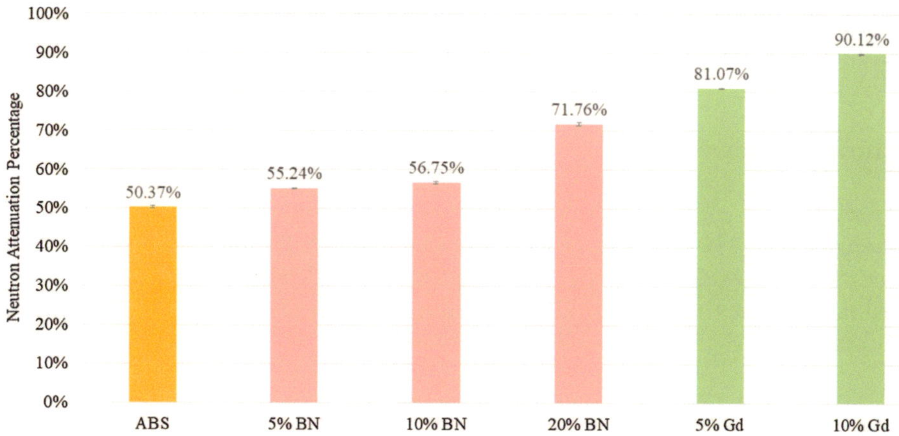

Figure 13.
Neutron attenuation for ABS, BN/ABS, and gadolinium/ABS composite 3D prints.

amount of hydrogen atoms, provides some limited shielding capability at high concentrations. 5% and 10% ABS/BN composite samples showed slightly higher neutron attenuation capability, between 55% and 57%. Around 71.76% incident neutron attenuation was exhibited in 20% ABS/BN 3D printed composites. In the case of ABS/gadolinium composites, 5% and 10% of gadolinium showed very significant attenuation for incident neutrons at 81.07% and 90.12%, respectively. The theoretical calculations for attenuation coefficients to further understand radiation shielding of these 3D printed composites are reported elsewhere [20].

4. Conclusions

In summary, in this work, we have shown that incorporation of functional additives, carbon black, nickel, boron nitride or gadolinium particles, into FDM printable ABS mixture is feasible. Various filament nanocomposite materials for use in FDM printers were successfully created by using simple solution mixing approach and desktop extruder. Even dispersion of additives in the ABS polymer matrix was confirmed using optical, electron microscopic, and thermogravimetric analysis. Furthermore, thermal analysis confirmed that thermal properties were only altered minimally compared with pure ABS. Even though, mechanical measurements revealed some variations in polymer strength in the composite filaments, compared with pure ABS, the printability on regular 3D FDM printers was not significantly affected.

Finally, functional testing of the 3D printed nanocomposite samples for properties such as conductivity, magnetic response, and neutron attenuation showed marked to very significant increase in functional properties, particularly in the case of radiation shielding capability of 3D printed composite with gadolinium particles. The neutron radiation shielding capability was enhanced from 50% with pure ABS samples to up to 90% with 10% ABS/gadolinium particles and around 72% with 20% ABS/boron nitride nanocomposites. This enhancement in radiation shielding capability is extremely valuable to reduce damage from neutron sources for aerospace, UAV, and satellite applications. In conclusion, the ability to manufacture functional FDM components is expected to assist the additive manufacturing community to move beyond just “form-and-fit” prototyping into the production of complete end-use parts with functional attributes.

Acknowledgements

This work was conducted at the Joint School of Nanoscience and Nanoengineering (JSNN), a member of the Southeastern Nanotechnology Infrastructure Corridor (SENIC), and a site for National Nanotechnology Coordinated Infrastructure (NNCI), with partial support from the National Science Foundation (EECS-1542174 and ECCS-2025462).

Conflict of interest

The authors declare no conflict of interest.

Author details

Smith Woosley and Shyam Aravamudhan*
Department of Nanoengineering, Joint School of Nanoscience and Nanoengineering, Greensboro, NC, USA

*Address all correspondence to: saravamu@ncat.edu

References

[1] Ortiz-Acosta D, Moore T. Functional 3D printed polymeric materials. Functional Materials. 2019;**9**:1-5

[2] Wickramasinghe S, Do T, Tran P. FDM-based 3D printing of polymer and associated composite: A review on mechanical properties, defects, and treatments. Polymers. 2020;**12**(7): 1529

[3] Awasthi P, Banerjee SS. Fused deposition modeling of thermoplastic elastomeric materials: Challenges and opportunities. Additive Manufacturing. 2021;**46**:102177

[4] Liao G, Li Z, Cheng Y, Xu D, Zhu D, Jiang S, et al. Properties of oriented carbon fiber/polyamide 12 composite parts fabricated by fused deposition modeling. Materials & Design. 2018;**139**:283-292

[5] Ning F, Cong W, Qiu J, Wei J, Wang S. Additive manufacturing of carbon fiber reinforced thermoplastic composites using fused deposition modeling. Composites Part B: Engineering. 2015;**80**:369-378

[6] Love LJ, Kunc V, Rios O, Duty CE, Elliott AM, Post BK, et al. The importance of carbon fiber to polymer additive manufacturing. Journal of Materials Research. 2014;**29**(17): 1893-1898

[7] Dickson AN, Barry JN, McDonnell KA, Dowling DP. Fabrication of continuous carbon, glass and Kevlar fiber reinforced polymer composites using additive manufacturing. Additive Manufacturing. 2017;**16**:146-152

[8] Masood SH, Song WQ. Development of new metal/polymer materials for rapid tooling using fused deposition modelling. Materials & Design. 2004;**25**(7):587-594

[9] Dul S, Fambri L, Pegoretti A. Fused deposition modelling with ABS–graphene nanocomposites. Composites Part A: Applied Science and Manufacturing. 2016;**85**:181-191

[10] Singh R, Singh N, Amendola A, Fraternali F. On the wear properties of Nylon6-SiC-Al2O3 based fused deposition modelling feed stock filament. Composites Part B: Engineering. 2017;**119**:125-131

[11] Turner BN, Strong R, Gold SA. A review of melt extrusion additive manufacturing processes: I. process design and modeling. Rapid Prototyping Journal. 2014;**20**:192-204

[12] Leigh SJ, Bradley RJ, Purssell CP, Billson DR, Hutchins DA. A simple, low-cost conductive composite material for 3D printing of electronic sensors. PLoS One. 2012;**7**(11):e49365

[13] Dorigato A, Moretti V, Dul S, Unterberger SH, Pegoretti A. Electrically conductive nanocomposites for fused deposition modelling. Synthetic Metals. 2017;**226**:7-14

[14] Christ JF, Aliheidari N, Ameli A, Pötschke P. 3D printed highly elastic strain sensors of multiwalled carbon nanotube/thermoplastic polyurethane nanocomposites. Materials & Design. 2017;**131**:394-401

[15] De Santis R, Gloria A, Russo T, Ronca A, D'Amora U, Negri G, et al. Viscoelastic properties of rapid prototyped magnetic nanocomposite scaffolds for osteochondral ti'ssue regeneration. Procedia CIRP. 2016;**49**:76-82

[16] Bollig LM, Hilpisch PJ, Mowry GS, Nelson-Cheeseman BB. 3D printed magnetic polymer composite transformers. Journal of Magnetism and Magnetic Materials. 2017;**442**:97-101

[17] Sandler N, Salmela I, Fallarero A, Rosling A, Khajeheian M, Kolakovic R, et al. Towards fabrication of 3D printed medical devices to prevent biofilm formation. International Journal of Pharmaceutics. 2014;**459**(1-2):62-64

[18] Muwaffak Z, Goyanes A, Clark V, Basit AW, Hilton ST, Gaisford S. Patient-specific 3D scanned and 3D printed antimicrobial polycaprolactone wound dressings. International Journal of Pharmaceutics. 2017;**527**(1-2):161-170

[19] Teo EY, Ong SY, Chong MS, Zhang Z, Lu J, Moochhala S, et al. Polycaprolactone-based fused deposition modeled mesh for delivery of antibacterial agents to infected wounds. Biomaterials. 2011;**32**(1):279-287

[20] Woosley S, Galehdari NA, Kelkar A, Aravamudhan S. Fused deposition modeling 3D printing of boron nitride composites for neutron radiation shielding. Journal of Materials Research. 2018;**33**(22):3657-3664

[21] Kennedy AR. Biological effects of space radiation and development of effective countermeasures. Life Sciences in Space Research. 2014;**1**:10-43

[22] Stiegler JO, Mansur LK. Radiation effects in structural materials. Annual Review of Materials Science. 1979;**9**(1):405-454

[23] Abbott AC, Tandon GP, Bradford RL, Koerner H, Baur JW. Process-structure-property effects on ABS bond strength in fused filament fabrication. Additive Manufacturing. 2018;**19**:29-38

[24] Kim J, Lee BC, Uhm YR, Miller WH. Enhancement of thermal neutron attenuation of nano-B4C,-BN dispersed neutron shielding polymer nanocomposites. Journal of Nuclear Materials. 2014;**453**(1-3):48-53

[25] Harrison C, Weaver S, Bertelsen C, Burgett E, Hertel N, Grulke E. Polyethylene/boron nitride composites for space radiation shielding. Journal of Applied Polymer Science. 2008;**109**(4):2529-2538

[26] Neutron Cross Section of the elements [Internet]. 1999. Available from: https://periodictable.com/Properties/A/NeutronCrossSection.html [Accessed: 2022-01-28]

[27] Reich S, Ferrari AC, Arenal R, Loiseau A, Bello I, Robertson J. Resonant Raman scattering in cubic and hexagonal boron nitride. Physical Review B. 2005;**71**(20):205201

[28] Colthup N. Introduction to Infrared and Raman Spectroscopy. New York, NY: Elsevier; 2012

Chapter 8

Quality Control of Metal Additive Manufacturing

Bojie Sheng, Jamil Kanfoud and Tat-Hean Gan

Abstract

Metal Additive Manufacturing (AM) is an emerging technology for rapid prototype manufacturing, and the structural integrity of printed structures is extremely important and should meet the specifications and high standards of the above industries. In several metal AM techniques, residual stresses and micro-cracks that occur during the manufacturing procedure can result in irreversible damage and structural failure of the object after its manufacturing. Thus effective quality control of AM is highly required. Most Non-Destructive Testing (NDT) techniques (X-Ray, Computed Tomography, Thermography) are ineffective in detecting residual stresses. Bulk, cost, and resolution are limitations of such technologies. These methods are time consuming both for data acquisition and data analysis and have not yet been successfully integrated into AM technology. However two sets of NDT techniques: Electromagnetic Acoustic Transducers (EMAT) and Eddy Current (EC) Testing, can be applied for residual stress detection for AM techniques. Therefore a crucial and novel extension system incorporation of big data collection from sensors of the both techniques and analysis through machine learning (ML) can estimate the likelihood of the AM techniques to introduce anomalies into the printed structures, which can be used as an on-line monitoring and detection system to control the quality of AM.

Keywords: metal, additive manufacturing, quality control, NDT, non-destructive testing, eddy current, electromagnetic acoustic transducers, inspection, monitoring, machine learning

1. Introduction

Additive manufacturing (AM) is a process to build complex 3D parts from computer-aided design (CAD) models through layer-by-layer or drop-by- drop deposition of materials [1]. It is an emerging technology due to its capability to build complex-shaped products with less tooling and production time [2, 3]. AM provides significant advantages over traditional subtractive (machining) and formative (casting, moulding) manufacturing processes, such as reducing material waste, eliminating specialised tooling cost, and enabling the creation of intricate and free-form geometries. Lower prices of AM technologies make it more accessible to industries [4]. Reducing costs of material, higher level of automatization with less intervention of humans and increasing of manufacturing quality are also impulses to expand AM in further sectors of industry. AM industries have grown rapidly since 2000 [5], and have shown almost six times the growth during the 2000s as compared with the growth during the 1990s [5]. It was reported that in 2013 AM system sales revenue of industrial products, consumer products, automotive, medical, aerospace and military are 19%, 18%, 17%, 14%, 12% and 5% separately. Automotive, medical, aerospace and military, which require high precision and reliability, lead 48% in total [6]. It has been estimated that the global market for AM processes and services will reach around $50 billion by the year of 2031 [7]. Nowadays, industries invest 10 times more on end-part production than on prototyping [8]. Therefore, final and functional part productions and relevant researches are growing faster than the general market.

Although AM technology has been dramatically developed and it brings high feasible applications into real industry, there are still many obstacles in the adoption of AM for reliable production, such as part quality inconsistency, repeatability, and absence of material process standard [9]. For example, process variations and uncertain factors significantly impact the microstructure and mechanical properties of AM builds, which will further lead to internal defects deteriorating the build hardness, strength, and residual stress [10]. Process monitoring is a big challenge in AM as different factors affect the monitoring from materials to geometries to hardware and software limitations [11]. AM even with consistent process parameters is affected by little variations [12] in air flow, melting pool, etc. Due to its high-quality variability particularly for critical industries such as automotive and aerospace, high precision and mechanical properties certification is required [5].

For quality control improvement, numerous researchers have adopted different analyses focused on quality-approached technologies. For example, the effect of printing parameters on precision and internal cavity of fabricated parts using fused deposition modelling 3D printer was studied in [13]. Another research [14] investigated the effects of layer printing delay on physical and mechanical properties. Yang et al. [15] identified that process monitoring with closed-loop process control allows to achieve consistent quality of manufactured parts. Due to the complexity of AM processes, more and more complex real-time monitoring technologies are investigated. A non-exhaustive list includes Thermal camera, high-speed optical camera, photodetector, pyrometer, acoustic emission. Advanced sensing leads to the generation of big and complex data. The full exploitation of the data is critical to understanding the quality variability during the printing. Advanced signal processing/deep learning frameworks are required to achieve near-real-time defect detection. As a result, there are increasing interests and rapid development of sensor-based models for the characterisation and estimation of defects in the past few years [10]. Paper [15] reviewed several experimental configurations and adopted the vision-based and thermal sensing metrology approaches for the in situ process monitoring of typical material variation and failure modes. A spectral-graph approach was proposed to study the photo-detector sensor signature for the identification of defects caused by material cross-contamination in LPBF AM process [16].

Depending on the target material and involved energy source, an AM system (phase 4 of **Figure 1a**) entails different types of processes. One of the most popular in industry is powder bed fusion (PBF) which typically uses a high source of thermal energy to melt powder-based materials (such as metal) into desired structures and shapes [18]. For applications targeting the production of metal components, the most diffused PBF technique exploits a laser as the heat source to either melt or sinter the metal powder together (see a schematic representation in **Figure 1b**). In several metal

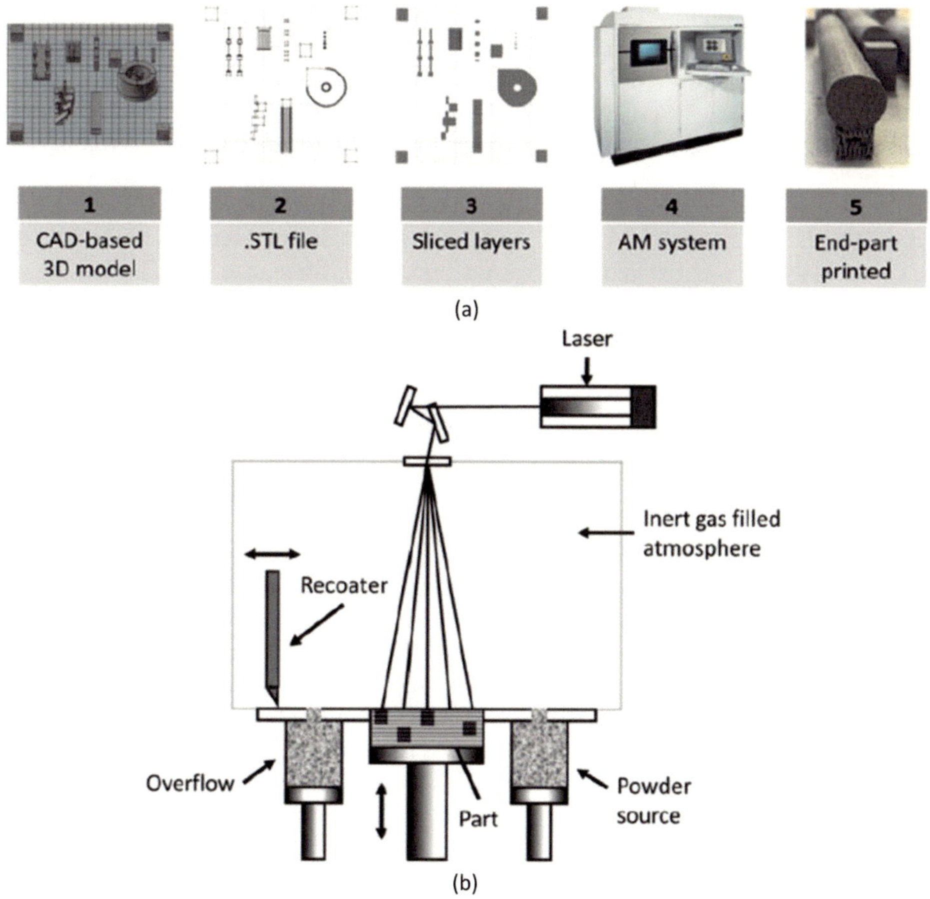

Figure 1.
Additive manufacturing in industry. (a) Main phases of a typical process. (b) Scheme of laser powder bed fusion [17].

AM techniques, e.g. selective laser melting (SLM), electron beam additive manufacturing (EBAM) and wire arc additive manufacturing (WAAM), residual stresses and micro-cracks that occur during the manufacturing procedure can result in irreversible damage and structural failure of the object after its manufacturing. Repetitive faults which occur during manufacturing due to incorrect estimation of appropriate operating conditions of the printer should be eliminated, as any waste is undesirable and costly for a company. Thus effective QC of metal AM is highly required. Currently there are many different monitoring techniques in detecting residual stresses and defects for QC of metal AM, for example, thermography, X-ray computed tomography (CT scan), eddy current (EC) and electromagnetic acoustic transducers (EMAT). Therefore, this chapter discussed and investigated different techniques for the QC of metal AM technology separately. Sections 2 and 3 reviewed temperature monitoring and CT scan techniques while Sections 4 and 5 investigated EC testing and EMAT testing methods.

2. Temperature monitoring

Kim et al. [19] studied mechanical property improvement of Ti-6Al-4 V alloy part printed by EBM. Increasing bed temperature or post-processing the printed parts (e.g. peening) allows improving the mechanical properties. Although useful, this approach requires modification of the process or adding another process which increases the cost. The complexity of the geometry makes it also not easily generalisable.

It is much more agile to dynamically optimise the process parameters thanks to quality control sensors [19]. A video microscopy system was used to observe sintering and flow behaviour in real-time, which can evaluate different sintering characteristics of the materials [20]. An infrared light thermal sensor was developed to control laser power for better uniform sintering performance [21]. Paper [22] reviewed literatures on the thermal modelling method in SLS and SLM techniques. It was summarised that uniform temperature distribution of fields during printing processes leads to better quality; there is a need to provide information from monitoring temperature of the melt pool to be able to control process parameters for part quality. As a temperature monitoring system, pyrometers and thermocouples were used for monitoring its temperature [22].

Another way of monitoring temperature is through emitted thermal radiation (pyrometry).

This is achieved using either photodiodes or CMOS/CCD digital cameras which convert radiation and light into electric signal. Schürmann et al. [23] developed a novel laser cladding head with photodiodes and digital camera (CCD) to enable real time condition monitoring. A high resolution CCD camera was used for surface error detection through image processing, as shown in **Figure 2** [24]. Research [25] analysed images of melt pool size to adjust laser output power by using a CMOS camera with an additional illumination source required for high scanning velocities, resolution, as well as photodiodes. A thermal model was developed that the temperature evolution and sintering formation can be simulated by a 3D FEA to predict thermal properties (i.e. thermal conductivity and specific heat) [26].

Unlike pyrometers, thermocouples are a contact measuring technique thus inconvenient for additive manufacturing where the part is changing over time. Nonetheless, thermocouples were used to monitor the temperature on the powder bed [27]. In addition, research [28] used thermocouple attached to the bottom of the base plate with a strain gauge to record residual stresses, and paper [29] also used thermocouples positioned under the bed's surface to monitor energy absorption and powder effective conductivity for better understanding of heat transfer in metal powder during laser processing of the powder bed. A thermocouple control system was invented to improve uniform distribution of temperature on powder bed during part build. The thermocouple was attached inside of the powder bed along with IR sensor and communicates with temperature transmitter through circuitry in real-time [30].

Besides, research was focused on the IR thermography based monitoring and control system incorporated in the EBM system. Research [31] demonstrated the feasibility of using a near IR thermal camera for temperature measurement in hatch melting, preheating and contour melting events during the EBS process. A galvanometric scanner system was used for temperature distribution monitoring of melt pool of Ti6Al4V alloy in the SLM system [32]. Paper [33] developed continuous data capturing method using the IR camera to demonstrate feasible work to detect porosities inside materials and understand thermal phenomena such as it happens when beam and powder interact with each other. A temperature feedback control system was developed by positioning photodiodes and CMOS digital camera on laser beam to stabilise melt pool temperature distribution in SLM system [34]. In 2013, Mireles developed automatic IR camera monitoring system for defect detection allowing process stoppage when porosity exceeds a threshold. In addition, the work allowed to stablise temperature and optimise the process automatically using image processing algorithms [35]. In 2015, Mireles developed an in-line inspection

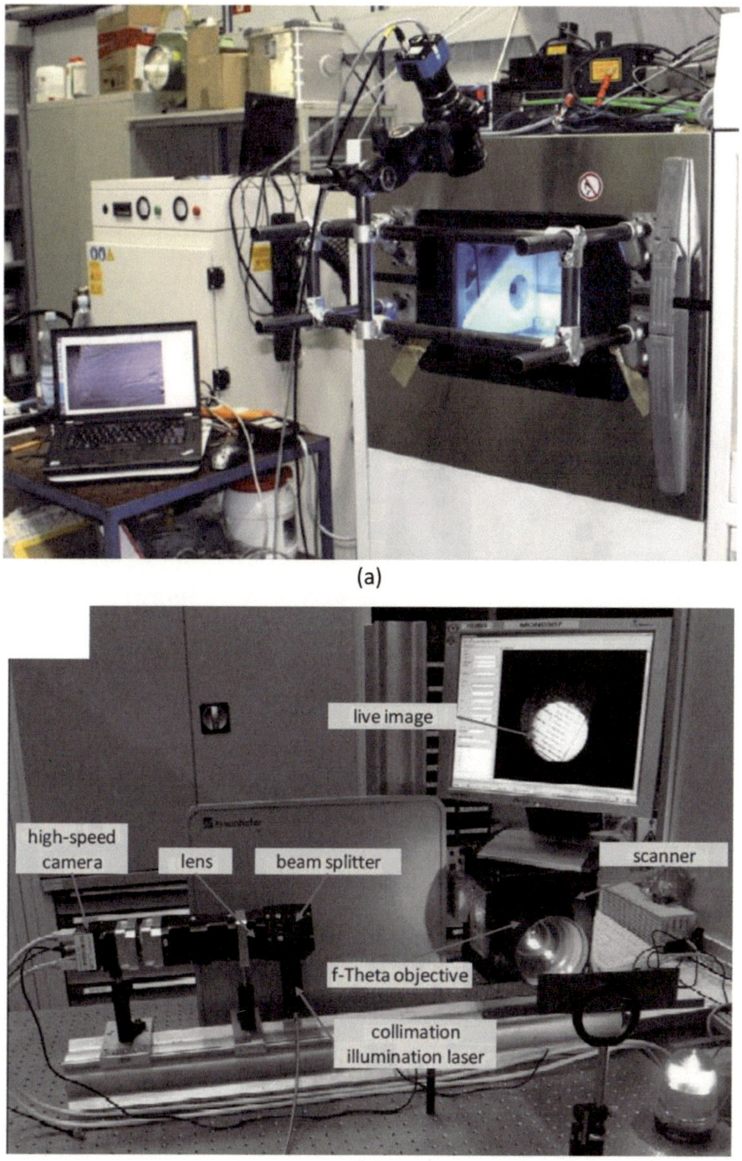

(b)

Figure 2.
(a) Setup of the CCD camera system in front of machine window [24] and (b) setup of CMOS camera system with illumination source and photodiode for high scanning velocities and resolution [25].

system using IR thermography applied to an EBM machine. This allowed measuring defects geometry and locating it. A re-melt allows porosity defects repair as shown in **Figure 3** [36]. Compared to mechanical processes such as hot isostatic pressing, the in-situ monitoring and repair allows saving total manufacturing time and avoiding microstructure alteration.

3. CT scan

Micro X-ray computed tomography (CT) is an advanced measuring technique that can be used for characterising AM components [37, 38]. Examples of an XCT

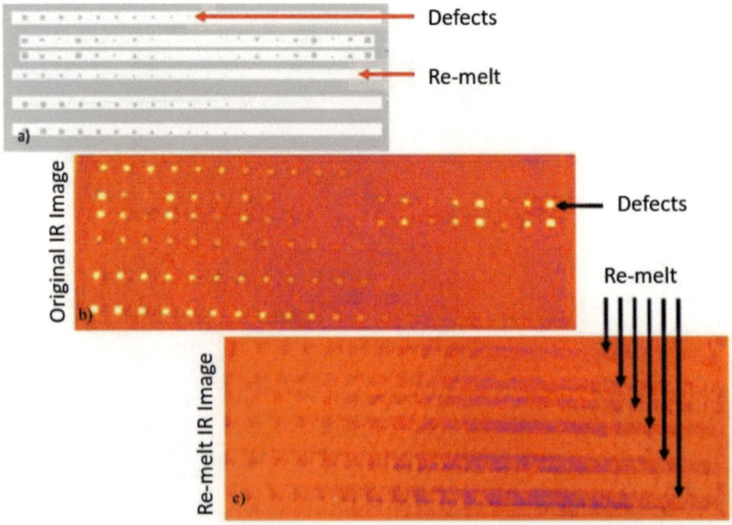

Figure 3.
Correction of un-melted powder through layer re-melt shown by (a) defects detection and localisation, (b) defect labelling, (c) repair after re-melting process [36].

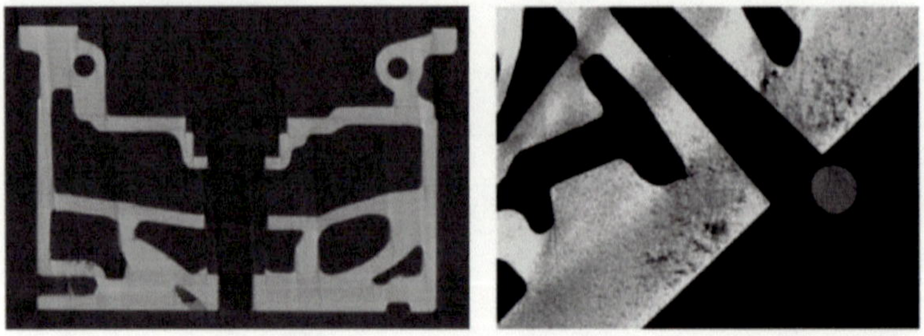

Figure 4.
Examples of an XCT slice of an aluminium sand cast engine cylinder head and of porosities [37].

slice of an aluminium sand cast engine cylinder head and of porosities are shown in [37]. Metrological CT systems allow accurate dimensional measurements [39] as well as internal and surface defect detection sizing and localisation (e.g. voids and inclusions) [40]. CT is an excellent technique allowing topographical measurements at micro-scale achieving 100% volume coverage [41, 42]. Due to the cost, radiation hazard and bulk and lengthy scanning time it is not a monitoring technique but more applicable as a post printing quality control technique (**Figure 4**).

Zanini et al. [43] has carried out experimental investigations on different SLM parts made of Ti6Al4V to address the accuracy of CT-based evaluations of AM. Based on samples with calibrated defects, it was proven that CT porosity measurements evaluate internal porosity with measurement errors below 5 μm for diameter and below 5% for volume measurements. However CT scan is not suited for residual stress detection and is time consuming both for data acquisition and data analysis due to slice by slice visual interpretation requirements as shown in **Figure 5**.

4. Eddy current monitoring

Eddy current non-destructive testing (ECNDT) as shown in **Figure 5** is currently used for surface defects detection in metals. It is efficient and reliable. The

ECNDT research is focusing on improving eddy current transducers (ECT) and the development of new methods for transducer signals processing [44, 45]. Previous work in this field has shown that metal AM parts can be tested with commercially available eddy current testing equipment (**Figure 5**) [46].

Paper [48] has further discussed the applicability of EC testing with magnetoresistive sensors for laser PBF parts using giant magnetoresistance arrays in combination with a single wire excitation coil. To evaluate the influence of the powder used in the manufacturing process on EC testing and vice versa, a laser PBF mock-up made from stainless steel powder (316L) is used with artificial surface defects down to 100 μm. This laser PBF specimen was then examined using eddy current testing and the underlying principles.

4.1 Case study I

The first case study was based on an EC dataset collected from a sample with defects through in-line scanning tests in laboratory. The in-line scanning data was pre-processed and manually labelled, then two different ML algorithms (time series data processing and image data processing) for anomaly detection were investigated for the dataset.

4.1.1 Pre-processing

As each raw data file saved a lot of in-line scanning data as shown in **Figure 6**, a program of using a sliding window to automatically split the raw data into multiple sub datasets was developed to pre- process the raw data. Then each subset was labelled manually while examples of each labelled class are shown in **Figure 7**. The number of each classes was summarised in **Table 1**. Finally the labelled data was prepared for the following ML model training and testing.

4.1.2 ML algorithms

Based on the above labelled data, two different ML algorithms (time series raw data processing and image data processing) were investigated separately and each ML algorithm was implemented for two objectives: anomaly detection and defect type classification. The anomaly detection is able to classify two classes: Normal class which comprises 'edge' and 'normal' data and abnormal class including

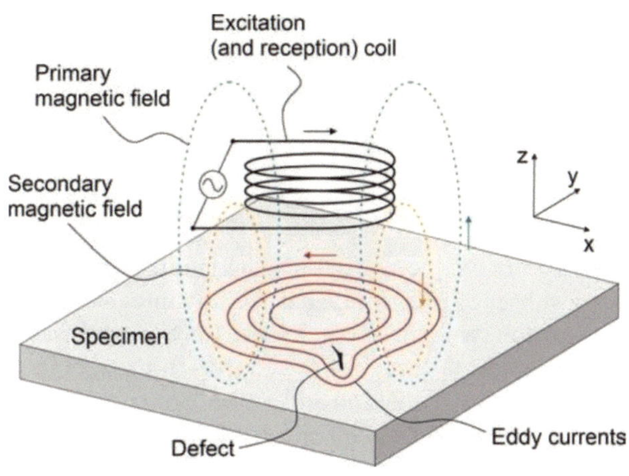

Figure 5.
Eddy current non-destructive testing [47].

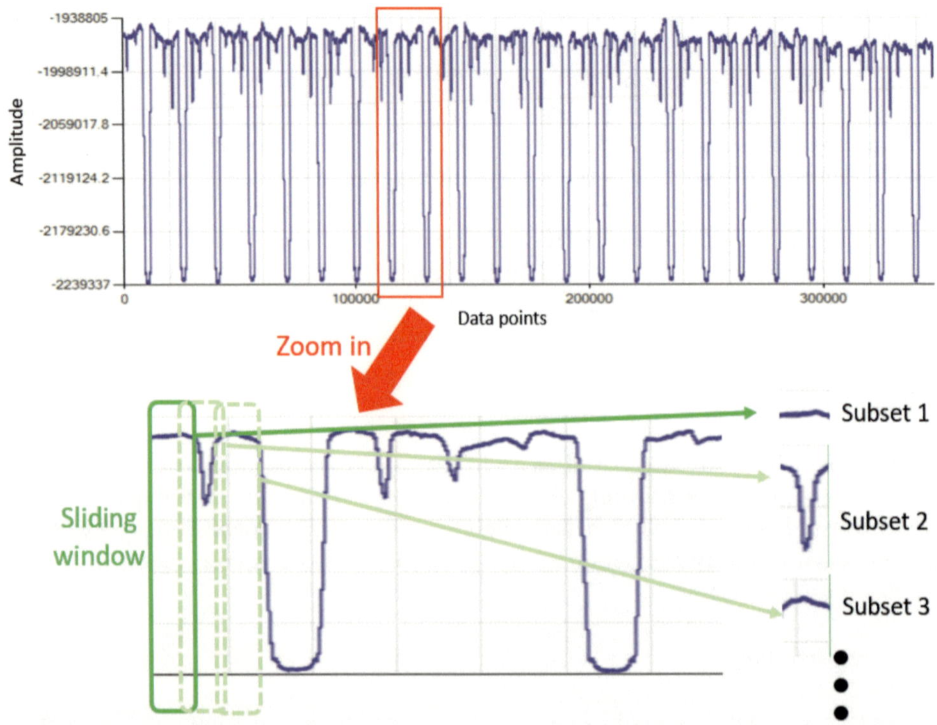

Figure 6.
Data pre-processing.

'0.2 mm', '0.5 mm' and '1.0 mm' defects. The defect type classification aims to identify all the different classes in **Table 1**.

4.1.2.1 Time series raw data processing

As the EC data is time series data, it was feed directly into training and testing the ML algorithms based on XGBOOST model. Results of the confusion matrix, accuracy, precision, recall and f1 score for anomaly detection and defect type classification are shown in **Figure 8**. In the images, scale number means the number of images and the number in each class means the image number been classified as the class. For example, in **Figure 8a**, 575 and 1994 images were correctly predicted as "Abnormal" and "Normal" separately while only 21 and 4 images were predicted incorrectly for the two classes. The results indicate good performance of the both models due to 99% accuracy for anomaly detection and 98% accuracy for defect type classification.

4.1.2.2 Image data processing

The EC time series data was converted into images so that convolution neural network (CNN) as a subset of ML, which has the advantage of image classification can be investigated for the two objectives. Resnet CNN model was trained and tested based on the image data. Results of the confusion matrix, accuracy, precision, recall and f1 score for anomaly detection and defect type classification are shown in **Figure 9**. For example, in **Figure 9a**, 288 and 1000 images were correctly predicted as "Abnormal" and "Normal" separately while only 5 and 6 images were predicted incorrectly for the two classes. Both results demonstrate very good performance of the models resulting from 99% accuracy for both objectives.

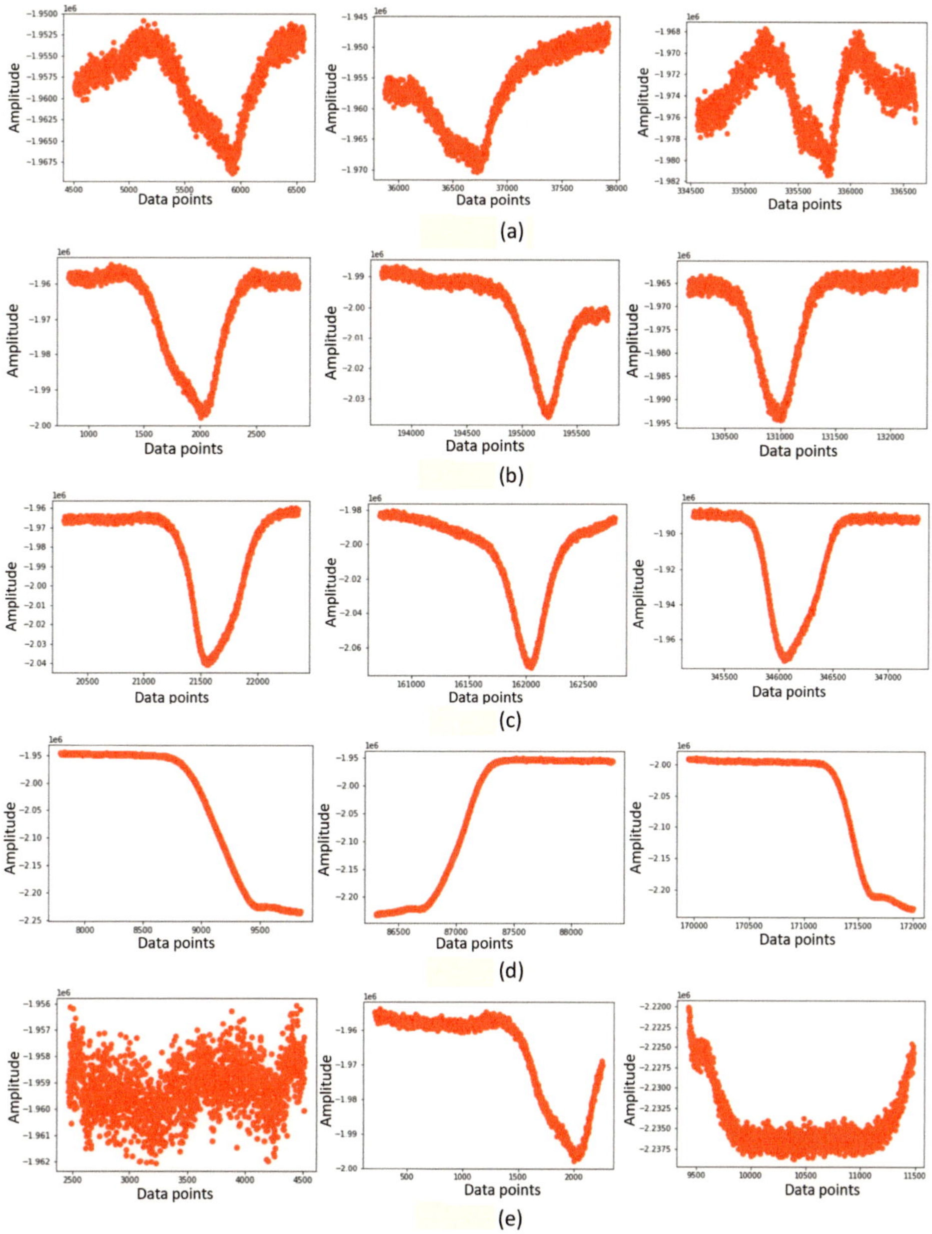

Figure 7.
Examples of labelled data. (a) 0.2 mm, (b) 0.5 mm, (c) 1 mm, (d) edge and (e) normal.

Class	Number
0.2 mm	956
0.5 mm	979
1.0 mm	985
Edge	2118
Normal test	7930

Table 1.
Data number of in-line scanning dataset.

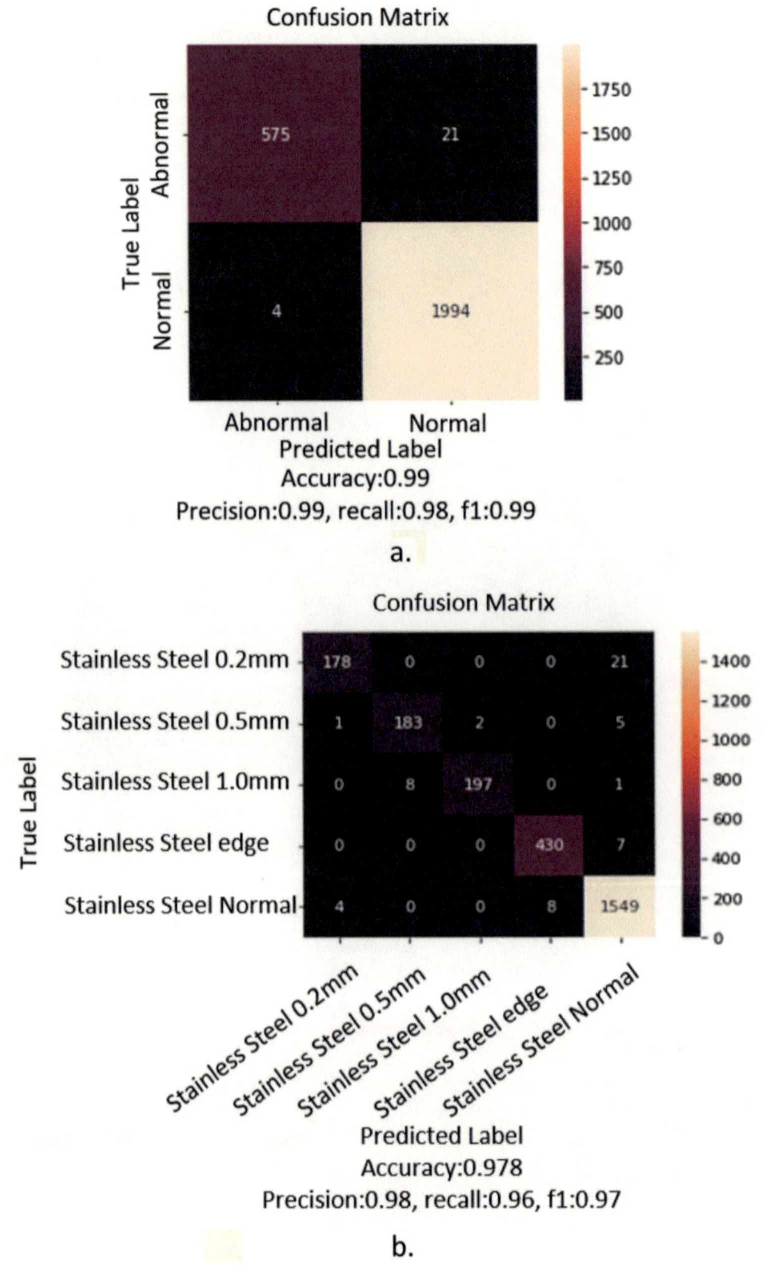

Figure 8.
ML results based on time series data format. (a) Anomaly detection, (b) defect type classification.

Although both of the ML results based on the two data formats has excellent performance (above 98% accuracy), Resnet model based on image data format has slightly better performance due to 99% accuracy for both objectives (anomaly detection and defect type classification).

4.2 Case study II

The second case study was based on an EC dataset collected from a sample with defects through 2D scanning tests in laboratory. The 2D scanning EC data was

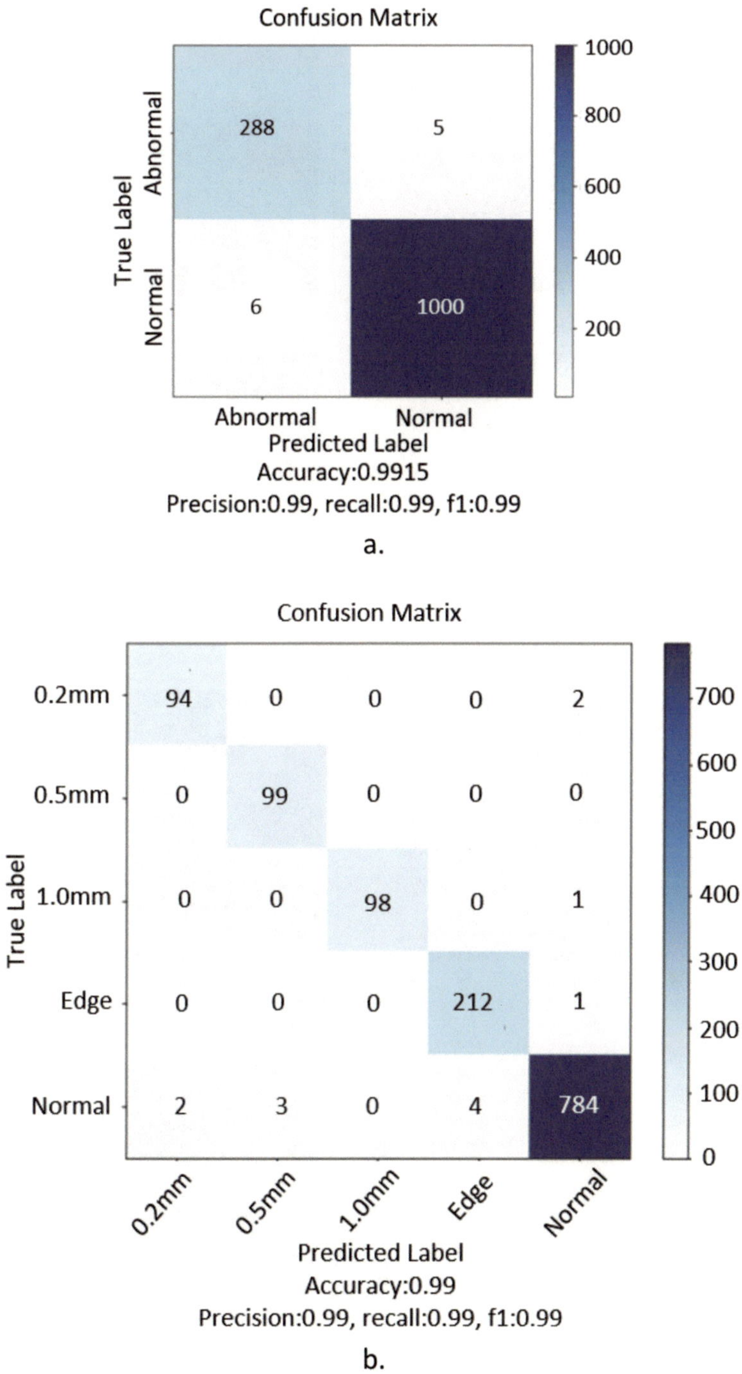

Figure 9.
ML results based on image data format. (a) Anomaly detection, (b) defect type classification.

pre-processed and manually labelled before feed into ML algorithm for anomaly detection.

4.2.1 Pre-processing

The 2D scanning data is a large dataset, which was based on tests on a 100 mm by 100 mm plate sample with different testing directions as shown in **Figure 10a**.

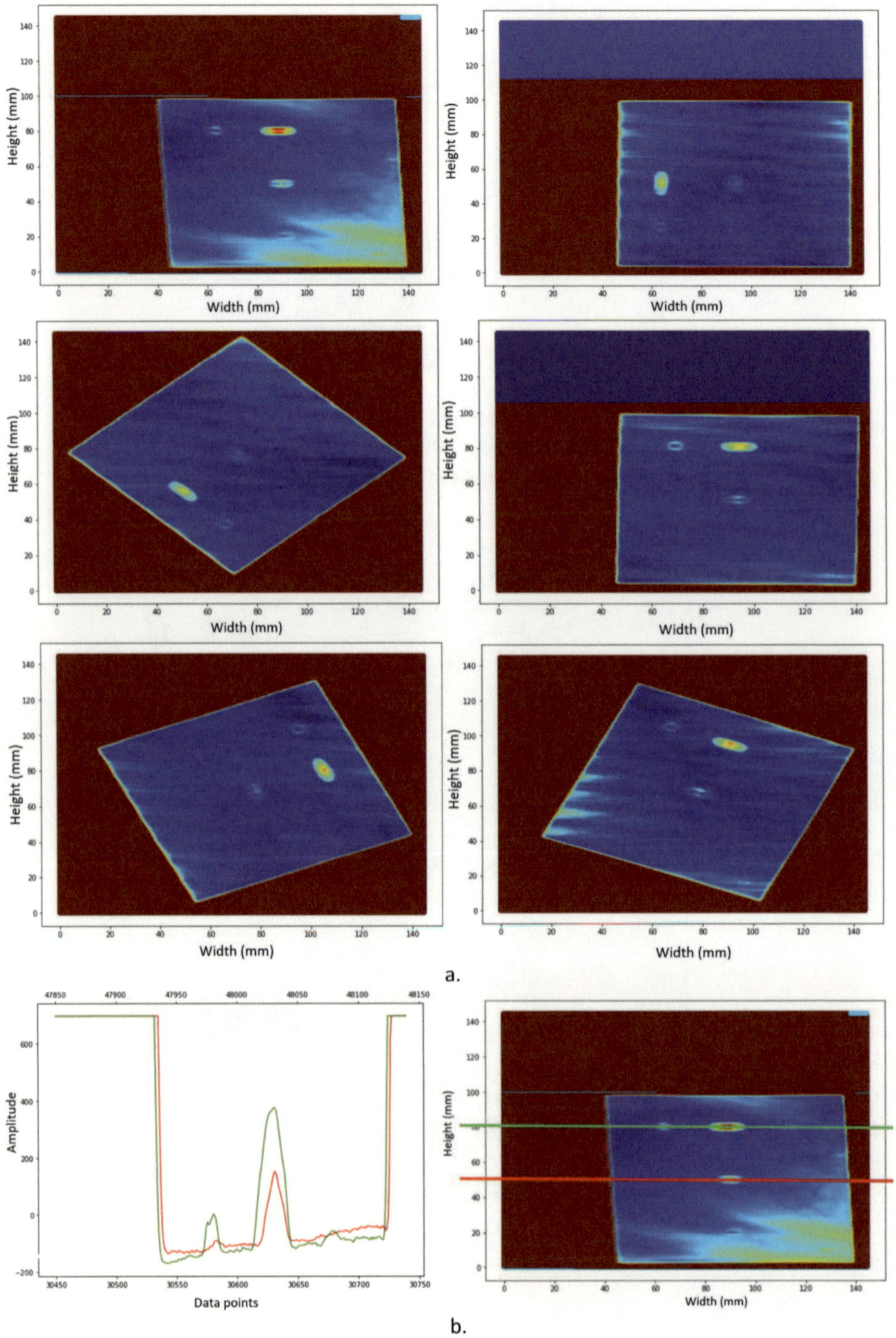

Figure 10.
2D scanning data. (a) Testing in different directions, (b) each 2D scan has multiple in-line scanning.

Each 2D scan has multiple in-line scanning as shown in **Figure 10b**, for example, the green line means the in-line scanning at the height of 82 while the red line means the in-line scanning at the height of 52. A program of using a sliding window to automatically split the raw data into multiple sub datasets was developed to pre-process the 2D scanning data. Then each subset was converted into images before labelled manually while examples of each labelled class are shown in **Figure 11**. The number of each classes was summarised in **Table 2**. Finally the labelled data was prepared for the following ML model training and testing.

4.2.2 ML algorithms

Based on the above labelled data, ML algorithm (Resnet model) based on image data was applied for anomaly detection. The anomaly detection aims to classify two classes: Normal class which comprises 'edge' and 'normal' data and abnormal class including different defects. Resnet model was trained and tested based on the image data. Results of the confusion matrix, accuracy, precision, recall and f1 score for anomaly detection are shown in **Figure 12**, for example, there were 309 and 8012 images correctly predicted as "Abnormal" and "Normal" separately while only 28 and 37 images were predicted incorrectly as the both classes. These results indicate great performance of the model due to 99% accuracy.

4.3 Summary

Above two case studies based on EC testing which collected two datasets with different formats: in- line scanning and 2D scanning, have been investigated. With respect to the in-line scanning data, two different ML algorithms (XGBOOST model for time series data processing and Resnet model for image data processing) have been applied for anomaly detection. Results indicate both ML models have good

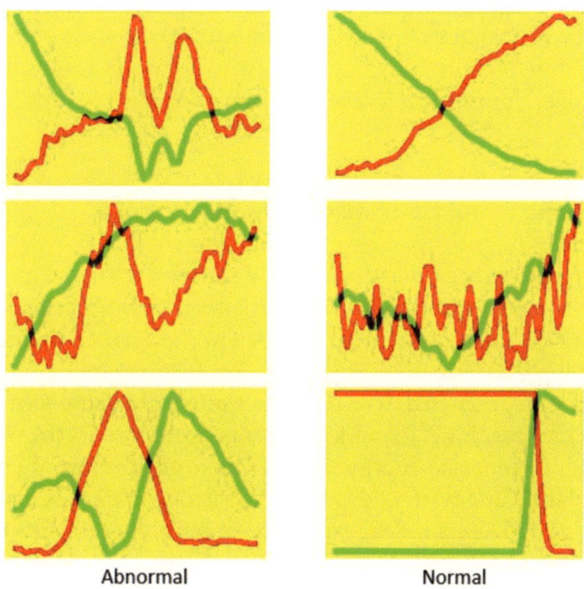

Figure 11.
Examples of labelled image data.

Class	Number
Abnormal	2237
Normal	53,653

Table 2.
Number of 2D scanning dataset.

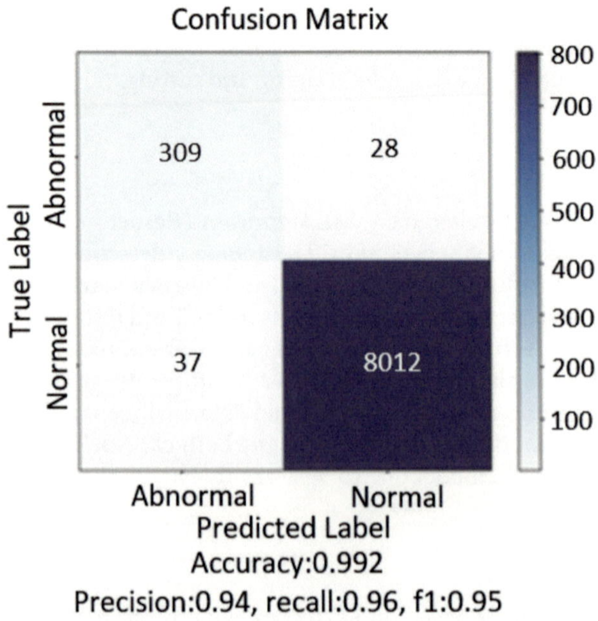

Figure 12.
ML results for anomaly detection.

performance due to 99% accuracy. With respect to the 2D scanning data, the raw data was converted into images before fed into ML training and testing for anomaly detection, which results of accuracy above 90% demonstrate the good performance of the ML models. Therefore both high accuracy anomaly detection results indicate that EC testing can be applied for residual stress detection to control the quality of metal AM.

5. Electromagnetic acoustic transducers monitoring

EMATs are devices made up of coils fed by a large dynamic current (a pulse or a tone-burst are commonly used) and a magnet or electromagnet providing a static magnetic field [49–53] as shown in **Figure 13**. They are used on metallic specimens based on Lorenz force as shown in **Figure 13**. For ferromagnetic metals, both Lorenz force and a magneto-strictive transduction mechanism occur.

Although contactless, EMAT require extreme proximity to the tested specimen (mm to few mms) to generate energy. EMATs have good repeatability since there is no external couplant. Common modes for non-ferrous materials include shear waves [49] making EMATs an ideal candidate for studying in- situ the interaction of shear ultrasonic waves with AM components (**Figure 13**).

5.1 Case study I

A case study was based on an EMAT data through testing in a CNC machine with samples in laboratory. Then signal processing algorithms and ML algorithms were developed for different tasks: automatic signal processing algorithm was developed based on filtering and peak detection methods to calculate acoustic birefringence results while different ML models based on EMAT time series plotting data and EMAT short-time Fourier transform (STFT) image data were developed for good/substandard measurement detection.

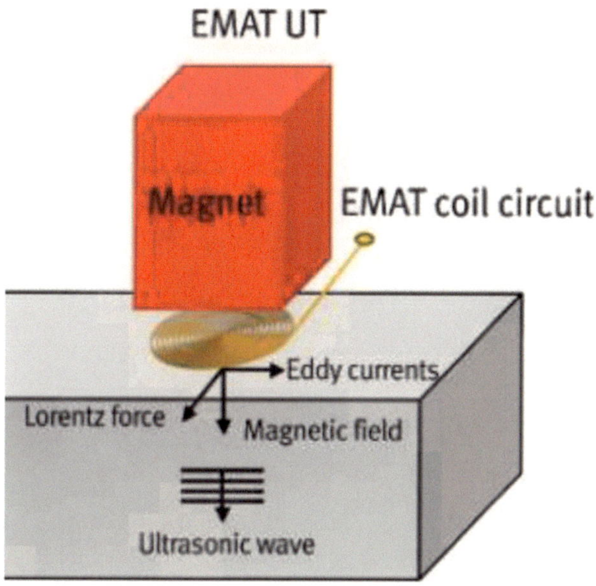

Figure 13.
EMAT operation inspection [54].

5.1.1 Signal processing

EMAT data has different measurements before and after additive printing while it also has good and substandard measurements, as summarised in **Table 3**. Each EMAT data file has two channels records while the beginning of the waveform has large variation, as example plot shown in **Figure 14**.

Based on the FFT analysis of the raw data as shown in **Figure 15**, it indicates that there are a lot of noise which frequency is higher than around 4/5 MHz. Thus automatic pre-processing program was developed through applying low pass filters (either 4 MHz filter or 5 MHz filter), as shown in **Figure 16**. This figure demonstrate that both of the filters have similar results as the red curve and black curve are close to each other. Then peak detection algorithm was developed to detect the peaks automatically (shown in **Figure 17**). The pattern of each peak difference between two channels for different measurements before and after additive printing are plotted in **Figure 18**. The pattern shows that all the measurements have the similar increasing rate of the peak difference. This is because of the increased time difference along the peaks resulting from different velocities of both waves after more reflections.

File name	File number
Steel pipe after printing left of new material (type 1 measurement)	9
Steel pipe after printing on new material (type 2 measurement)	46
Steel pipe after printing on new material2 (type 3 measurement)	45
Steel pipe before printing—bad data	45
Steel pipe before printing full axial scan (type 4 measurement)	45
Steel pipe before printing—poor data	47

Table 3.
EMAT data files.

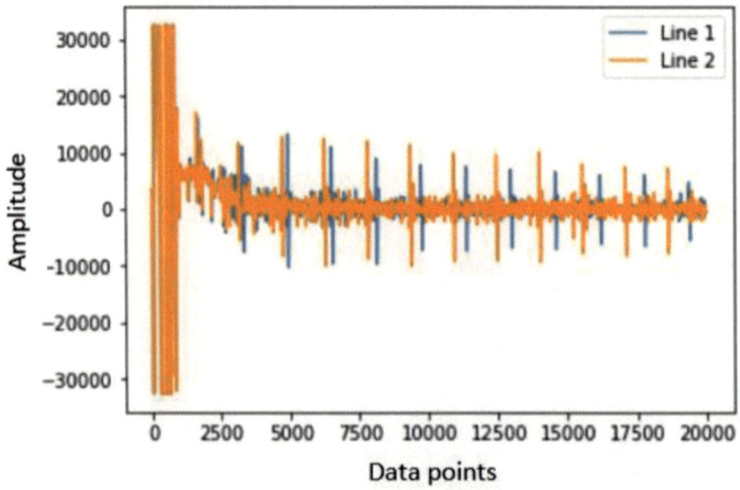

Figure 14.
EMAT data plot.

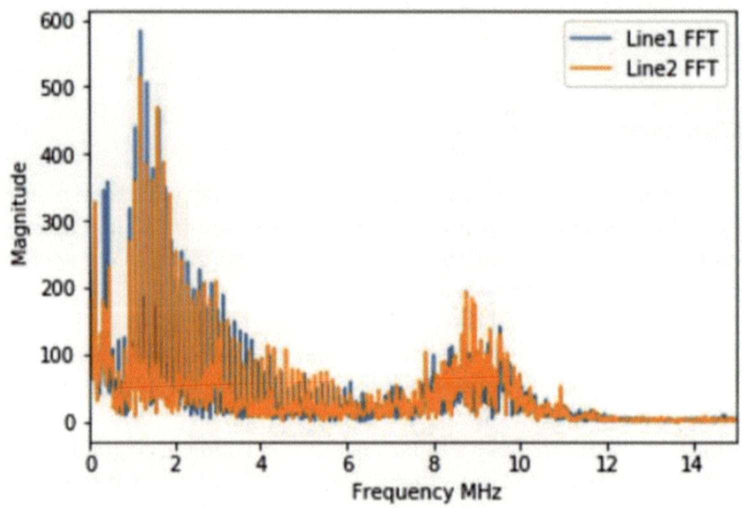

Figure 15.
EMAT data FFT analysis.

After pre-processing the data, acoustic birefringence (AB) was calculated through following function:

$$AB = \frac{t_s - t_f}{\left(\frac{t_s + t_f}{2}\right)} \tag{1}$$

ts = arrival time of 'slow angle' wavetf = arrival time of 'fast angle' wave

The arrival time is the time at maximum point of each peak. The time difference is a direct indication of change in velocity (assuming sample is the same thickness across the probe area). It is assumed that a change in velocity is an indication of residual stress change.

AB results for all the different measurements before and after printing are shown in **Figure 19**, which indicates that AB results decrease monotonically alone the peaks. The boxplots of AB results of the four types of measurements before and

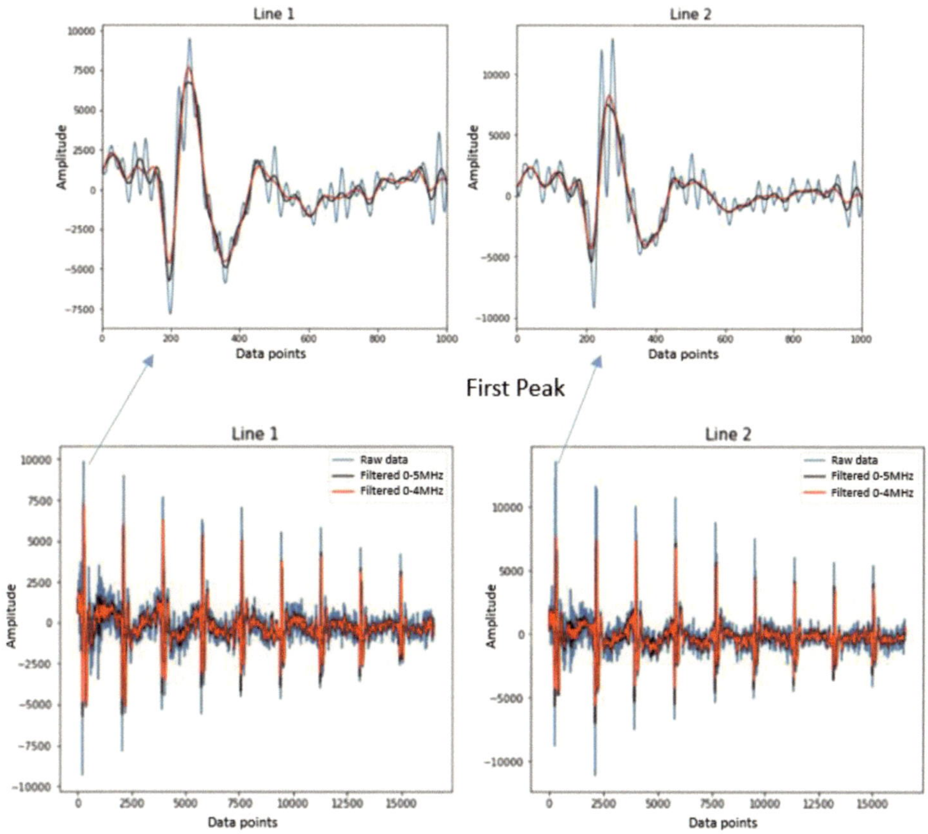

Figure 16.
EMAT data pre-processing through low path filters.

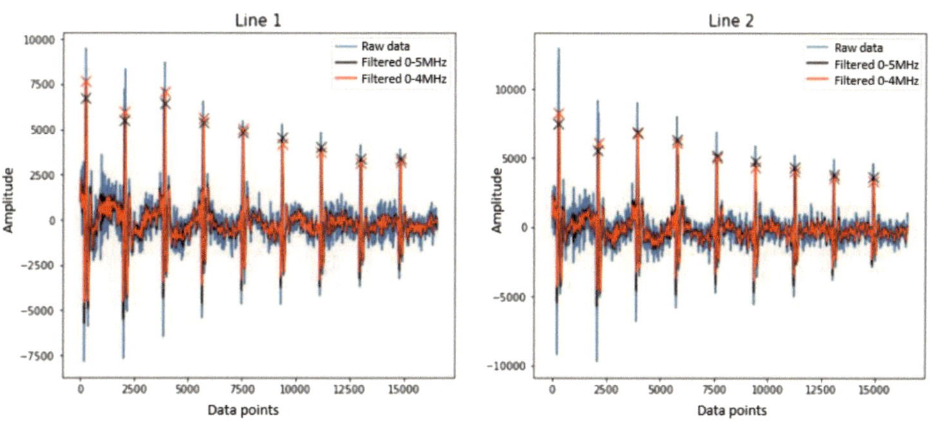

Figure 17.
Peak detection.

after printing for all the six peaks are shown in **Figure 20**, which demonstrates that all six peaks have the similar pattern: AB results of type 2&3 measurements have the same level and lower than that of type 1&4 measurement. These results are consistent with the facts that type 2&3 measurements were tested on the similar new materials after printing, which are different with type 1&4 measurements as type 1 measurements were tested before printing and type 4 measurements were tested at

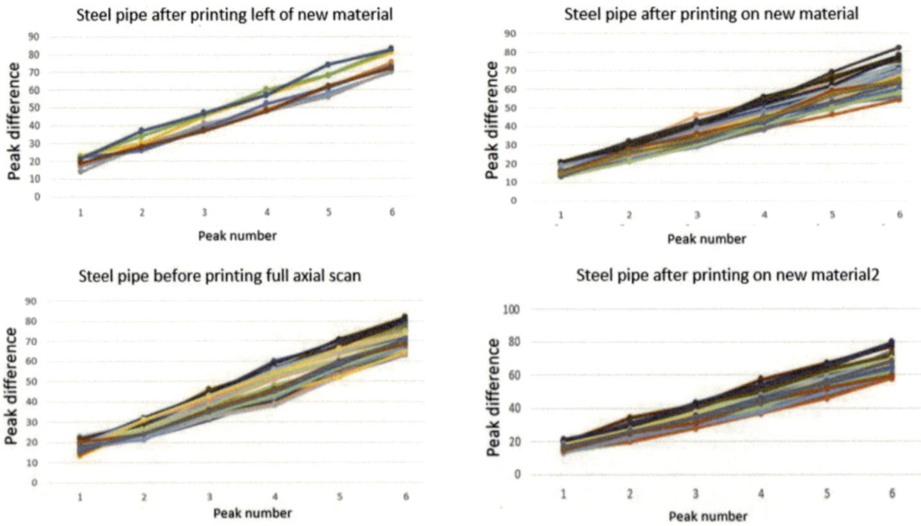

Figure 18.
Pattern of peak difference for different measurements.

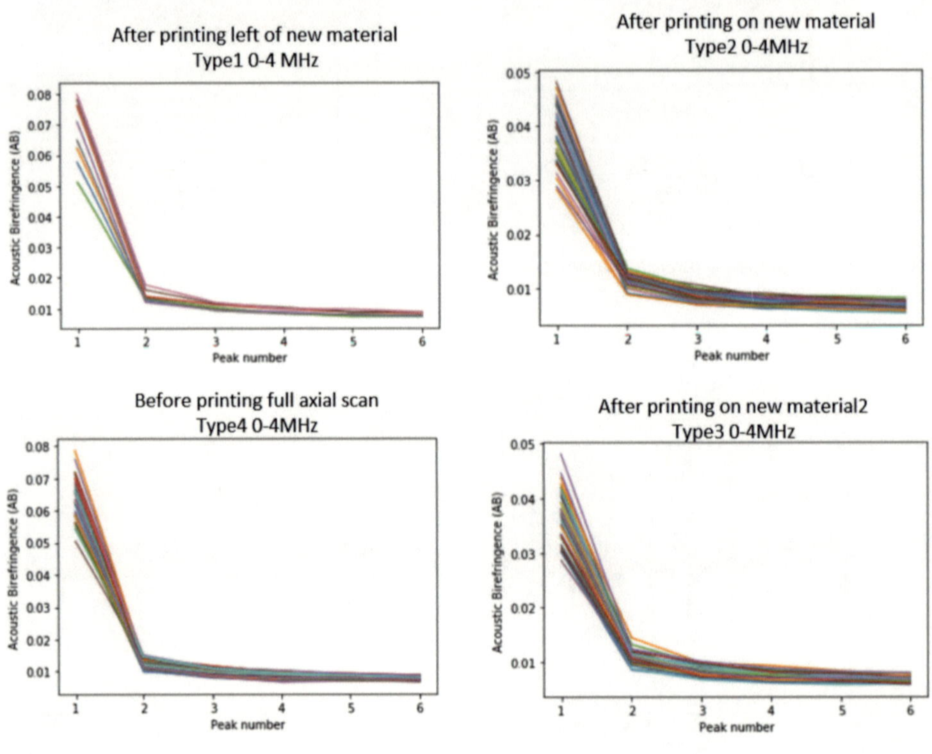

Figure 19.
AB results for all the different measurements before and after printing.

the edge after printing new material. This indicates that AB results can indicate the different residual stresses in a certain level thus as an indicator of QC of metal AM.

5.1.2 ML algorithms

Based on the good and substandard EMAT data, the raw data was converted into images through two pre-processing methods, then different ML algorithms based

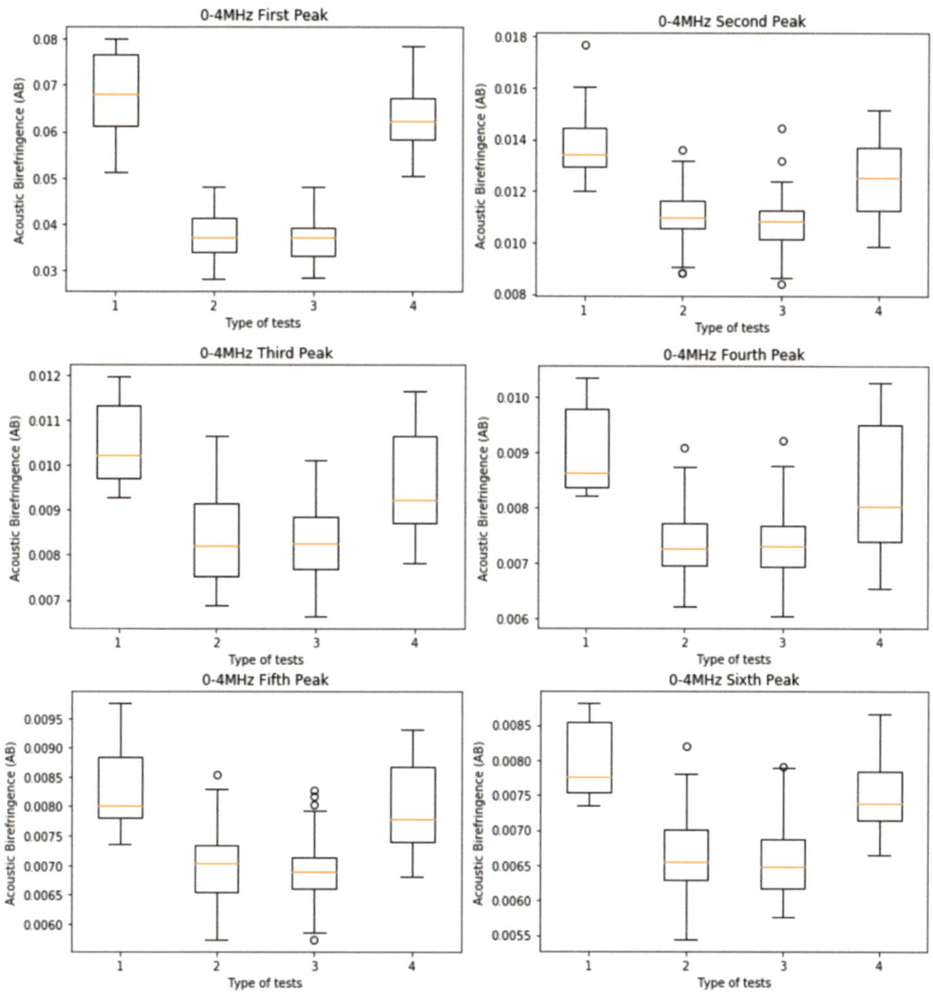

Figure 20.
AB results of the four types of measurements before and after printing for all the six peaks.

on the two image datasets were applied for good/substandard classification. **Table 4** lists the number of each class. The first pre-processing method is plotting the time series data directly in a figure as shown in **Figure 21a** while the second

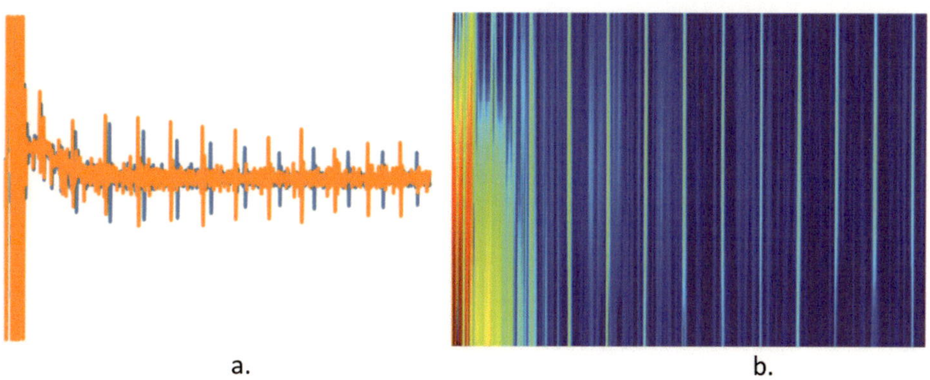

Figure 21.
EMAT image data. (a) Time series plotting image, (b) STFT image.

pre-processing method is convert the time series data into STFT images as shown in **Figure 21b**.

STFT is a frequency-time transform of a time signal and an effective tool to analyse the non- stationary signals because of the avoiding of severe interference by the cross-terms [55]. The basic principle for STFT is as follows: divide these characteristic signal into small time intervals, use the Fourier transformation to analyse separately each time interval. The definition of STFT for signal s(t) is:

$$S_t(\omega) = \frac{1}{\sqrt{2\pi}} \int s(\tau)h(\tau - t)e^{j\omega\tau}d\tau \tag{2}$$

The two formats of datasets were used to train and test ML algorithms based on Resnet model. Results of the ML models show in **Figure 22**, for example, there were 23 and 15 images correctly predicted as "Good" and "Substandard" measurements separately while no image was predicted incorrectly. The results indicate excellent performance due to 100% accuracy for both of data formats (**Table 4**).

5.2 Summary

Based on above case study of MEAT testing, automatic signal processing algorithm was developed based on filtering and peak detection methods to calculate acoustic birefringence, which results of type 2&3 measurements have the same level and lower than that of type 1&4 measurement. These results are consistent with the facts that type 2&3 measurements were tested on the similar new materials after printing, which are different with type 1&4 measurements as type 1 measurements were tested before printing and type 4 measurements were tested at the edge

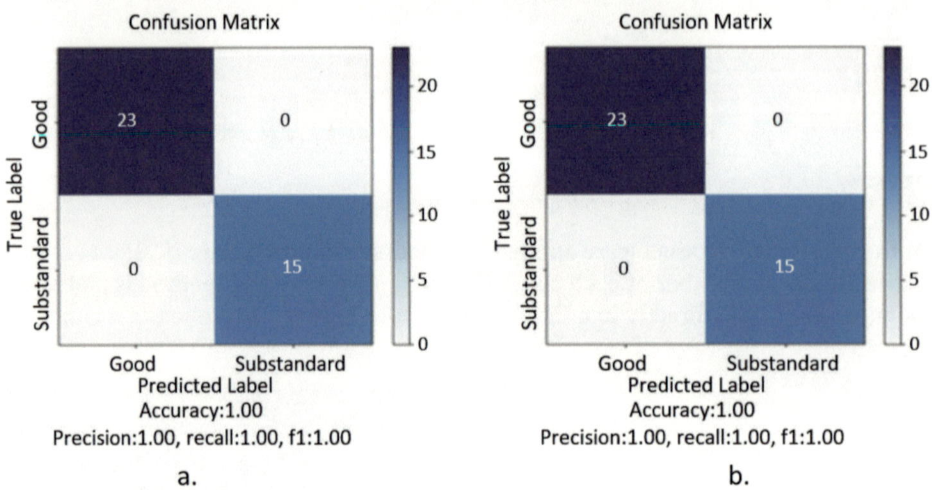

Figure 22.
ML results. (a) Based on time series plotting image, (b) based on STFT image.

Class	Number
Good data	217
Substandard data	134

Table 4.
Total good/substandard data.

after printing new material. This indicates that AB results can indicate the different residual stresses in a certain level thus as an indicator of QC of metal AM.

Additionally, different ML models based on EMAT time series plotting data and EMAT STFT image data were applied for good/substandard measurement detection, which results demonstrate excellent performance due to 100% accuracy.

6. Conclusion

Additive manufacturing technology has progressed a lot allowing to achieve higher parts quality and improved consistency although a lot of challenges remain with regards to consistency of quality, sensitivity to tiny process tolerances. Many techniques have been developed to enhance the quality: the application of post-processing of 3D printed parts, the use of monitoring sensors for early defect detection and the use of monitoring sensors with feedback loop for continuous process optimisation. In this chapter, many sensing techniques have been investigated for process monitoring. Thermography and CT scan are limited by the resolution of images, they are bulky and costly, and not suited to residual stress detection. These methods are time consuming both for data acquisition and data analysis (CT requires slice by slice visual interpretation) and have not yet been successfully integrated into AM technology due to health and safety standards and its poor resolution at the edges of AM structures. However investigation of the applications through EMAT and EC testing techniques shows that both of the methods can be applied for residual stress detection for meatal AM techniques.

Based on the case study of two sets of EC data which have two formats: in-line scanning and 2D scanning, different ML algorithms were applied for anomaly detection. With respect to the in-line scanning data, results indicate both of the two different ML algorithms (time series data processing and image data processing) have excellent performance due to 99% accuracy. With respect to the 2D scanning data, ML models have good performance resulting from accuracy above 90%. Therefore both high accuracy anomaly detection results indicate that EC testing can be applied for residual stress detection to control the quality of metal AM.

Based on above case study of MEAT testing, automatic signal processing algorithm was developed based on filtering and peak detection methods to calculate acoustic birefringence, which results of type 2&3 measurements have the same level and lower than that of type 1&4 measurement. These results are consistent with the facts that type 2&3 measurements were tested on the similar new materials after printing, which are different with type 1&4 measurements as type 1 measurements were tested before printing and type 4 measurements were tested at the edge after printing new material. This indicates that AB results can indicate the different residual stresses in a certain level thus as an indicator of QC of metal AM. Additionally, different ML models based on EMAT time series plotting data and EMAT STFT image data were applied for good/substandard measurement detection, which results demonstrate excellent performance due to 100% accuracy.

Above all, a crucial and novel extension system incorporation of big data collection from sensors of the both EC and EMAT techniques and analysis through ML can estimate the likelihood of the metal AM techniques to introduce anomalies into the printed structures before the beginning of the manufacturing, thus the system can be used as an on-line monitoring and detection system to control the quality of metal AM.

Author details

Bojie Sheng[1], Jamil Kanfoud[1] and Tat-Hean Gan[1,2*]

1 Brunel Innovation Centre, Brunel University London, London, United Kingdom

2 TWI Ltd, Great Abington, United Kingdom

*Address all correspondence to: tat-hean.gan@brunel.ac.uk

References

[1] Conner BP, Manogharan GP, Martof AN, Rodomsky LM, Rodomsky CM, Jordan DC, et al. Making sense of 3-D printing: Creating a map of additive manufacturing products and services. Additive Manufacturing. 2014;**1-4**: 67-76

[2] Ngo TD, Kashani A, Imbalzano G, Nguyen KTQ, Hui D. Additive manufacturing (3D printing): A review of materials, methods, applications and challenges. Composites Part B: Engineering. 2018;**143**:172-196

[3] Wang X, Jiang M, Zhou Z, Gou J, Hui D. 3D printing of polymer matrix composites: A review and prospective. Composites Part B: Engineering. Feb. 2017;**110**:442-458

[4] Sculpteo. The state of 3D PRINTING. Edition 2018. France; 2018

[5] Wong KV, Hernandez A. A review of additive manufacturing. ISRN Mechanical Engineering. 2012;**2012**(2012):1-10

[6] Wohlers Report 2013 (ISBN 0-9754429-9-6)

[7] Thomas D. Costs, benefits, and adoption of additive manufacturing: A supply chain perspective. The International Journal of Advanced Manufacturing Technology. 2016;**85** (5-8):1857-1876

[8] Cotteleer M, Joyce J. 3D opportunity: Additive manufacturing paths to performance, innovation, and growth. Deloitte Review. 2014;**14**:5-19

[9] Yang H, Huang C, Adnan M, Hsu C, Lin C, Cheng F. An online AM quality estimation architecture from pool to layer. IEEE Transactions on Automation Science and Engineering. 2021;**18**(1): 269-281

[10] Yao B, Yang H. Constrained Markov decision process modeling for sequential optimization of additive manufacturing build quality. IEEE Access. 2018;**2**(1): 1-9

[11] Gao W, et al. The status, challenges, and future of additive manufacturing in engineering. Computer-Aided Design. 2015;**69**:65-89

[12] Frazier WE. Metal additive manufacturing: A review. Journal of Materials Engineering and Performance. 2014;**23**(6):1917-1928

[13] Kaveh M, Badrossamay M, Foroozmehr E, Etefagh AH. Optimization of the printing parameters affecting dimensional accuracy and internal cavity for HIPS material used in fused deposition modeling processes. Journal of Materials Processing Technology. 2015;**226**:280-286

[14] Farzadi A, Waran V, Solati-Hashjin M, Rahman ZAA, Asadi M, Osman NAA. Effect of layer printing delay on mechanical properties and dimensional accuracy of 3D printed porous prototypes in bone tissue engineering. Ceramics International. 2015;**41**(7):8320-8330

[15] Everton SK, Hirsch M, Stravroulakis P, Leach RK, Clare AT. Review of in-situ process monitoring and in-situ metrology for metal additive manufacturing. Materials and Design. 2016;**95**:431-445

[16] Montazeri M, Yavari R, Rao P, Boulware P. In-process monitoring of material cross-contamination defects in laser powder bed fusion. Journal of Manufacturing Science and Engineering. 2018;**140**(11):111001

[17] Cannizzaro D, Varrella AG, Paradiso S, Sampieri R, Chen Y, Macii A, et al. In-situ defect detection of

metal additive manufacturing: An integrated framework. IEEE Transactions on Emerging Topics in Computing. 2021;**10**(1):1-12

[18] Stucker B. Additive manufacturing technologies: Technology introduction and business implications. In: Frontiers of Engineering: Reports on Leading-Edge Engineering From the 2011 Symposium. Washington, DC: National Academies Press; 2012. pp. 19-21

[19] Edwards P, O'conner A, Ramulu M. Electron beam additive manufacturing of titanium components: Properties and performance. Journal of Manufacturing Science and Engineering. 2013;**135**: 061016

[20] Melvin LS III, Das S, Beaman S Jr. Video microscopy of selective laser sintering. In: Proceedings of the Solid Freeform Fabrication Symposium, DTIC Document. Austin, TX, USA; 1994. pp. 34-41

[21] Benda J. Temperature controlled selective laser sintering. In: Proceedings of the 1994 Solid Freeform Fabrication Symposium. Austin, TX, USA: DTIC Document. 1994. pp. 277-284

[22] Zeng K, Pal D, Stucker B. A review of thermal analysis methods in laser sintering and selective laser melting. In: Proceedings of 2012 Solid Freeform Fabrication Symposium; Austin, TX, USA. 2012

[23] Bi G, Schürmann B, Gasser A, Wissenbach K, Poprawe R. Development and qualification of a novel laser-cladding head with integrated sensors. International Journal of Machine Tools and Manufacture. 2007;**47**:555-561

[24] Kleszczynski S, Zur Jacobsmühlen J, Sehrt J, Witt G. Error detection in laser beam melting systems by high resolution imaging. In: Proceedings of the 2012 Solid Freeform Fabrication Symposium. Austin, TX, USA; 2012

[25] Lott P, Schleifenbaum H, Meiners W, Wissenbach K, Hinke C, Bültmann J. Design of an optical system for the in situ process monitoring of selective laser melting (SLM). Physics Procedia. 2011;**12**:683-690

[26] Kolossov S, Boillat E, Glardon R, Fischer P, Locher M. 3D FE simulation for temperature evolution in the selective laser sintering process. International Journal of Machine Tools and Manufacture. 2004;**44**:117-123

[27] Shishkovsky I, Scherbakov V, Morozov Y, Kuznetsov M, Parkin I. Surface laser sintering of exothermic powder compositions. Journal of Thermal Analysis and Calorimetry. 2008;**91**:427-436

[28] Van Belle L, Vansteenkiste G, Boyer JC. Investigation of residual stresses induced during the selective laser melting process. Key Engineering Materials. 2013;**554/557**:1828-1834

[29] Taylor C, Childs T. Thermal experiments in direct metal laser sintering. In: Proceedings of the 2001 Solid Freeform Fabrication Symposium, Austin, TX, USA; 2001

[30] Low SC, Ake BE. Thermocouple Control System for Selective Laser Sintering Part Bed Temperature Control. United States Patent Application 20030222066, USA; 2004

[31] Price S, Cooper K, Chou K. Evaluations of temperature measurements by near-infrared thermography in powder-based electron-beam additive manufacturing. In: Proceedings of the Solid Freeform Fabrication Symposium. Austin, TX: University of Texas; 2012. pp. 761-773

[32] Yadroitsev I, Krakhmalev P, Yadroitsava I. Selective laser melting of Ti6Al4V alloy for biomedical applications: Temperature monitoring and

microstructural evolution. Journal of Alloys and Compounds. 2014;**583**:404-409

[33] Dinwiddie RB, Dehoff RR, Lloyd PD, Lowe LE, Ulrich JB. Thermographic in-situ process monitoring of the electron-beam melting technology used in additive manufacturing. In: SPIE Defense, Security, and Sensing, International Society for Optics and Photonics. Baltimore, Maryland, United States; 2013. pp. 87050K-1-87050K-9

[34] Kruth J-P, Duflou J, Mercelis P, van Vaerenbergh J, Craeghs T, De Keuster J. On-line monitoring and process control in selective laser melting and laser cutting. In: Proceedings of the 5th Lane Conference, Laser Assisted Net Shape Engineering. Erlangen, Germany; 2007. pp. 23-37

[35] Mireles J, Terrazas C, Medina F, Wicker R, Paso E. Automatic feedback control in electron beam melting using infrared thermography. In: Proceedings of the 2013 Solid Freeform Fabrication Symposium. Austin, TX, USA; 2013

[36] Mireles J, Ridwan S, Morton PA, Hinojos A, Wicker RB. Analysis and correction of defects within parts fabricated using powder bed fusion technology. Surface Topography: Metrology and Properties. 2015;**3**:034002

[37] Thompson A, et al. X-ray computed tomography for additive manufacturing: A review. Measurement Science and Technology. 2016;**27**(7):072001

[38] Khademzadeh S, et al. Precision additive anufacturing of NiTi parts using micro direct metal deposition. International Journal of Advanced Manufacturing Technology. 2018

[39] Carmignato S, et al. Industrial X-ray Computed Tomography. Midtown Manhattan, New York City: Springer International Publishing; 2018. ISBN 978-3-319-59571-9

[40] Wits WW, et al. Porosity testing methods for the quality assessment of selective laser melted parts. CIRP Annals-Manufacturing Technology. 2016;**65**(1): 201-204

[41] Thompson A, et al. Topography of selectively laser melted surfaces: A comparison of different measurement methods. CIRP Annals-Manufacturing Technology. 2017;**66**:543-546

[42] Townsend A, et al. Areal surface texture data extraction from x-ray computed tomography reconstructions of metal additively manufactured parts. Precision Engineering. 2017;**48**: 254-264

[43] Zanini F, Sbettega E, Carmignato S. X-ray computed tomography for metal additive manufacturing: Challenges and solutions for accuracy enhancement. In: 15th CIRP Conference on Computer Aided Tolerancing. Milan, Italy; 2018; **75**:114-118

[44] Teterko A, Nazarchuk Z. Selektyvna Vykhrostrumova Defektoskopija [Selective Eddy Current Non-Destructive Testing]. Lviv: Karpenko Physico-Mechanical Institute of the National Academy of Sciences of Ukraine (in Ukraine); 2004. p. 247

[45] Bazhenov V, Protasov A, Gloinik K. Increasing of operation speed of digital eddy current defectoscopes based on frequency synthesizer. In: MRRS 2017—2017 IEEE Microwaves, Radar and Remote Sensing Symposium, Proceedings 8075051. Kiev, Ukraine. pp. 155-158

[46] Todorov EI. Non-destructive evaluation of additive manufacturing components using an eddy current array system and method. U.S. Patent 2016/ 034 921S A1. 2016

[47] Elsenbarth D, Stoll P, Klahn C, Heinis T, Meboldt M, Wegener K. Unique coding for authentication and

anti-counterfeiting by controlled and random process variation in L-PBF and L-DED. Additive Manufacturing. 2020; **35**:101298

[48] Ehlers H, Pelkner M, Thewes R. Heterodyne eddy current testing using magnetoresistive sensors for additive manufacturing purposes. IEEE Sensors Journal. 2020;**20**(11):1

[49] Hirao M, Ogi H. EMATs for Science and Industry: Non-Contacting Ultrasonic Measurements. Midtown Manhattan, New York City: Springer Science & Business Media; 2013

[50] Hirao M, Ogi H. An SH-wave EMAT technique for gas pipeline inspection. NDT & E International. 1999;**32**:127-132

[51] Hirao M, Ogi H, Yasui H. Contactless measurement of bolt axial stress using a shear-wave electromagnetic acoustic transducer. NDT & E International. 2001;**34**: 179-183

[52] Gao H, Lopez B. Development of single-channel and phased array electromagnetic acoustic transducers for austenitic weld testing. Materials Evaluation. 2010;**68**:821-827

[53] Ribichini R. Modelling of electromagnetic acoustic transducers [PhD dissertation]. London: Imperial College; 2011

[54] Available from: https://www.arorandt.com/emat-solutions/

[55] Zhang Z, Ren Y. Time–frequency analysis of echoes signal in ultrasonic testing of adhesion based on short-time Fourier transformation. In: 2010 International Conference on Measuring Technology and Mechatronics Automation; Changsha, China. 2010: 1023-1026

Chapter 9

Modeling of LPBF Scanning Strategy and its Correlation with the Metallic 316 L, 321, and Alnico Magnets Samples Structure

Pavel Kuznetsov, Anna Mozhayko, Ivan Shakirov, Vitaliy Bobyr, Mikhail Staritsyn and Anton Zhukov

Abstract

This chapter presents the influence of powder bed laser scanning strategy on the crystallographic structure of the fused specimens 316 L, 321 stainless steel, and Alnico magnets. The main parameters affecting structure are as follows—laser power, stripe width, number of repeated passes with different power, and type of scanning (circle, bidirectional or interlaced, etc.). Changes in the crystallographic structure are studied with regard to melt pool geometry, surface temperature, and surface heat transfer. The correlation is shown between stripe width and laser beam focal spot diameter. Depending on the ratio between stripe width and laser beam focal spot diameter one can see growth elongated and oriented grains or quasi-equiaxed non-oriented grains. The influence of the energy input on the melt pool size and the microstructure of the sample is studied. The influence of the scanning mode (bidirectional and circular) on the temperature distribution in the sample and the microstructure of the sample made of Alnico alloy is considered. All these experimental and model examples clearly demonstrate that it is possible to produce a controllable structure during LPBF process building for advanced additive manufacturing.

Keywords: Laser powder bed fusion, finite element modeling, scanning strategy, melting pool geometry, crystallographic structure

1. Introduction

Laser powder bed fusion (LPBF) is one of the additive manufacturing (AM) techniques in which a metal powder bed is selectively melted by a laser beam using a layer-by-layer method based on the designed scanning strategy to produce final parts [1, 2]. In general, AM enables to control the microstructure of metals by changing various values of the process parameters that cannot be achieved by traditional metal parts manufacturing technologies due to their inability to control heat transfer conditions on a very limited scale [3].

The microstructure of the sample obtained by the LPBF method depends on many parameters, such as laser power, scanning speed, hatch spacing, stripe width, and scanning strategy. Alteration of scanning parameters leads to a change in the geometry of the melt pool. In turn, when the geometry of the melt pool changes, the thermodynamic conditions of crystallization will change, which will lead to the formation of various structures. Thus, thanks to the ability to select various scanning parameters, it becomes possible to control the microstructure of products [4].

Currently, different numerical models of LPBF are available. Some of them evaluate and predict the temperature distribution and melt pool size during the selective laser melting process [5–12]. Others investigate the influence of various LPBF parameters on temperature distribution [13–16] and the sample's microstructure [3, 17, 18]. For instance, Yingli Li et al. developed a model, which incorporates a phase function to differentiate powder phase, melting liquid phase, dense solid phase, and vaporized gas phase to determine the melt pool size [6]. Zhichao Dong et al. investigated the effects of hatch spacing on the temperature field, microstructure and melt pool size, overlap rate, surface quality, and relative density during the selective laser melting of 316 L SS [16]. They found that increasing the hatch spacing reduced maximum temperature and heat accumulation. The change in the melt pool size when the laser beam moves from the center of the first layer to the center of the second layer was studied by Yali Li et al. [13].

Besides, the research studies the influence of process parameters on the structure of fused steels or alloys. For example, as shown in [19, 20], various scanning strategies can make for a clear epitaxial growth along the heat flow with an orientation <100>, which can enhance isotropy. It was also shown that the change in yield strength was mainly due to different grain sizes, and the improved plasticity was associated with a changed grain structure based on different scanning [20].

Despite a large number of studies in the field of LPBF, the relationship between fabrication and microstructure has not been fully studied. There are many process parameters whose influence on temperature fields and microstructure needs to be studied.

In this chapter, we focus on understanding the effect of LPBF parameters on the microstructure of the fused specimens. The chapter discusses the influence of energy input, ratio of stripe width, and various scanning strategies (circle, bidirectional or interlaced, and so on) on the microstructure of the sample.

Austenitic stainless steels AISI 316 L and AISI 321 and hard magnetic alloy Alnico were selected as investigated materials. Section 2 contains a detailed description of the LPBF model and the physical properties of the austenitic steel AISI 316 L selected as the modeled material since this steel is widely used in additive manufacturing. Sections 3 and 4 provide an investigation of the stripe width and generalized energy input effect on microstructure. Thermal fields and microstructure of austenitic steels AISI 316 L and AISI 321 are studied in these sections. Austenitic steels do not have phase transitions in the entire temperature range under study, therefore, they make it possible to study directly the influence of LPBF parameters on the structure being formed. Research on the influence of the scanning strategy on the metal quality is necessary for such labor-intensive materials in additive manufacturing as the two-phase Alnico alloy. The high-stress level leading to cracking creates many difficulties in the way of additive production of permanent magnets. It is shown that with the help of a favorable change in the distribution of thermal fields by means of applying an optimized scanning strategy, it is possible to manufacture Alnico alloy permanent magnet with a good structure.

2. Modeling

2.1 Model description

In general, heat transfer can be described by the heat conduction equation:

$$\rho C_p\left(\frac{\partial T}{\partial t}+\left(\vec{u}\bullet\vec{\nabla}T\right)\right)=\vec{\nabla}\bullet\left(k\vec{\nabla}T\right), \tag{1}$$

where ρ is the density of the material, C_p is the heat capacity, T is the temperature, $\vec{u}$ is the velocity vector of the liquid, and k is the thermal conductivity.

It is assumed that the distribution of the surface heat flux Q_{in} over the powder layer represents Gaussian distribution, which mathematically can be rendered as:

$$Q_{in}=\frac{2PA}{\pi r^2}\exp\left(-\frac{2(x-vt)^2+y^2}{r^2}\right), \tag{2}$$

where P is the laser power, A is the laser energy absorption coefficient, r is the radius of the laser beam, and v is the speed of the laser beam. The scanning direction is set by replacing of x by $(x-vt)$.

The initial conditions of the model include a uniform temperature field along the full sample before the heat source is supplied, which can be described as:

$$T(x,y,z,t)_{t=0}=293\ K \tag{3}$$

Heat exchange between the sample and ambient occurs on the upper surface of the model.

$$-k\frac{\partial T}{\partial z}=Q_{in}-\alpha(T_w-T_0)-\sigma\varepsilon\left(T_w^4-T_0^4\right)-Q_{vap}, \tag{4}$$

where α is the heat transfer coefficient, T_w is the surface temperature, and T_0 is the ambient temperature, σ is the Stefan-Boltzmann constant, ε is the surface absorption coefficient, and Q_{vap} is the vapor flow.

On the right side of Eq. (4) all the input and output heat fluxes on the upper surface of the model are described. Heat losses due to convection, radiation, and evaporation are described in the succession.

When the evaporation temperature is reached, a vapor flow appears in the model, which is given in the formula [21]:

$$Q_{vap}=L_v(1-\beta)\sqrt{\frac{M}{2\pi RT}}P_{amb}\,exp\left[\frac{ML_v}{RT_v}\left(1-\frac{T_v}{T}\right)\right], \tag{5}$$

where L_v is the specific heat of evaporation, β is the rate of re-condensation, R is the universal gas constant, M is the molar mass, P_{amb} is the ambient pressure, and T_v is the evaporation temperature.

β is the fraction of evaporated particles that re-condenses upon the interaction with the surrounding gas.

The melting phase transition is taken into account by the enthalpy method [22]. Latent heat of fusion is taken into account in the heat capacity equation.

$$C_p^{eq}=C_p+D_mL_m, \tag{6}$$

where L_m is the latent heat of fusion, C_p is the heat capacity, D_m is the Gaussian function normalized around the melting temperature T_m:

$$D_m=\frac{exp\left(-\frac{(T-T_m)^2}{\Delta T}\right)}{\sqrt{\pi\Delta T^2}}, \tag{7}$$

where ΔT is the smoothing interval equal to 50 K.

2.2 Numerical methods

Numerical modeling is widely used to determine the optimal scanning strategies in the LPBF process. FEM is the most commonly used numerical approach to analyze the temperature profile and the size of the melt pool during the melting and solidification of materials. Commercial software Comsol Multiphysics 5.4. was used to create models of the LPBF process. The present model involves "Heat Transfer" module that is solved with a time-depended solver. The mesh elements are triangular in shape. The upper part is of relatively higher importance where many phenomena are involved at relatively high temperature. The meshing of this part is done with extra fine mesh with a maximum element size of 25 microns. The mesh is relatively coarse at the remaining parts of the model with a maximum element size of 0.2 mm. The results change by less than 1% when a smaller grid is used. Such size of the grid elements was chosen as optimal because it provides maximum accuracy at the lowest calculation speed.

The temperature dependence of the heat capacity of 316 L steel is shown in **Figure 1**. The physical properties of the modeling material are shown in **Table 1**.

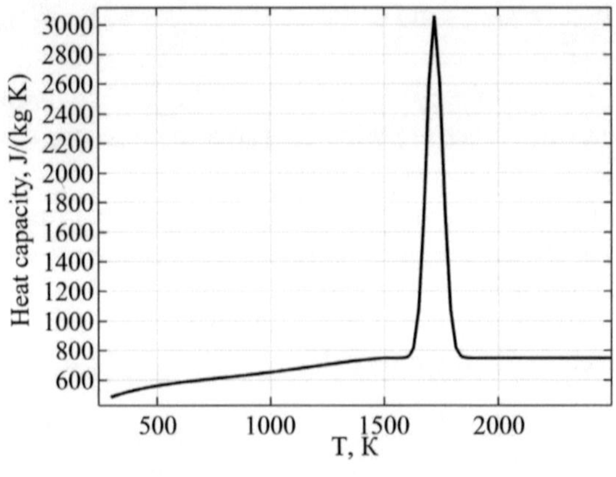

Figure 1.
Equivalent heat capacity of 316 L steel.

Parameter (Designation)	Value [Dimension]
Melting temperature (T_m)	1720 [K]
Evaporation temperature (T_v)	3200 [K]
Laser energy absorption coefficient (A)	0.8
Thermal conductivity (k)	24.9 $[\frac{W}{mK}]$
Density (ρ)	7750 $[kg/m^3]$
Surface absorption coefficient (ε)	0.8
Latent heat of fusion (L_m)	205 [kJ/kg]
Latent heat of evaporation (L_v)	6000 [kJ/kg]

Table 1.
Physical properties of 316 L steel [14, 16].

3. Investigation of stripe width effect on the microstructure

LPBF technology is conditioned that the sample in its volume is built from a set of melt pools. Consequently, the geometry of the melt pool and the rate of metal crystallization has a decisive influence on the formed structure of the sample. The geometry of the melt pool and the crystallization rate directly depend on all parameters of fusing the surface of the powder layer with a laser beam—the described beam trajectory, the speed, the laser power, the geometry of a specific section of the part when scanning a specific layer. The stripe width is one of the key parameters that affect the geometry of the melt pool and the conditions of metal crystallization.

The printing process on most LPBF installations is carried out by fusing the part section in the form of blocks with a certain width. The thermal pattern of the distribution of temperature fields depends on the width of the block, therefore, this has a direct effect on the rate of metal crystallization in the melt pool and the structure created.

It is of interest to study the influence of such paint parameters as the stripe width on the formation of the melt pool and the metal structure.

The main objective of the present study was to get a deeper understanding of the correlation between LPBF process parameters and microstructure evolution. The temperature field close to the solid–liquid interface determines the morphology and grain size, as well as the crystallographic structure of each track of the material obtained by LPBF. The solidified microstructure depends on local solidification conditions at the trailing edge of the melt pool. Hence, the melt pool geometry (i.e., size and shape) and thermal profile need to be first predicted as a function of the process parameters. Further, taking the solidification theory as a basis and calculating the profile of the melt pool on the basis of numerical modeling, it is possible to establish a connection with the microstructure of the experimentally obtained product.

3.1 Modeling of the LPBF process with different stripe widths

Three modes of the LPBF process with various stripe widths (1 mm, 2.5 mm, and 5 mm) were considered. The model parameters are shown in **Table 2**. The diameter of the focal beam of the laser was equal to 100 microns and did not change. Thus, in the first case, 10 focal spot diameters fit the stripe, 25 in the second, and 50 in the third (**Figure 2**).

No	Power, W	Speed, mm/s	Hatch, μm	Stripe width, mm
1, 2, 3	190	800	100	1; 2, 5; 5

Table 2.
The model parameters.

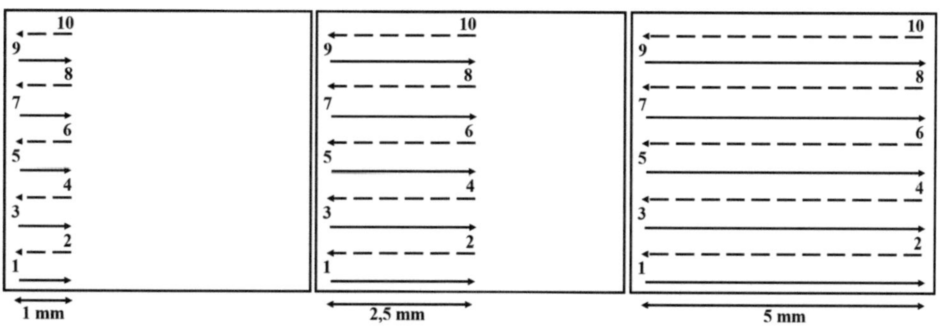

Figure 2.
Schematic representation of different stripe widths (1 mm, 2.5 mm, and 5 mm).

Based on the simulation results, the dimensions of the melt pool were measured in three directions. The melt pool was measured at the midpoint of the track with a step equal to the hatch. The melt pool dimensions (length, width, and depth) were observed from the tracks temperature distribution results and considered from the melting point to the peak temperature along the scanning direction.

Figures 3–5 show the dependences of the width, length, and depth of the melt pool on the number of the laser beam track.

It can be seen in **Figures 3–5** that for the first pass of the laser, the melt pool dimensions differ insignificantly. This is explained by the fact that the initial

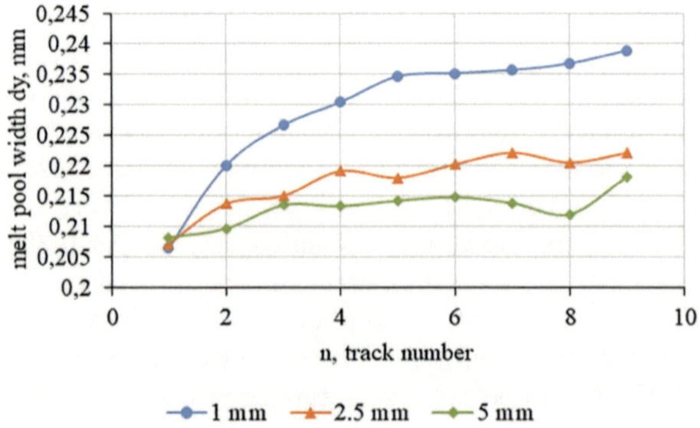

Figure 3.
Dependence of the melt pool width on the track number.

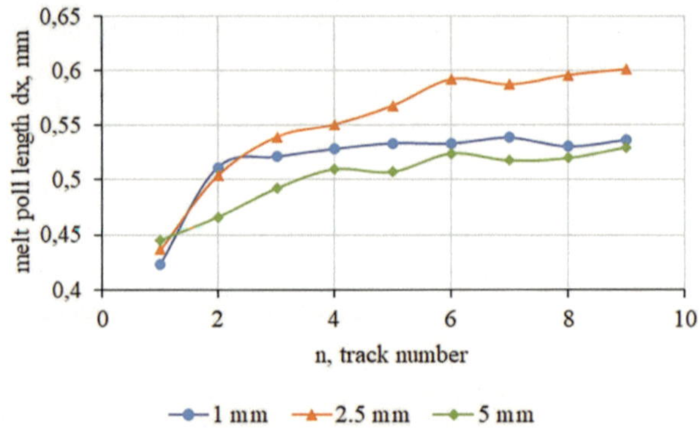

Figure 4.
Dependence of the melt pool length on the track number.

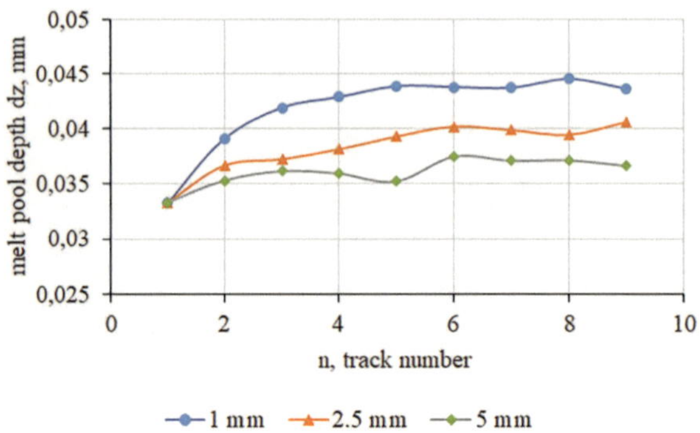

Figure 5.
Dependence of the melt pool depth on the track number.

parameters for the three modes are identical (the substrate temperature is equal to room temperature), as a result of which heating occurs at the same temperature. With an increase of the stripe width, the width and depth of the melt pool decrease. Since the rest of the process parameters remain unchanged, the time of one laser pass increases, with an increase of the stripe width, but the cooling rate does not change. Thus, the next pass of the laser starts after a longer period, and heating occurs at a lower temperature.

Melt pool profiles were described for 1, 5, and 9 laser passes at the midpoint of the track (**Figures 6–8**). The melt pool profiles differ insignificantly for the first laser pass. However, the melt pool rotates already at the fifth pass for a stripe width of 1 mm, the center of the melt pool shifts toward the beginning of the hatching (**Figures 6–8**). The side of the sample from which the hatching starts is heated to a higher temperature for all passes except for the first one than the other side (**Figure 9**). This leads to an increase in the melt pool width from the beginning of the hatching. Therefore, the melt pool becomes asymmetric about the straight line along laser movement, with rotation around its symmetry axis. The most significant rotation of the molt pool about the beam path is observed for a stripe width of 1 mm.

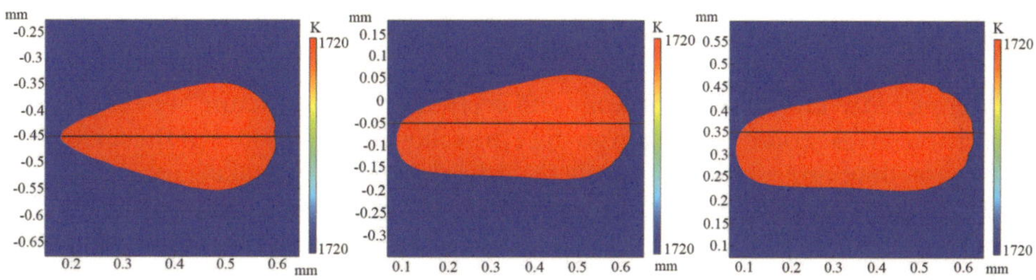

Figure 6.
Melt pool for 1, 5, and 9 track numbers (1 mm).

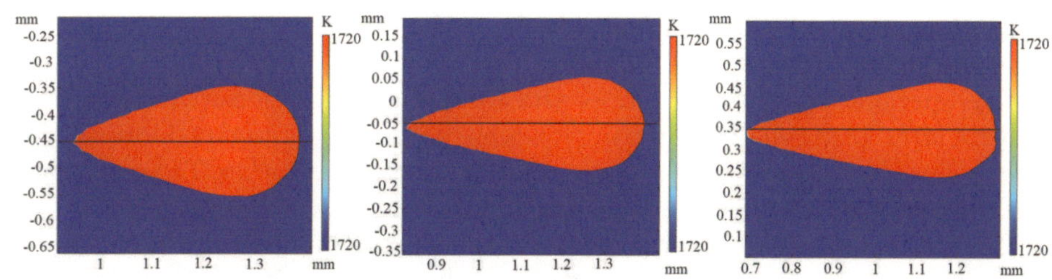

Figure 7.
Melt pool for 1, 5, and 9 track numbers (2.5 mm).

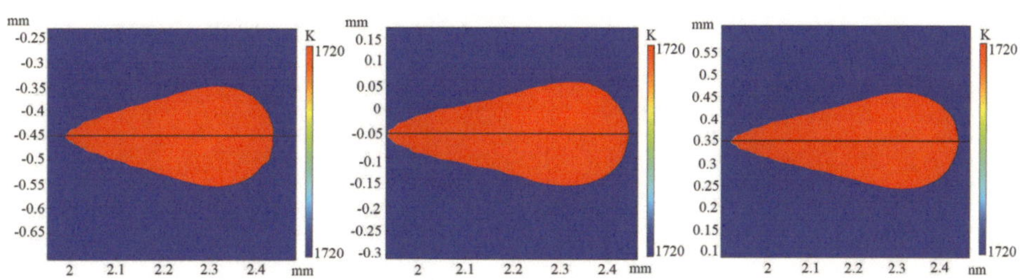

Figure 8.
Melt pool for 1, 5, and 9 track numbers (5 mm).

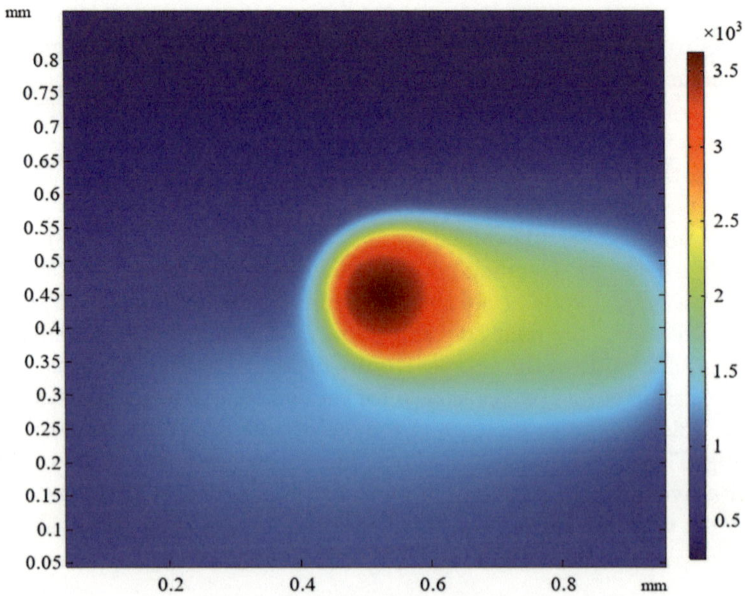

Figure 9.
Temperature profile for 10 laser track (1 mm stripe width, 12.35 ms).

During LPBF the grain morphology and crystallographic structure are determined by a characteristic epitaxial grain growth at the solid–liquid interface of the melt pool. Grain growth occurs mainly in specific material-dependent crystallographic directions, along the maximum temperature gradient [17, 23]. Grains grow epitaxially from the previously deposited layer to be partially re-melted. A decrease in the stripe width increases the molten pool depth and penetrates deeper into the previous layers. Such conditions may lead to an increased grain elongation degree, considerable epitaxial growth, and thus an increased morphological and crystallographic texture.

3.2 Investigation of LPBF samples

The experimental samples are cubes with a side of 10 mm, made of 316 L steel. To study the effect of the width of the paint block on the resulting structure, LPBF samples were produced in accordance with the previously selected modes (see **Table 2**), which were modeled. LPBF samples were manufactured on the EOSint M270.

Figures 10–12 show the microstructure of the samples in the vertical plane (in the direction of sample growth). The boundaries of the former melt pools and grains grown epitaxially through several layers of deposited metal are clearly visible.

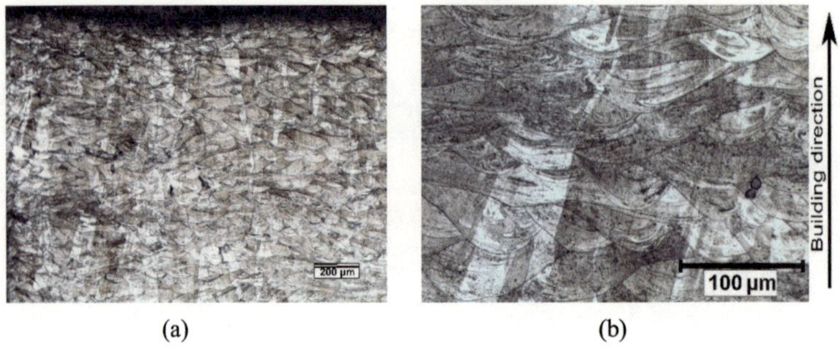

Figure 10.
Microstucture of sample 1: (a) is the general plan, (b) is the view of the melt pool and single grains (the direction of growth of the sample is shown by an arrow).

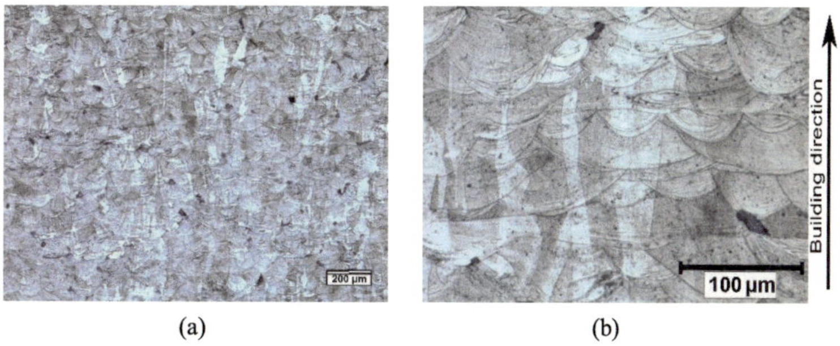

Figure 11.
Microstucture of sample 2: (a) is the general plan, (b) is the view of the melt pool and single grains (the direction of growth of the sample is shown by an arrow).

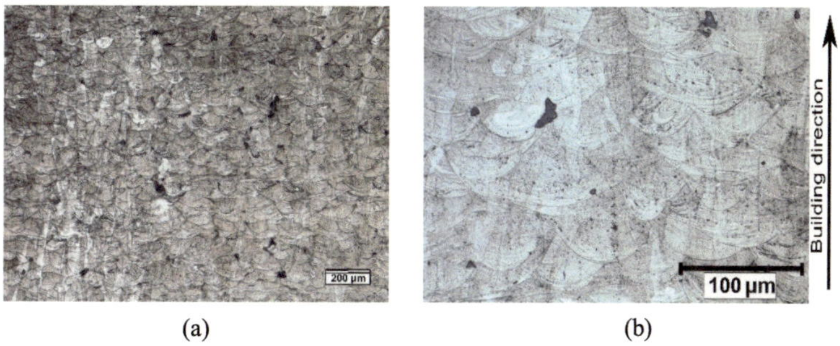

Figure 12.
Microstucture of sample 3: (a) is the general plan, (b) is the view of the melt pool and single grains (the direction of growth of the sample is shown by an arrow).

According to the study results obtained by optical microscopy methods, the structure of all the three samples at a first glance is identical with no obvious differences in the modes.

Based on the EBSD analysis of the vertical plane of samples 1–3, maps of crystallite orientations were obtained (see **Figure 13**).

It should be noted that the morphology of the crystallites of the first sample (**Figure 13a**) is characterized by epitaxial growth. The contours of the crystallites freely extend over the several printed layers of metal, the outlines of the boundaries of the melt pools are not traced on the EBSD map. The phenomenon of epitaxial growth of crystallites is also observed in the second sample, but at the same time, crystallites boundaries take the form of the boundaries of the printed layers.

The morphology of the crystallites of sample 3 indicates that epitaxial growth is strongly reduced and the structure is fragmented, the boundaries of the printed layers are easily traced, while sample 1 has a pronounced line structure.

For the purpose of quantitative analysis of the detected EBSD structure, pole figures of orientation densities were constructed (**Figure 14**).

Figure 14 shows the pole figures of the orientation densities of the crystallographic planes {100}, {110}, and {111}. The polar figures of samples 1 and 2 have a similar appearance, that is, a symmetrical intensity distribution (**Figure 14a** and **b**), whereas in sample 3 (**Figure 14c**) the pole figure is not symmetrical.

Based on the numerical modeling results (**Figure 6**), a feature of the geometry of the melt pool is established for the stripe width of 1 mm—a significant rotation of the melt pool with relation to the trajectory of the beam.

The crystal lattice of individual grains is characterized by preferential orientation. Since the pole figures show a set of orientation densities of all crystallographic planes in the examined area, the observed symmetrical image of these areas is

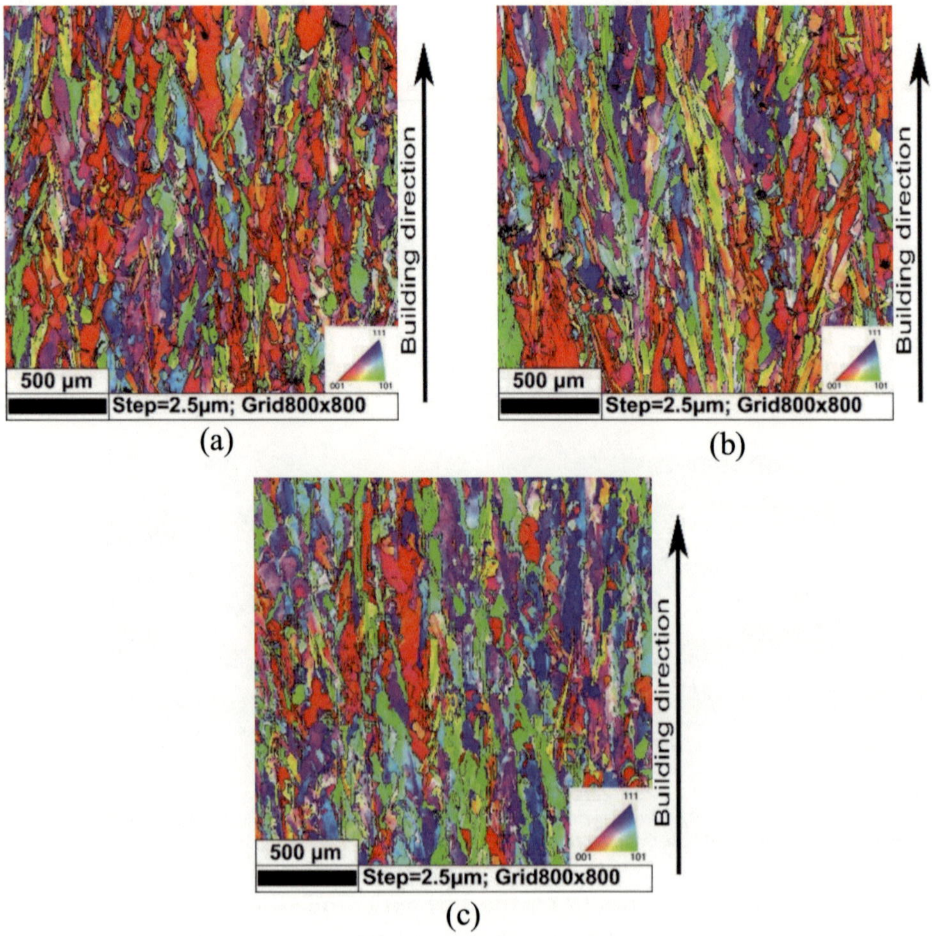

Figure 13.
EBSD sample cards no.: (a) – 1, (b) – 2, (c) – 3.

associated with the characteristic morphology of the grains revealed in **Figure 13a** and **b**. The structure of sample 3 is represented by smaller vertically oriented grains shown in the pole figure (**Figure 14c**) as a difference between the structure of samples 1 and 2.

The structure of samples 1 and 2 with different stripe widths of 1 mm and 2.5 mm, respectively, is represented by a set of symmetrical crystallites elongated in the growth direction of the sample and rotated at different angles around the normal to the plane. The structure of sample 3 does not have a pronounced symmetry of the crystallographic in the vertical plane. Thus, it was found that the stripe width affects the morphology of the structure of the generated LPBF sample.

4. The generalized energy input

In this section, to analyze the influence of LPBF parameters on thermal fields, microstructure and mechanical properties of AISI 321 austenitic steel, the LPBF process was simulated and a series of samples was created to combine various combinations of LPBF technological modes, such as scan speed, laser power, and scanning strategy. The possibility of controlling the structure formation of steel in the LPBF process in order to obtain a specific crystallographic texture, grain size, and morphology is evaluated. The relationship between the resulting anisotropic

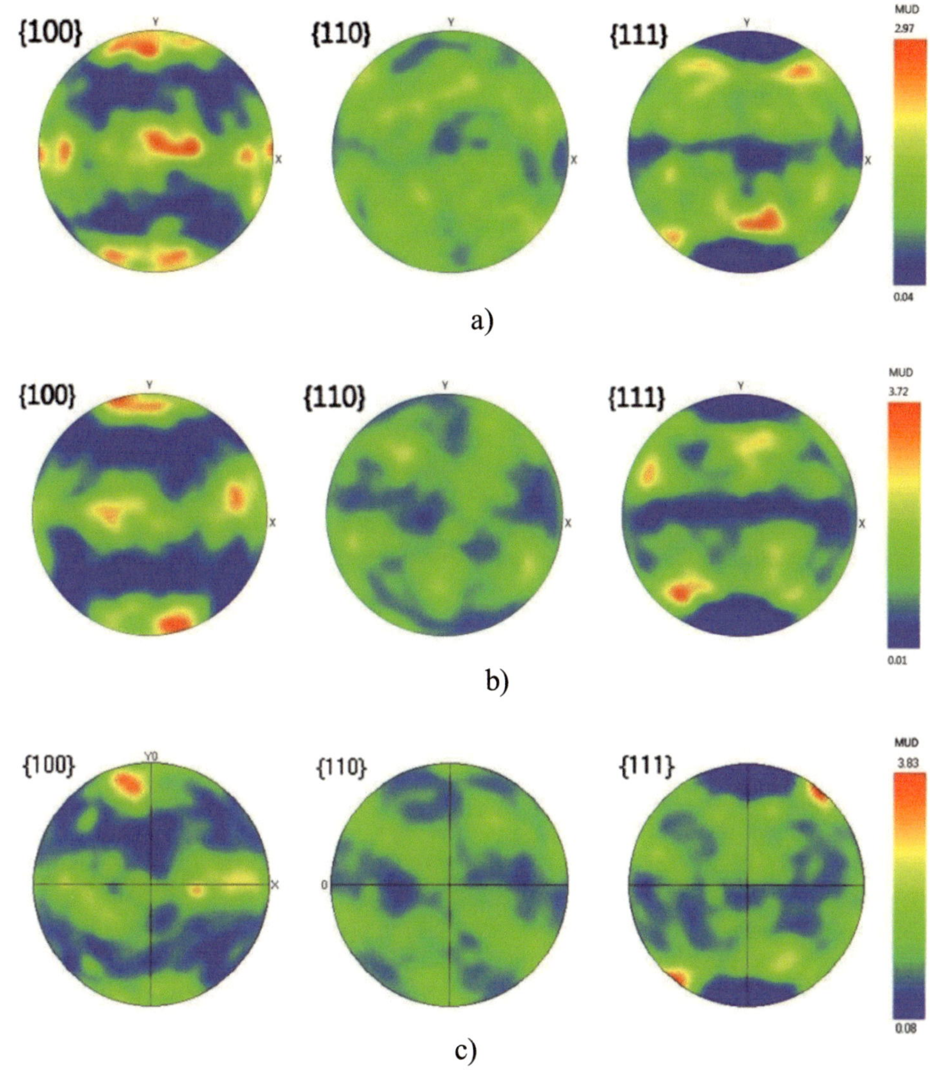

Figure 14.
Pole figures of samples no.: (a) – 1, (b) – 2, (c) – 3.

structure and mechanical properties is investigated. The aim is to understand the influence of the scan speed on the microstructure with the same energy input. The generalized energy input parameter is calculated as a ratio of laser power to the scan speed on the irradiated surface. In addition, the study is aimed at understanding whether there is additivity of power with the number of passes.

In Section 2, the effect of the stripe width on the microstructure of the samples was studied. The stripe width is selected 5 mm and remains constant for all modes studied in this section. To study the dependence of the melt pool size and microstructure on the energy input, nine modes of the LPBF process were selected. The laser speed varies in the range of 800–1013 mm/s, the laser power varies in the range of 75–190 W. The modes were chosen in such a way to study three models with an energy input of 0.188 W·s/mm, taken as 100%, three more ones with an energy input of 0.141 W·s/mm (50%), and the others with an energy input of 0.094 W·s/mm (75%), with each mode being different in power and scanning speed. In addition, various scan modes are used: the first three models go with one scan, the rest use double scan (**Table 3**).

Mode No.	Power, W	Speed, mm/s	Energy input, %	Energy input, W·s/mm	Scanning
1	150.0	800	100	0.188	Single scan
2	170.0	906	100	0.188	
3	190.0	1013	100	0.188	
4	75.0	800	50	0.094	Double scan
5	85.0	906	50	0.094	
6	95.0	1013	50	0.094	
7	112.5	800	75	0.141	
8	127.5	906	75	0.141	
9	142.5	1013	75	0.141	

Table 3.
LPBF Parameters.

The experimental samples are cylinders with a diameter of 4 mm and a height of 7 mm. The thickness of the powder layer was 40 microns.

4.1 Modeling

The melting pool size (length, width, and depth) was observed from the tracks temperature distribution results considered from the melting point to the peak temperature along the scanning direction.

The dependences of the length, width, and depth of the melt pool on the laser power are shown in **Figures 15–17**.

Figures 15–17 show that the increase in energy input extends the width, length, and depth of the melt pool. Comparing 1 and 4 scanning modes, where the input power was doubled, it was found that the melt pool width increased from 160 μm to 210 μm, the melt pool length increased from 190 μm to 530 μm, and the melt pool depth increased from 20 μm to 54 μm. For modes 4–6, the melt pool depth is less than the depth of the powder layer (40 μm). Thus, lack of fusion can be suggested with these process parameters. The melt pool width increased 1.3 times, while the length and depth increased 2.8 and 2.7 times, respectively. Therefore, it can be concluded that a change in energy input has a greater effect on changes in length and depth of the melt pool, while the width changes insignificantly.

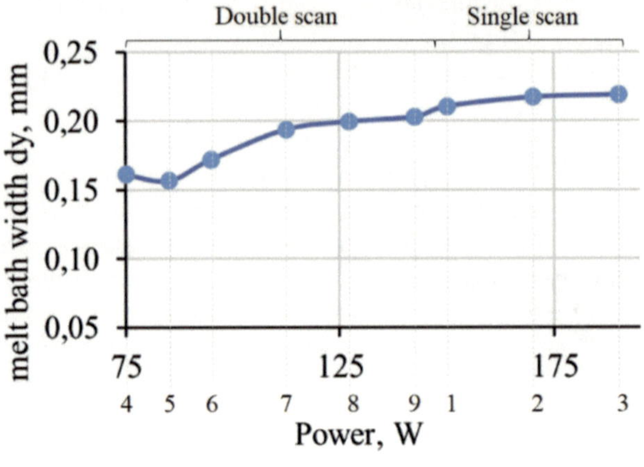

Figure 15.
Dependence of the melt pool width on the power.

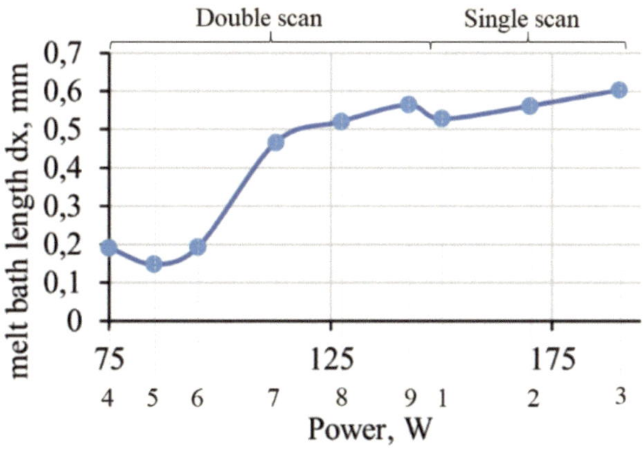

Figure 16.
Dependence of the melt pool length on the power.

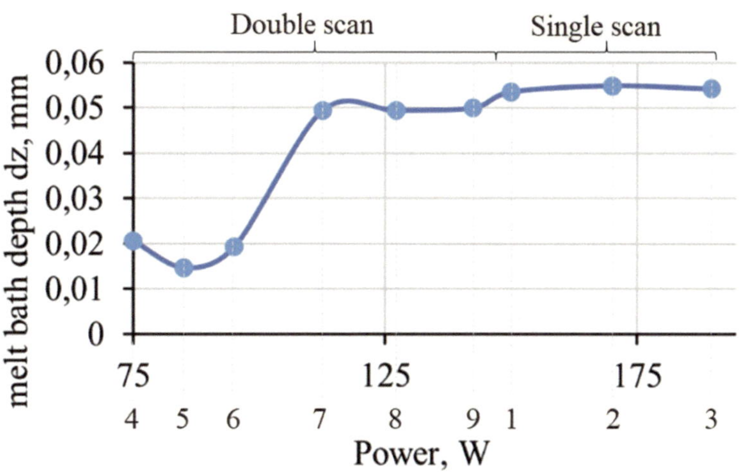

Figure 17.
Dependence of the melt pool depth on the power.

Comparing the modes with the same energy input (laser's speed and power may vary, but their ratio remains unchanged). For example, in modes 1, 2, and 3, the melt pool width and depth practically remain invariable, while the length grows with raising speed and power. To assume this case the laser scanning speed affects the change in the melt pool length largely. A high scanning speed leads to a longer tail of the melt pool in the x-y plane, thereby increasing its length. Modes 1–3 correspond to the greatest melt pool depth (**Figure 17**) with deeper penetration into the previous layers. These conditions might represent a more considerable epitaxial growth. Modes 7–9 correspond to the smallest melt pool depth (**Figure 17**), which might favor the grinding of the microstructure, since upon multiple scans, previous layers of the sample will not be melted.

4.2 Investigation of LPBF samples

4.2.1 Density

The results of density measurement depending on the LPBF modes are shown in **Figure 18**. It demonstrates that at the values of energy input of 100 and 75% with

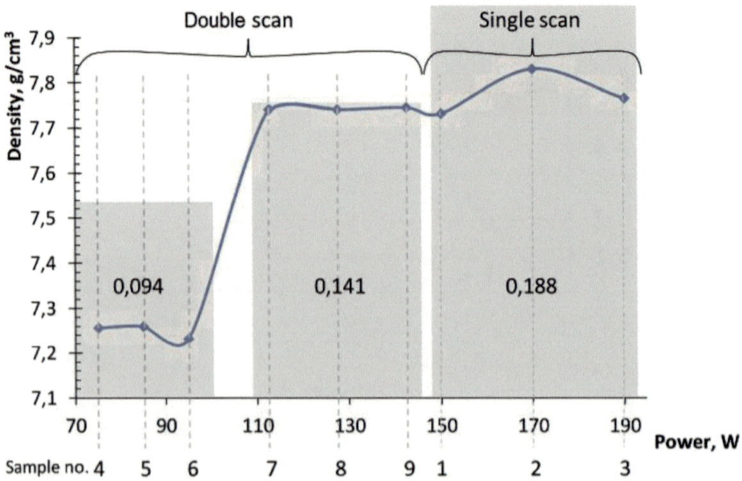

Figure 18.
Dependences of the sample density on the laser power.

single or double scanning, respectively, it is possible to achieve a density of 7.8 g/cm^3, which corresponds to porosity of 1% or less. At an energy input of 50%, the sample density decreases to 7.2 g/cm^3, which corresponds to a porosity of about 10%.

Thus, we can conclude that the energy input at the level of 100 and 75% for single or double scanning is sufficient to melt the powder layer and form a high-quality melt pool. This enables to achieve a high density.

4.2.2 Vickers hardness

The measured values of the samples hardness on a plane parallel to the direction of construction are shown in **Figure 19**. It can be seen that with a single scan at 100% of energy input, the hardness values are a bit less than with a double scan at 75% of energy input and are 230 and 250 HV, respectively.

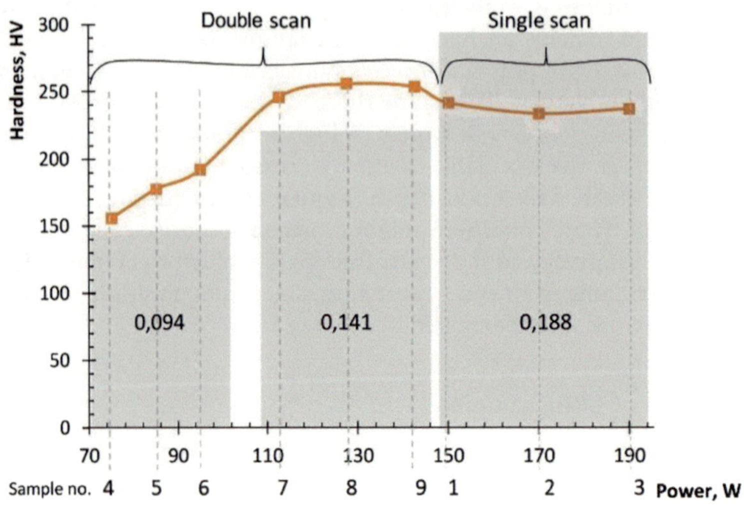

Figure 19.
Samples hardness dependences on laser power.

The measured hardness values enable to evaluate strength properties of the samples. Thus, for the samples obtained at 100% of the energy input, the time resistance is 800 MPa, whereas for the samples obtained at 75% of the energy input the time resistance is 750 MPa. This indicates a significant hardening of the samples, which is consistent with other authors' works [16], where such hardening factors as dislocation, dimensional, and dispersed particles are mentioned.

4.2.3 Microstructure

Figure 20 shows the microstructure obtained by optical metallography of all nine samples.

A large number of pores is observed on samples 4–6, which corresponds to reduced density values (**Figure 20d–f**).

The micrographs in **Figure 20** show the effect of various modes (**Table 3**) on the solidification structure. The merge lines are clearly visible. Grains consist of colonies of hardened cells with the same orientation. The grains grow parallel to the construction direction and pass through several layers of powder melted by a laser. According to the simulation results, the melt pool depth was obtained for samples No. 4–6 with power values of 75, 85, and 95 W (see **Figure 17**). As you can see, the estimated melt pool depth is not enough to melt a powder layer of 40 microns.

The structure of samples 1–3 in **Figure 21** is formed by grains elongated in the direction of sample growth extending through several layers that indicates their epitaxial growth. From the first to the third sample, a proportional increase in power and speed occurred. In general, this accelerates the thermokinetic process of structure formation, hence a change is observed from large columnar grains, similar to those

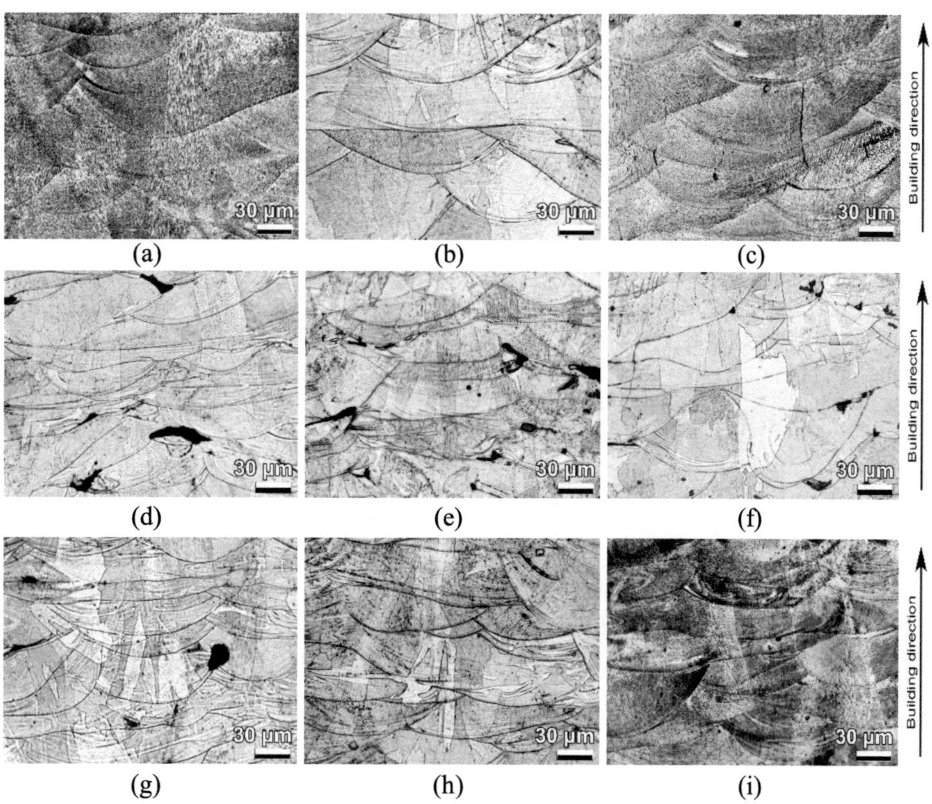

Figure 20.
Optical microscopy. Sample no.: (a) – 1, (b) – 2, (c) – 3, (d) – 4, (e) – 5, (f) – 6, (g) – 7, (h) – 8, (i) – 9.

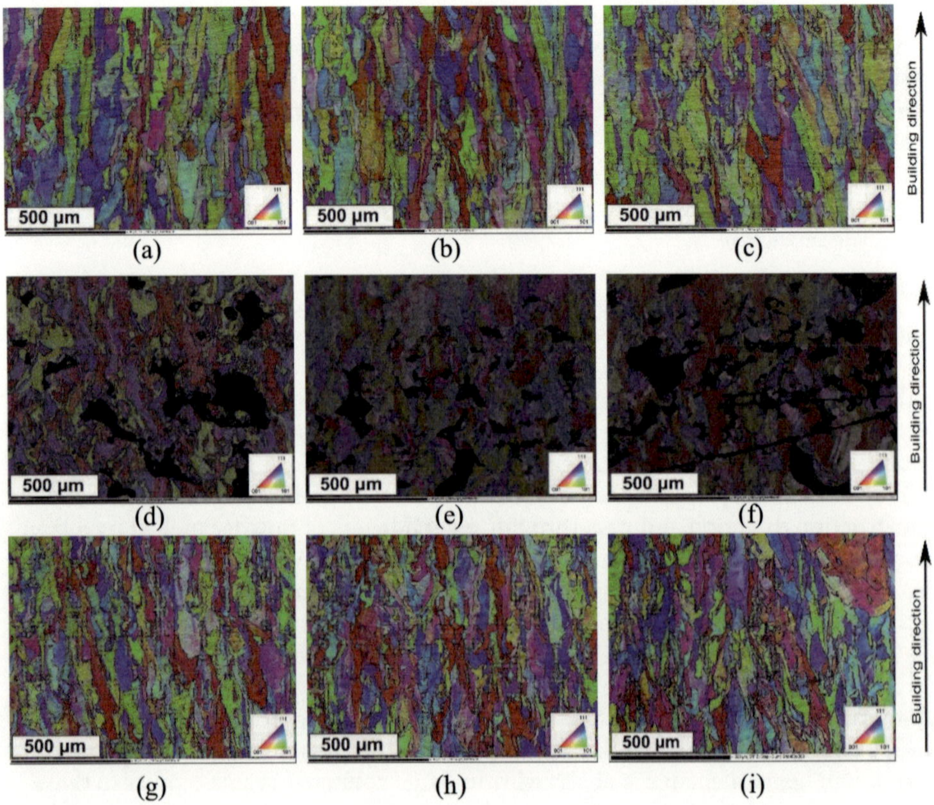

Figure 21.
EBSD maps for sample numbers: (a) – 1, (b) – 2, (c) – 3, (d) – 4, (e) – 5, (f) – 6, (g) – 7, (h) – 8, (i) – 9.

formed at casting with directional solidification, to grains with morphology that is similar to the outlines of melt pools. Lower rates of formation of the deposited metal contribute to the growth of crystallites, the appearance of grain boundaries changes with increasing speed, the shape of the grains acquires the appearance of crystallized melt pools; the grain structure is crushed since the crystallization conditions do not favor epitaxial growth. The grains are mainly oriented along the growth axis (building direction) and the length of these grains approaches 1 mm, which is much larger than the thickness of the layer used in the construction (40 microns).

Double scanning leads to the structure grinding. Samples 7–9 (see **Figure 21g–i**), in comparison with the structure of samples 1–3, have a structure with smaller grains of 100 µm in length order, with a drop-shaped morphology. In the future, double scanning can be applied for the structure grinding to obtain a favorable one with functional gradient properties of the part as a whole, or on a separate site.

In this section, the change in the parameters of the LPBF process, like scanning speed and the number of repeated scans was evaluated in terms of the effect on melt pool size, density, hardness, and microstructure. The use of experimental methods with computer modeling methods contributed to a better fundamental understanding of the correlation between process, microstructure, and properties.

5. Scanning strategy

The laser beam scanning strategy in the LPBF process has a direct impact on the formed sample structure. The forming metal structure strongly depends on the thermokinetic conditions of crystallization. In the previous sections, the effect of stripe width (Section 3) and energy input (Section 4) on temperature fields and microstructure was considered. In this section, with a constant stripe width equal to 5 mm, as well as a fixed speed of 800 mm/s and a power of 190 W, the influence of the scanning strategy on temperature fields and microstructure is studied. The scanning strategy of the powder layer with a laser beam in each section of the part being created can change the nature of the distribution of temperature fields and, as a consequence, the formed structure. It is shown that it is possible to eliminate the process of cracking in the Alnico alloy by applying an optimized scanning strategy developed on the basis of modeling. The magnetic hard alloy alnico was chosen as the material, which is characterized by a difficult-to-process in the LPBF.

5.1 Modeling

In this section, several modes of selective scanning are investigated—1) bidirectional; 2) circular from the center; 3) circular to the center. The model parameters for different scanning modes are shown in **Table 4**. Here L is the sample length, W is the sample width, H is the sample height, and R is the radius of the cylindrical sample.

Using the created model, it is possible to obtain the temperature distribution at any time during the entire considered LPBF process within the entire volume of the sample. In order to study the history of temperature distribution under different scan modes, temperature distributions for all models were analyzed at the end of the LPBF process (**Figure 22**). The melting zone is shown in burgundy. It was observed that the nature of the scan mode has little effect on the size of the melt pool, and the temperature gradient is relatively high near the melt pool.

The scan strategy has a significant impact on the temperature distribution. As can be seen from **Figure 22**, different scan strategies lead to different temperature

No	Power, W	Speed, mm/s	Line hatch, μm	Sample size, mm	Mode
1	190	800	95	L = 5; W = 2; H = 1	Bidirectional
2	190	800	75	R = 1; H = 0.8	Circular from the center
3	190	800	75	R = 1; H = 0.8	Circular to the center

Table 4.
Model parameters for different scan modes.

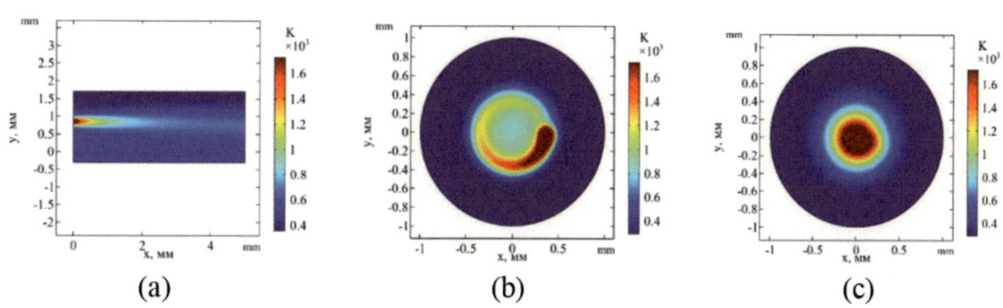

Figure 22.
Temperature distribution on the upper surface of the sample after the LPBF process for (a) bidirectional; (b) circular from the center; (c) circular to the center mode.

distributions in the sample. The circular scan mode from the center provides a temperature distribution with approximately diagonal symmetry (**Figure 22b**). The circular scan mode in the center provides a temperature distribution with circular symmetry (**Figure 22c**).

The stresses generated in the longitudinal direction of the laser beam are higher than in the transverse direction due to non-uniform compression during cooling [24–26]. Since the longitudinal direction of the laser does not change from pass to pass in bidirectional scan mode, the stresses in the longitudinal direction are higher than in the transverse one. This might lead to the formation of cracks in the sample. When the bidirectional scanning mode changes to a circular one, the residual stresses become more directional, which in turn can lead to a reduction in cracks in the manufactured sample.

Temperature dependences were plotted for 10 laser beam passes for 1–3 scanning modes. The cross-sectional distributions x = 2.5 mm (**Figure 22a**) and y = 0 mm (**Figure 22b** and **c**) were considered.

Figure 23 illustrates that in the bidirectional scan mode, heating occurs asymmetrically in cross-section. If we consider the first pass of the laser (indicated in blue in **Figure 23a**), it can be seen that the temperature on one side reaches about 2500 K, while the other side of the sample remains almost in equilibrium with a temperature of 300 K. During the circular scan mode from the center, heating occurs from the center along small radiuses, so the time of one laser pass is very short. This leads to the fact that the initial temperatures of the next pass will be greater. From the graph of the temperature distribution over the cross-section of the sample (**Figure 23b**), it can be observed that during the circular scan from the center, the temperatures are distributed more evenly than for the circular to the center and bidirectional modes.

Thus, the circular mode from the center is optimal, since the temperature distribution in this mode is more uniform. This corresponds to the residual stresses directions are pointed to the center, and, together with the cylindrical shape of the sample, contributes to the proper distribution of stresses and prevents cracking. These considerations will be verified in the next section.

5.2 Investigation of LPBF samples

In accordance with the scanning modes selected for modeling (**Figure 24**; **Table 4**), samples from the Alnico alloy were obtained. The samples were obtained using the equipment "Russian SLM Factory," which allows for more detailed configuration of scanning modes.

The standard technology of the bidirectional scanning strategy (mode 1) in LPBF with the microstructure is shown in **Figure 25**. A directional grid of cracks permeates the entire volume of the resulting sample, which changes its appearance from the center to the surface.

Since the morphology of multiple cracks has a clearly expressed anisotropy associated with the sample's geometry and the growth direction under LPBF, the crack formation process is apparently associated with the distribution of thermal stresses in the volume of the fused sample.

Based on the modeling data, when studying the structure of the obtained samples to rearrange the thermal stresses during the sample's construction, the samples were obtained by two types of ring scanning. The laser-scanned section of a cylindrical sample in each powder layer with a beam from the generatrix to the center and vice versa with 75 μm scanning step, 800 mm/s speed, and a power of 190 W (**Table 4**). This experiment allowed us to evaluate the effect of scanning strategy on

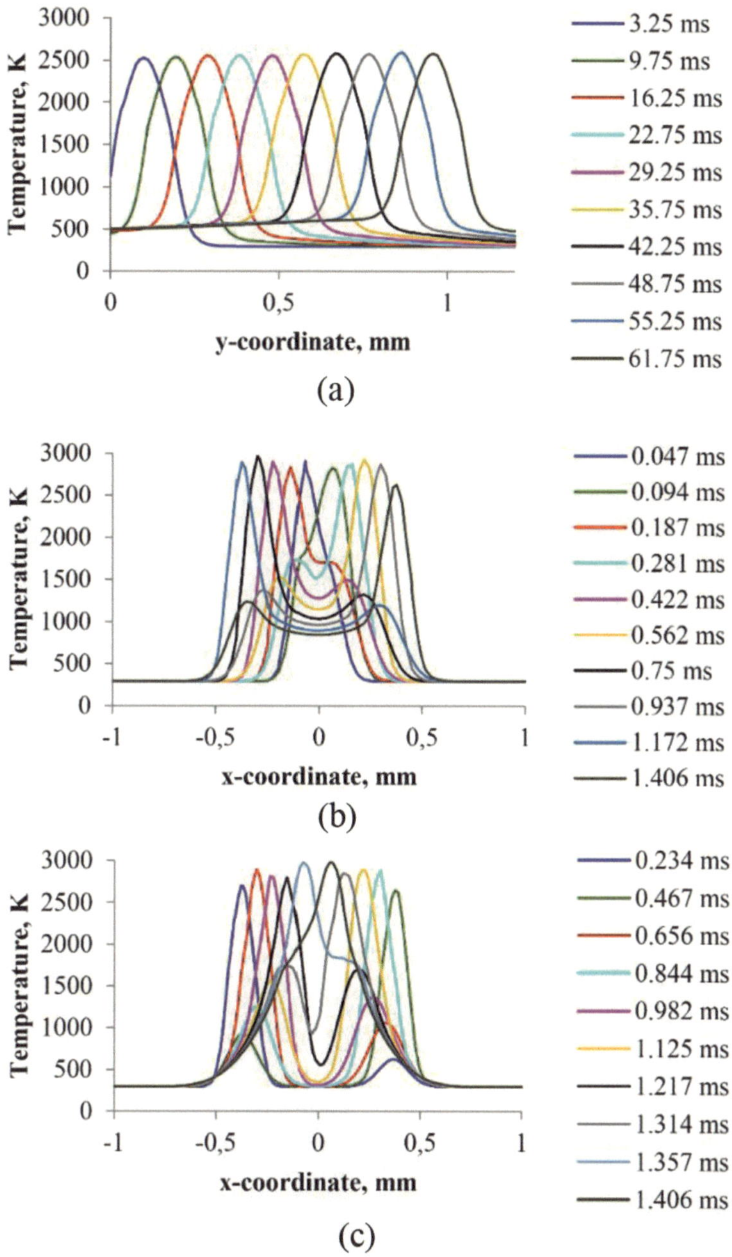

Figure 23.
Temperature distribution in cross-section: (a) bidirectional scan mode, $x = 2.5$ *mm; (b)* circular from the center mode, $y = 0$ *mm; (c)* circular to the center mode, $y = 0$ *mm.*

thermal stresses in LPBF samples and lead to cracks formation. The structure of the samples corresponding to mode 2 and mode 3 (**Figure 24c**; **Table 4**) is shown in **Figures 26** and **27**, respectively.

The structure of sample 3 has no cracks in its volume, with only a grid of cracks found in the radial direction into a depth of about 300 μm from the surface, and the presence of many pores distributed over the sample volume.

Based on numerical modeling, the distribution of temperature fields in cylindrical-shaped LPBF samples was estimated under various scanning modes of the sample—bidirectional, ring-centered. This experiment allowed us to evaluate

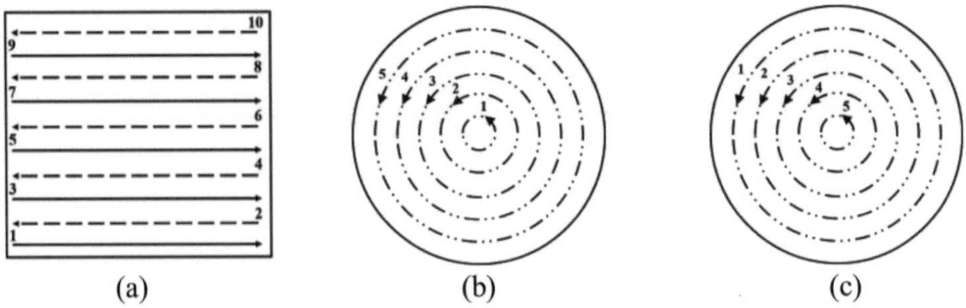

Figure 24.
Scanning modes: (a) bidirectional; (b) circular from the center; (c) circular to the center.

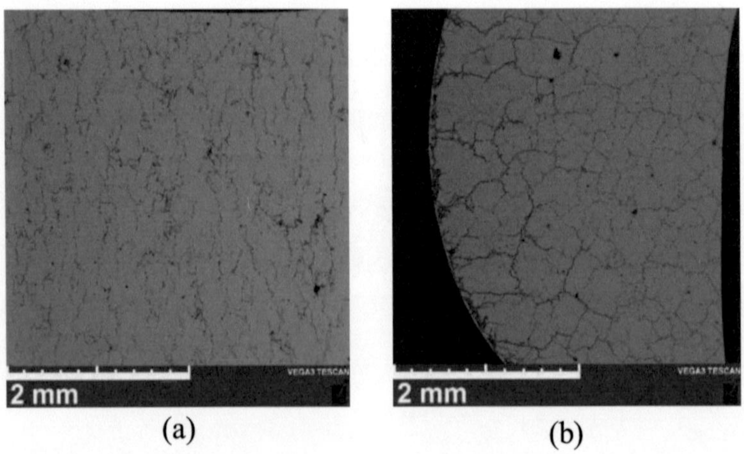

Figure 25.
Mode 1 (bidirectional). (a) – vertical direction, (b) – horizontal direction.

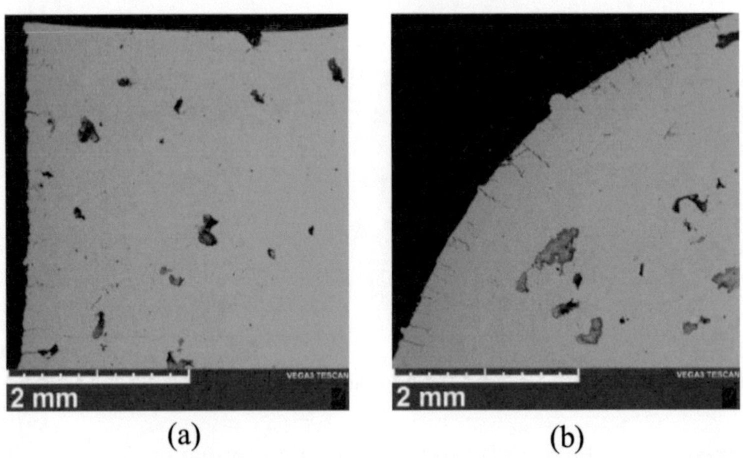

Figure 26.
Mode 2 (circular from the center). (a) – vertical direction, (b) – horizontal direction.

the influence of the scanning strategy under LPBF on the thermal stresses in the sample volume and lead to the formation of the crack.

Comparative analysis of temperature fields and microstructure of LPBF samples revealed that the scan strategy (bidirectional or circular) had an intense effect on the distribution of thermal fields and, consequently, on the microstructure of

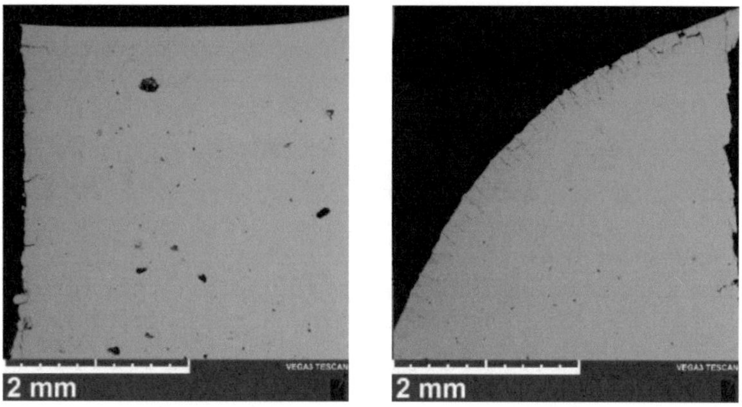

Figure 27.
Mode 3 (circular to the center). (a) – vertical direction, (b) – horizontal direction.

samples produced by LPBF. The structure of the specimens produced by the circular scan method had no cracks in the sample volume.

6. Conclusions

In this chapter by computer modeling and experimental investigations, we revealed that the type of scanning has an essential effect on microstructure. Energy input was constant 0.188 W·s/mm, but speed and power were changed in such a way: 800, 906, 1013 mm/s, 190, 170, 159 W, respectively. Additionally, energy input was reduced by 25 and 50% by decreasing laser power, but at the same time, there was double scanning to estimate the influence of repeated heat input.

1. Reducing the stripe width increases the melt pool depth and deepens the penetration of the melt pool into the previous layers. These conditions favor an increased grain elongation degree, pronounced epitaxial growth and thus, bring to an increased morphological and crystallographic texture. The samples structure with different stripe widths of 1 mm and 2.5 mm, respectively, is represented by a set of crystallites, elongated in the direction of sample growth, rotated at different angles around the normal plane to the studied one. The structure of the sample with 5 mm stripe width has isotropy in the vertical plane. Thus, it is established that the stripe width affects the morphology of the LPBF sample structure.

2. The influence of power and speed is estimated since the acceleration of the process from 800 to 1013 mm/s with a proportional rise in power leads to a change in the thermokinetic process of structure formation. This change is observed from large columnar grains, similar to those formed at casting with directional solidification, to grains with morphology repeating the outlines of the melt pools.

Double scanning does not have a summing effect, that influences the physicomechanical properties of the obtained specimens. Double scanning leads to shredding of the structure. In the future, double scanning can be used for structure grinding to obtain a favorable structure with functional gradient properties of the part as a whole, or on a separate site.

The low depth of penetration into the melt pool, high-temperature gradients, and high solidification rates stimulate the grinding of the microstructure and reduced intensity of texture due to a partial change in the solidification mode from columnar to equiaxed.

The density of 321 steel depends directly on the energy input as a ratio of laser power to scan speed at the level of 100 and 75% with single or double scanning, which is enough to melt the powder layer and form a high-quality melt pool.

3. Uneven temperature distribution under LPBF, a sharp temperature gradient in the melt pool and thermal influence zone lead to cracking. The fused metal and the heated part of the sample expand and cause stresses in the rest of the sample, which leads to characteristic cracks and their directed growth during layer-by-layer sample creation. The casting shrinkage of the molten Alnico alloy amount to 3%; this leads to internal stresses and cracking. Thus, a typical selection of modes for selective laser fusion of the Alnico alloy does not provide proper effect.

Ring scanning has a positive effect on the redirection of thermal stresses and cracking. Topological optimization of the sample's geometry scanning for thermal stress control enabled us to reduce cracking.

It was found that the LPBF fusion mode had a strong influence on the structure of the sample (especially the Alnico alloy); the scanning strategy is of great importance. Therefore, good sample's quality cannot be achieved merely by changing scanning power and speed. As stated above, the selection of LPBF modes for some materials refers to a scanning strategy for a specific 3D model of the sample.

Acknowledgements

This work was supported by the Russian Science Foundation under Grant No. 21-73-30019. Experimental studies were performed on the equipment in the Shared Use Center entitled "Composition, structure, and properties of structural and functional materials" of the NRC "Kurchatov Institute" – CRISM "Prometey" with the financial support of the state represented by the Ministry of Education and Science of the Russian Federation under the Agreement No. 13.CKP.21.0014 (075-11-2021-068). The unique identifier is RF——2296.61321X0014.

Author details

Pavel Kuznetsov*, Anna Mozhayko, Ivan Shakirov, Vitaliy Bobyr, Mikhail Staritsyn and Anton Zhukov
NRC "Kurchatov Institute" – CRISM "Prometey", St. Petersburg, Russia

*Address all correspondence to: prometey_35otdel@mail.ru

References

[1] Cao L, Li J, Hu J, Liu H, Wu Y, Zhou Q. Optimization of surface roughness and dimensional accuracy in LPBF additive manufacturing. Optics and Laser Technology. 2021;**142**:107246. DOI: 10.1016/j.optlastec.2021.107246

[2] Schmidt M, Merklein M, Bourell DL, Dimitrov D, Hausotte T, Wegener K, et al. Laser based additive manufacturing in industry and academia. CIRP Annals. 2017;**66**: 561-583. DOI: 10.1016/j.cirp.2017.05.011

[3] Liu J, To AC. Quantitative texture prediction of epitaxial columnar grains in additive manufacturing using selective laser melting. Additive Manufacturing. 2017;**16**:58-64. DOI: 10.1016/j.addma.2017.05.005

[4] Wang Y, Yu C, Xing L, Li K, Chen J, Liu W, et al. Processing technology, grain structure and texture of the SLM single track. Journal of Materials. 2020; **281**:116591. DOI: 10.1016/j. jmatprotec.2020.116591

[5] Kuznetsov P, Shakirov I, Mozhayko A, Zhukov A, Bobyr V. Comparison of sequential and circular scanning thermal fields and their influence on microstructure of Alnico alloy produced by laser powder bed fusion. Journal of Physics: Conference Series. 2021;**1967**:012064. DOI: 10.1088/ 1742-6596/1967/1/012064

[6] Li Y, Zhou K, Tor S, Chua C, Leong K. Heat transfer and phase transition in the selective laser melting process. International Journal of Heat and Mass Transfer. 2017;**108**:2408-2416. DOI: 10.1016/j.ijheatmasstransfer.2017.01.093

[7] Liu B, Fang G, Lei L. An analytical model for rapid predicting molten pool geometry of selective laser melting (SLM). Applied Mathematical Modelling. 2021;**92**:505-524. DOI: 10.1016/j.apm.2020.11.027

[8] Ilin A, Logvinov R, Kulikov A, Prihodovsky A, Xu H, Ploshikhin V, et al. Computer aided optimization of the thermal management during laser beam melting process. Physics Procedia. 2014;**56**:390-399. DOI: 10.1016/j.phpro. 2014.08.142

[9] Krakhmalev P, Yadroitsava I, Fredriksson G, Yadroitsev I. In situ heat treatment in selective laser melted martensitic AISI 420 stainless steels. Materials and Design. 2016;**87**: 380-385. DOI: 10.1016/j.matdes.2015. 08.045

[10] Bayat M, Mohanty S, Hattel JH. Thermo-Fluid-Metallurgical Modelling of Laser-Based Powder Bed Fusion Process [Internet]. 2018. Available from: https://www.comsol.ru/paper/download/570031/bayat_paper.pdf [Accessed: January 19, 2022]

[11] Ansari MJ, Nguyen D-S, Park HS. Investigation of SLM process in terms of temperature distribution and melting pool size: Modeling and experimental approaches. Materials. 2019;**12**:1272. DOI: 10.3390/ma12081272

[12] Zhang T, Li H, Liu S, Shen S, Xie H, Shi W, et al. Evolution of molten pool during selective laser melting of Ti-6Al-4V. Journal of Physics D: Applied Physics. 2018;**52**(5):055302. DOI: 10.1088/1361-6463/aaee04

[13] Li Y, Gu D. Parametric analysis of thermal behavior during selective laser melting additive manufacturing of aluminum alloy powder. Materials and Design. 2014;**63**:856-867. DOI: 10.1016/ j.matdes.2014.07.006

[14] Antony K, Arivazhagan N, Senthilkumaran K. Numerical and experimental investigations on laser melting of stainless steel 316L metal powders. Journal of Manufacturing

Processes. 2014;**26**:345-355. DOI: 10.1016/j.jmapro.2014.04.001

[15] Yadroitsev I, Krakhmalev P, Yadroitsava I. Hierarchical design principles of selective laser melting for high quality metallic objects. Additive Manufacturing. 2015;7:45-56. DOI: 10.1016/j.addma.2014.12.007

[16] Dong Z, Liu Y, Wen W, Ge J, Liang J. Effect of hatch spacing on melt pool and as-built quality during selective laser melting of stainless steel: Modeling and experimental approaches. Materials. 2019;**12**(1):50. DOI: 10.3390/ma12010050

[17] Köhnena P, Létanga M, Voshageb M, Schleifenbaumb JH, Haasea C. Understanding the process-microstructure correlations for tailoring the mechanical properties of L-PBF produced austenitic advanced high strength steel. Additive Manufacturing 2019;**30**:100914. DOI: 10.1016/j.addma.2019.100914

[18] Sun Z, Tan X, Tor SB, Chua CK. Simultaneously enhanced strength and ductility for 3D-printed stainless steel 316L by selective laser melting. NPG Asia Materials. 2018;**10**:127-136. DOI: 10.1038/s41427-018-0018-5

[19] Guan J, Jiang Y, Zhang X, Chong X. Microstructural evolution and EBSD analysis of AlSi10Mg alloy fabricated by selective laser remelting. Materials Characterization. 2020;**161**:110079. DOI: 10.1016/j.matchar.2019.110079

[20] Liu C, Tong J, Jiang M, Chen Z, Xu G, Liao H, et al. Effect of scanning strategy on microstructure and mechanical properties of selective laser melted reduced activation ferritic/martensitic steel. Materials Science and Engineering. 2019;**766**:138364. DOI: 10.1016/j.msea.2019.138364

[21] Bruyere V, Touvrey C, Namy P. A Phase Field Approach to Model Laser Power Control in Spot Laser Welding [Internet]. 2014. Available from: https://www.comsol.com/paper/download/199279/touvrey_paper.pdf [Accessed: January 19, 2022]

[22] Bonacina C, Comini G, Fassano A, Primicerio M. Numerical solution of phase change problems. International Journal of Heat and Mass Transfer. 1973; **16**(10):1825-1832. DOI: 10.1016/0017-9310(73)90202-0

[23] Thijs L, Verhaeghe F, Craeghs T. A study of the microstructural evolution during selective laser melting of Ti–6Al–4V. Acta Materialia. 2010;**58**:3303-3312. DOI: 10.1016/j.actamat.2010.02.004

[24] Hussein A, Hao L, Yan C, Everson R. Finite element simulation of the temperature and stress fields in single layers built without-support in selective laser melting. Materials and Design. 2013;**52**:638-647. DOI: 10.1016/j.matdes.2013.05.070

[25] Li C, Liu JF, Fang XY, Guo YB. Efficient predictive model of part distortion and residual stress in selective laser melting. Additive Manufacturing. 2017;**17**:157-168. DOI: 10.1016/j.addma.2017.08.014

[26] Liu Y, Yang Y, Wang D. A study on the residual stress during selective laser melting (SLM) of metallic powder. International Journal of Advanced Manufacturing Technology. 2016;**87**:647-656. DOI: 10.1007/s00170-016-8466-y

Chapter 10

Plasma Metal Deposition for Metallic Materials

Enrique Ariza Galván, Isabel Montealegre Meléndez, Cristina Arévalo Mora, Eva María Pérez Soriano, Erich Neubauer and Michael Kitzmantel

Abstract

Plasma metal deposition (PMD®) is a promising and economical direct energy deposition technique for metal additive manufacturing based on plasma as an energy source. This process allows the use of powder, wire, or both combined as feedstock material to create near-net-shape large size components (i.e., >1 m) with high-deposition rates (i.e., 10 kg/h). Among the already PMD® processed materials stand out high-temperature resistance nickel-based alloys, diverse steels and stainless steels commonly used in the industry, titanium alloys for the aerospace field, and lightweight alloys. Furthermore, the use of powder as feedstock also allows to produce metal matrix composites reinforced with a wide range of materials. This chapter presents the characteristics of the PMD® technology, the welding parameters affecting additive manufacturing, examples of different fabricated materials, as well as the challenges and developments of the rising PMD® technology.

Keywords: additive manufacturing, plasma metal deposition

1. Introduction

Additive layer manufacturing (ALM) is an emerging manufacturing process that produces parts close to their final shape or ready to use. It is getting a rising interest in the present and future production technologies due to the open range of opportunities. ALM uses an energy source to melt or sinter material, depositing it layer by layer.

There is a wide variety of classifications of the different variants of the ALM process. One of the most popular is considering the energy source to perform the specimens. Hence, depending on the energy source, the ALM technologies can be divided into plasma-based (Gas Tungsten Arc Welding, GTAW; Gas Metal Arc Welding, GMAW; Plasma Arc Welding, PAW; Wire Arc Additive Manufacturing, WAAM), laser-based (Selective Laser Sintering, SLS), or electron beam-based systems (Electron Beam Welding, EBW; Wire Electron Beam Welding, WEBW) [1, 2].

The main advantage of the plasma-based techniques over the laser-based or the electron beam-based ones is the cost of the equipment. Moreover, in this last group, there is a necessity to work in a vacuum chamber, where the manufacturing atmosphere has to be clean of fumes and particles; otherwise, these could divert the electron beam. A disadvantage of the plasma-based techniques is the low resolution of the final parts produced; with the other techniques, it is possible to control the

IntechOpen

energy supplied and to have a smaller welding pool that allows a better geometric resolution.

Figure 1a presents an example of the devices employed to develop plasma metal deposition (PMD) parts. Moreover, **Figure 1b** shows a typical specimen manufactured *via* ALM, which is employed by the aerospace industry. There is a schematic representation of the arrangement of the layers in **Figure 1c**.

Moreover, building up parts with different materials is possible due to the development in powder metallurgy, whose advances have made it possible to produce powders of any material. Additive manufacturing (AM) is not only fed by powder, but also with wire shape material, which has the advantage of introducing the material directly into the heat source, although it has the disadvantage of being limited by commercially available materials. In this regard, advances in the field of the raw materials for AM might contribute to developing new material combinations with outstanding properties and complex geometries.

In line, another named classification can be considered according to disposing of the raw material. The material can be injected or introduced directly into the energy source, blown-powder/wire-feed techniques (Direct Metal Deposition, DMD) [3–5], or disposed on a bed or platform in powder shape (Laser Powder Bed Fusion, LPBF) [6–8]. Such powder-bed based ALM equipment is typically limited in manufacturing large parts due to the size of their building platforms. However, DMD processes are suitable for large-scale parts. Both being close to the final geometry and working with any type of material, AM processes are a point of interest due to the significant reduction of energy, saving time in post-production processes, and saving raw material in the manufacturing of a product.

One big advantage is the possibility of building pieces close to the final geometry even with complex shapes. Thanks to the advances in computer science, additive manufacturing is helped by computer-aided design (CAD) programs to produce any part, even with a difficult geometry [9].

The technology described in this chapter, known as plasma metal deposition, belongs to the plasma-based processes, similar to PAW but with the additional advantage of using not only a wire as feedstock, but also powder, or a combination of both. The equipment has been developed at RHP-Technology GmbH (Seibersdorf, Austria), and this system can produce parts with various metallic materials, as well as metal matrix composites (MMCs), or gradient material structures.

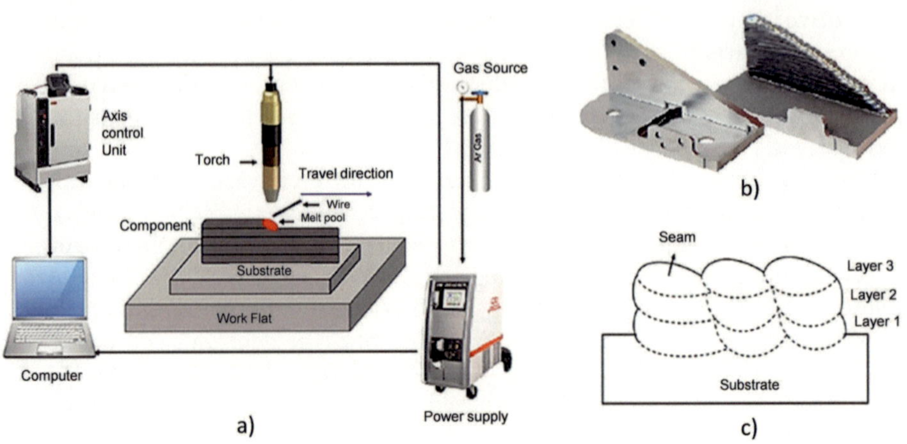

Figure 1.
a) Schematic of the PMD system, b) photograph of the sample as build vs. sample after post-processing, c) schematic of a cross section of the sample.

2. Plasma metal deposition

The PMD process is based on a torch where the plasma is produced for melting or sintering materials that are injected in powder shape or introduced as wire directly in the plasma focus (**Figure 2**) [10–12]. The gas used for producing the plasma is argon (pilot gas). It is introduced between the electrode and the copper torch. Due to a difference of potential of 20 V, the gas gets ionized, and a high-temperature plasma plume is created.

This plasma plume creates electrical contact between the electrode and the workpiece to ignite the main arc. Around the main plasma arc, an inert gas stream (shielding gas) is injected to protect locally the welding area from oxidizing during the manufacturing process or from other external agents entering the welding. The plasma torch is actively cooled with water through internal ducts, which prevents the torch and its components from being damaged due to high temperatures.

The PMD process can supply a DC mode, where the electrons flow from the electrode to the substrate, and AC mode, which alternates the flow direction of electrons. The AC mode is needed for manufacturing materials with high-oxidation surface alloys such as aluminum or magnesium. When the electrons go from the workpiece to the electrode, they are used for breaking the oxide layer and later on the flow changes to melt the material below.

The powder feedstock is injected with the help of a metric wheel feeder and pressurized argon (powder gas) through pipes directly into the plasma focus, where a welding pool of molten material is produced.

In the case of using wire feedstock, two wires are driven into the plasma focus with feeder wheels. Before introducing them into the plasma focus, the wires pass through a wire heater system that can be activated, to preheat them in advance and use less energy from the plasma to melt them. The heater works by inducing a current through both wires warming them. As it can be seen in **Figure 3**, the wire feeder is set up on a specific holder where the XYZ position of the wires can be adjusted, as well as the relative angle of the wires to the weld pool.

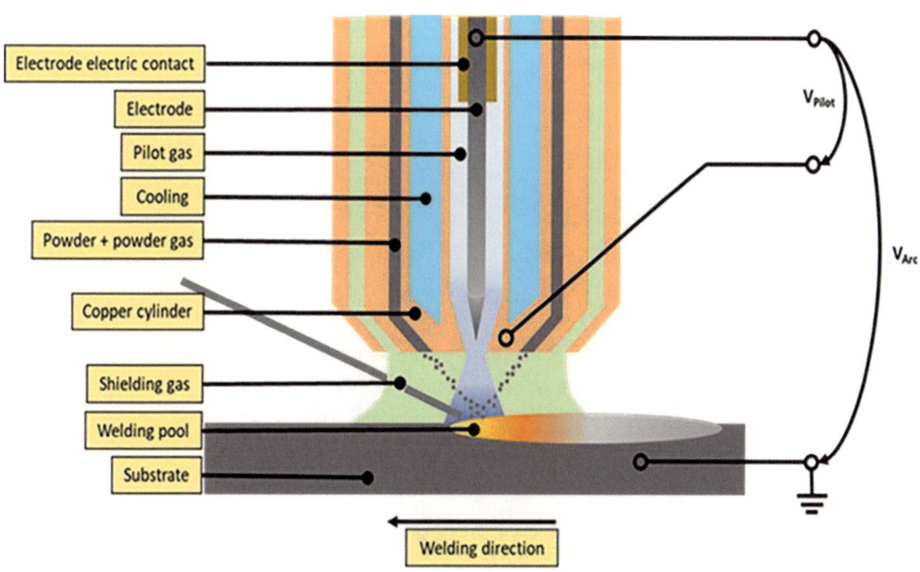

Figure 2.
PMD torch scheme.

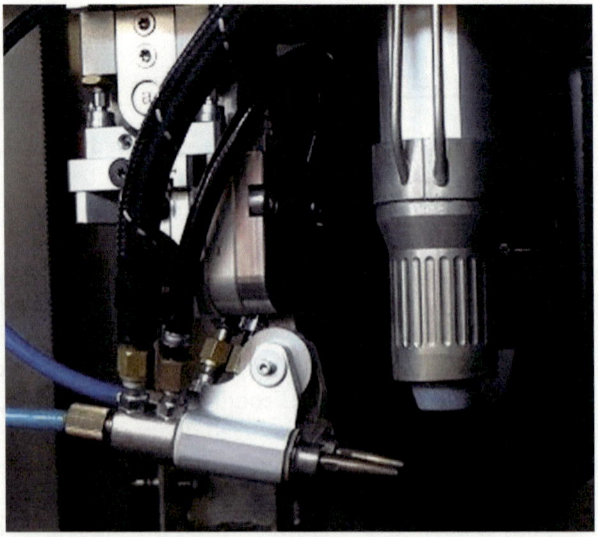

Figure 3.
Image of the hot-wire wire feeder next to the PMD torch.

The plasma torch is fixed to a gantry system that allows its movement in three dimensions over a working table (X-Y-Z axes) (**Figure 4**). This working table, at the same time, is chilled to avoid damaging the mechanical components of the equipment. The size of the gantry system and the working table will determine the maximum size of the produced part, being possible the manufacturing of large-size components. Thus, this technique offers more advantages regarding greater flexibility of specimen size, in comparison with the LPBF technique.

Furthermore, a turning/tilting table is available for producing revolution components or parts with high complexity where a 5-axis system is needed: for example, on high-overhang angle parts, cantilevers, or the edges of flange-kind components.

Figure 4.
Picture of PMD machine.

Moreover, the torch and wire feeder system are installed on another turning axis that eases the direction formed between the plasma focus and the wires following the toolpath. Also, by turning 180°C the torch leads to two different welding configurations, which can be used by adding the wires in front or after the weld pool.

In particular, and in accordance with the previously commented, the PMD device in RHP-Technology GmbH facilities has a working volume of 2000 mm × 1000 mm × 600 mm. In this respect, that allows the manufacturing of large-size components. On the contrary, in powder-bed systems where the part size depends on the platform in which the powder is disposed of and on the load that it can hold, the size of the specimens is limited to the axis geometry.

The plasma torch could be installed on any mechanical axis. The installation of a PMD torch on a robotic arm is the most effective way for repairing parts with complex geometry. Advantages include the possibility of repairing large pieces with localized defects or even improving the properties of the surface by cladding a material with greater hardness.

The welding and axis systems are located inside a welding chamber that enables the manufacturing of components in a protective gas atmosphere during the whole welding process, not only locally in the surroundings of the welding pool, but also giving general protection to the part during the cooling. This fact is of high importance when the used raw material has high oxidation sensitivity, for example, titanium alloys.

The chamber is filled with inert gas, usually argon. Through pipes with multiple diffusers on the floor of the equipment, it is sought to have a laminar flow of inert gas in order to achieve the evacuation of oxygen from the bottom to the top. Its effect is very important to avoid undesired secondary reactions during manufacturing. Sensors placed along the height of the machine measure the oxygen content at any moment, while the filling of the chamber allows having a precise amount of oxygen inside. In this regard, there is a guaranty of the quality protection of the manufactured specimens.

In the case of the welding gases used, generally, the main plasma is created by ionizing argon gas as explained above, but to increase the energy input in aluminum alloys, mixtures of argon-helium can be employed. The shielding gas that locally protects the welding pool can influence the mechanical properties and welding conditions of some materials; for example, for some steel alloys, the use of argon-carbon dioxide mixture is recommendable [13–16].

Figure 5.
Image of the PMD weld pool by monitoring welding cameras.

One of the most important aspects to control the welding process is the weld pool. The geometry, viscosity of the deposited melted material, and the time the deposited material is in the liquid phase give an idea of the penetration and dilution with the substrate or layers below, and the surface quality of the weld seam. Monitoring the weld pool during the complete additive manufacturing process is of interest; therefore, two welding cameras are installed on the PMD equipment next to the plasma torch to control in real-time and record the behavior of the welding process at any moment (**Figure 5**).

During the deposition, the layer height can be not homogeneous due to the geometry of the workpiece itself. In order to make corrections to the workpiece during the welding process, the equipment consists of a system called "automatic height control". The PMD system keeps the welding current constant during the process to always maintain the same value, the voltage between the electrode and the part varies. This variation is directly linked with the distance between torch and component. Using this voltage variation and coupling it with the material feeding system, more material can be added instantaneously on the areas where the workpiece layers are lower and the other way around when the layers are higher than initially designed.

Every welding parameter is logged and, together with the welding process videos, allows the identification of possible building mistakes or pores if found on a post-processing analysis.

3. Material feedstock

In PMD, different kinds of feedstock can be used as powder, wire, or both at the same time. In the case of powder feedstock, since the powder is directly injected into the welding pool, the process is not as limited as selective laser melting (SLM) or other powder-bed fusion techniques. In these processes, the powder size range has a direct impact on the final quality of the produced specimens.

There are specific powders that can be employed in PMD, as spherical gas atomized powders, irregularly rounded powders, and sharp edge powders typical of ceramics feedstock. On this subject, there is a great range of raw powders that could be combined promoting the development of a wide variety of specimens. Moreover, powders produced experimentally as water atomized are also suitable, and even the shape and range size are in this case irregular (**Figure 6**).

Introducing wire feedstock in the welding process increases the deposition rate significantly versus the use of powder. Furthermore, powders are linked to an overspray that does not occur when a wire is used. Different ranges of wire diameters can

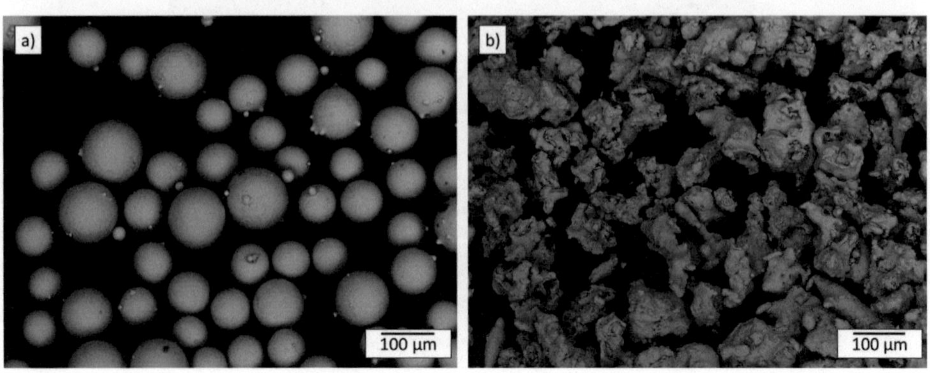

Figure 6.
Image of two different powders used in PMD: a) spherical plasma atomized powder, commercially available and b) irregular water atomized powder, experimentally manufactured [17].

be used from 0.6 to 3 mm. There are pure alloyed wires, filled wires, and rods that are suitable for the PMD process. In particular, some wire manufactures are introducing on their catalog wires specifically for additive manufacturing where the surface roughness of the wire is controlled and cleaned, with a minimized impurity content.

As commented previously, the use of wire feedstock facilitates a high deposition rate (~10 kg/h). Although not every alloy is available in a wire shape, powder metallurgy advances have made it possible to have commercially available powders of any alloy. Moreover, if an alloy is not available in powder shape, PMD enables the manufacturing of components by *in situ* alloying from elemental powders.

4. Plasma metal deposition limitations

In this subsection, some limitations of the PMD technique are reported. In order to be objective and concise, all of these limitations are briefly described comparing it with other AM techniques commented above.

4.1 Surface roughness

In contrast to Laser Powder Bed Fusion, where the component surface has a high resolution and exalted detail part can be manufactured, PMD components have a higher roughness due to the layer height (~1–3 mm), and the high deposition rate used could derive to metal droplets on the surface. These effects could affect the final behavior of the specimens fabricated.

PMD produces near-to-net-shape components that need to be post-machined at least on the functional areas. The post-machined process is also compromised with the droplets on the surface, which might contain a high-impurity content and be able to damage the cutting tools. However, despite this fact, the process is still economically beneficial because of the high raw material saved, and the manufacturing time reduction.

4.2 Shadowing effect

If wire feedstock is used, the position and the angle of how the wire is introduced into the plasma focus have a high impact on the final component geometry. This problem does not occur when powder blown is used where the powder is directly injected through the plasma torch and deposited on the weld pool from above (**Figure 7**).

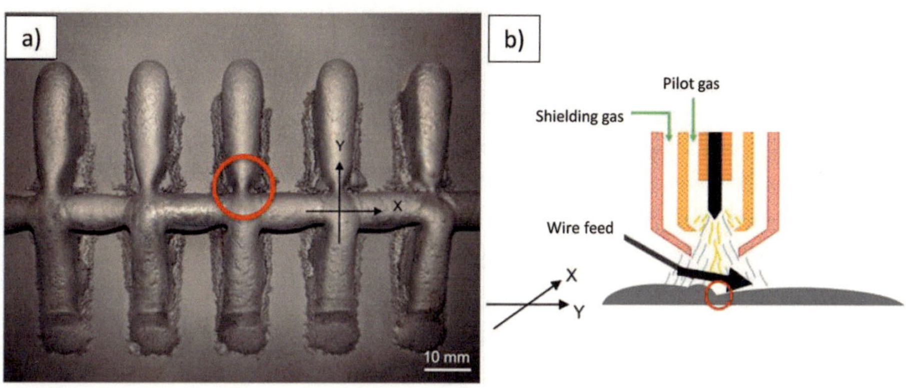

Figure 7.
Shadowing effect: a) caption of a shadowing effect mistake on a crossing points deposition and b) scheme of the shadowing effect.

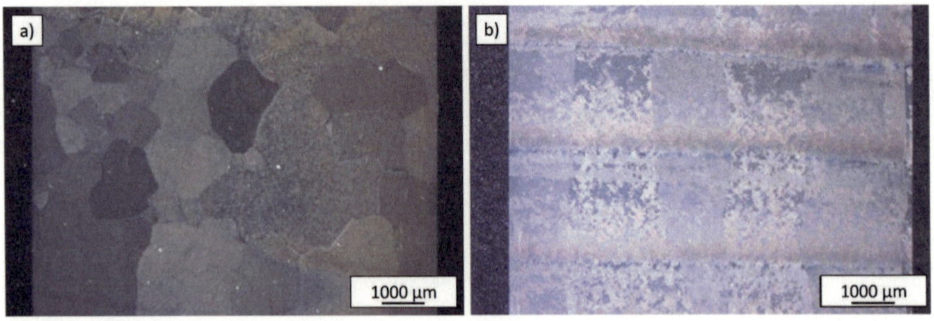

Figure 8.
Light microscope images from Ti6Al4V PMD manufactured sample, grain size detail: a) grain cross section, b) grain's length section.

4.3 Thermal cycle and grain size

The high energy input and thermal cycle from the layer-by-layer building up process provoke material mechanical stress that can induce cracks and deformation on the component. The substrate where the material is deposited is also suffering deformation and needs to be clamped. The stress on the substrate can produce that the deposited material gets detached.

Furthermore, the thermal cycles induce columnar big grains in the metallic matrix growing in the direction to the coolest area of the component that usually coincides with the vertical direction (**Figure 8**).

Incompatibility between the cladding material and the piece may appear as another disadvantage. Due to the high energy input of the PMD process, the geometry of the part can suffer distortions.

5. Results

In this section, several images of the produced specimens *via* PMD are presented classified according to the raw material employed, the kind of materials, also the processing parameters, as well as some of their properties.

5.1 Steels

The number of applications for steel is as numerous as the steel types available. Stainless steel, carbon steel, high-hardness steels, high-impact steels, etc., are feasible to be manufactured with PMD, as seen in **Figure 9** [18–20].

Also, reparations of broken tools or cladding with materials that give an added value to the existing component are possible with PMD.

In addition, when repairing some steels, it is necessary to preheat the piece before adding material; this can be previously done with the same plasma torch locally in the area that needs to be repaired, without adding material.

5.2 Nickel alloys

Nickel alloys are of high interest in the industry [20–22]. Their high corrosion resistance, high-temperature resistance, or as in the case of the alloy Invar 36 that suffers no deformation on a temperature range from −100 to 200°C, which make these alloys suitable for numerous high-performance applications.

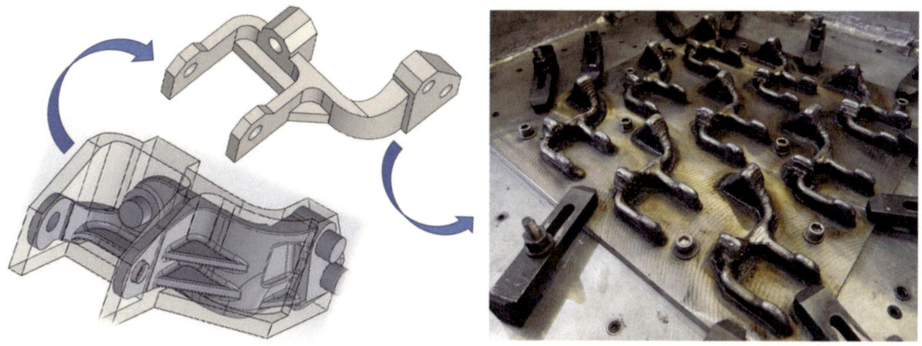

Figure 9.
Steel 17-4PH gearbox bracket. On the left, design adaption for PMD manufacturing from the casting model. On the right, gearbox bracket on a PMD production.

Figure 10.
Nickel alloy Invar 36 CFC tooling PMD manufactured. From left to right, PMD manufacturing, tooling as-deposited, and tooling as-milled.

Moreover, the nickel-based alloys have a high cost related to the raw material, and this cost is drastically reduced by using an additive manufacturing technique as PMD where near-net shape components are built, and the buy-to-fly ratio is significantly shortened (**Figure 10**).

5.3 Titanium alloys

The low density and high specific stiffness in titanium alloys as Ti6Al4V make the technique of interest for aeronautical and space applications [12, 23] **Figure 11** illustrates one example of fabrication with this material.

To avoid oxidation on titanium alloys, these materials need to be PMD manufactured under a protective atmosphere.

5.4 Aluminum alloys

Aluminum alloys have the characteristic that their oxides have a much higher melting temperature than the not oxidized material. When using a fixed current value to deposit aluminum, the oxide layer will not melt; therefore, the material will not deposit if the current is low. On the other hand, if the current applied is high, the material will overheat, and the weld pool cannot be controlled.

AC current is used to weld aluminum because its positive half cycle provides a "cleaning" action by breaking the oxide layer, and its negative half cycle provides

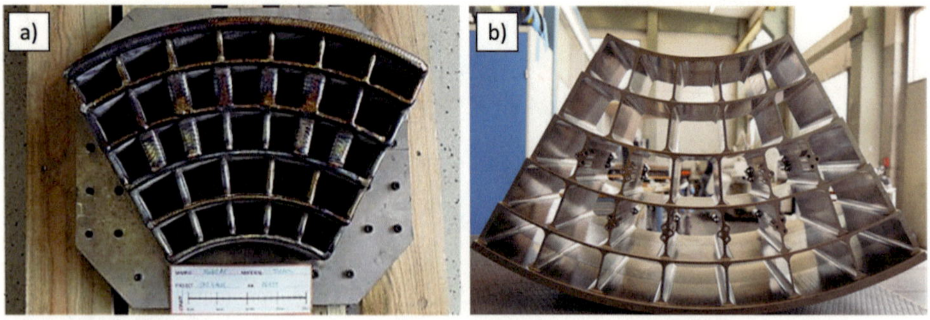

Figure 11.
Ti6Al4V optical bench demonstrator manufactured under the European Space Agency (ESA) activity SME4ALM: a) component as-deposited, b) component after post-process milling.

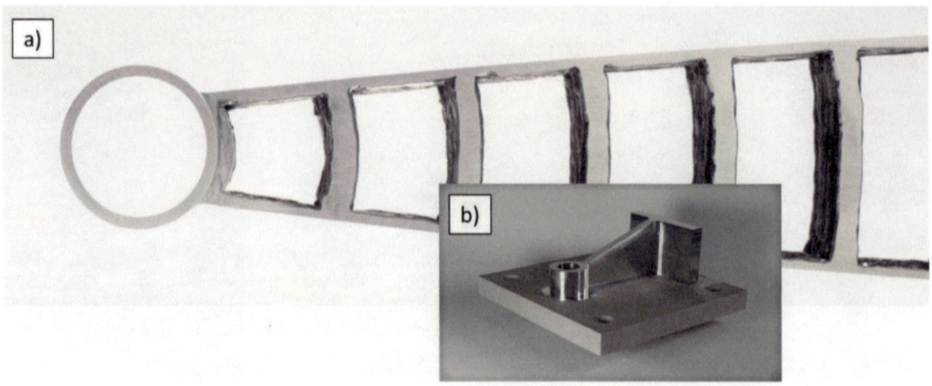

Figure 12.
Two examples of PMD manufactured aluminum alloys components: a) lightweight arm from AlMg5Cr, b) bearing bracket made of AW 5356.

energy input directly on the metal matrix that allows penetration and dilution of the deposited material. The time the positive and the negative half cycle are activated is controlled by the welding parameters known as frequency [Hz] and balance [%]. Two examples of PMD manufactured aluminum are shown in **Figure 12**.

5.5 Magnesium alloys

In addition for the aluminum alloys described above, magnesium alloys' oxides have a higher melting temperature than the matrix material; thus, AC current is needed for PMD deposition (see **Figure 13**) [24].

Figure 13.
PMD AZ91 magnesium alloy test specimen.

Furthermore, an extra risk appears if fine magnesium powder is used due to its high flammability and spontaneous firing.

5.6 On-going studies

This manufacturing technique is currently being developed for the processing of composite materials, in particular titanium-matrix composites, using powder material with reinforcement ceramic particles, mainly TiC and B_4C.

There is also work going on producing intermetallic, mixing the elemental component wires, since intermetallic ones do not still exist in the market.

6. Conclusions

To sum up, the PMD additive manufacturing process is an interesting and innovative technology that allows the following (**Figure 14**).

- Manufacturing large-size components: The size limitation is given by the structure where the torch is held.
- High deposition rate: Using welding wires of high diameters combined with the high energy input of the plasma allows the deposition of a considerable amount of material per hour.
- Flexibility: The possibility of using wire feedstock, as well as powder, as starting material gives to the user a large field of possibilities where *in situ* alloying, metal matrix ceramic composites, multi-material components, etc. are possible to manufacture.
- Scalability: The PMD equipment is set up as modular, and the equipment can be enlarged easily depending on the application requirements.
- Economically friendly: building near-net-shape components reduces the manufacturing times and material costs, especially on high-price alloys.

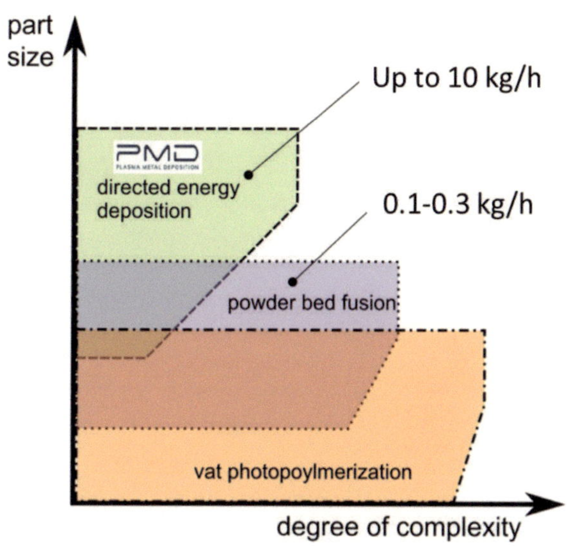

Figure 14.
Scheme of the application range of PMD vs. other additive manufacturing processes.

Conflict of interest

The authors declare no conflict of interest.

Notes/Thanks/Other declarations.

The authors warmly thank the European Space Agency (ESA), who fund the AZ91PMD (n°4,000,122,896/18/NL/BJ/gp) and SME4ALM (n°4,000,121,981/17/NL/BJ/gp) activities; and the European Commission (Horizon 2020) who fund the SUPREME (grant n° 768,612) project, within some of the presented components were carried out.

Author details

Enrique Ariza Galván[1], Isabel Montealegre Meléndez[2*], Cristina Arévalo Mora[2], Eva María Pérez Soriano[2], Erich Neubauer[1] and Michael Kitzmantel[1]

1 RHP-Technology GmbH, Seibersdorf, Austria

2 Universidad de Sevilla, Seville, Spain

*Address all correspondence to: imontealegre@us.es

References

[1] Molitch-Hou M. Overview of additive manufacturing process. In: Zhang J, Jung Y-G, editors. Additive Manufacturing: Materials, Processes, Quantifications and Applications. 1st ed. Oxford: Butterworth-Heinemann; 2018. pp. 1-38. DOI: 10.1002/9780128121559.ch1

[2] Ho A, Zhao H, Fellowes JW, Martina F, Davis AE, Prangnell PB. On the origin of microstructural banding in Ti-6Al4V wire-arc based high deposition rate additive manufacturing. Acta Materialia. 2019;**166**:306-323. DOI: 10.1016/j.actamat.2018.12.038

[3] Attar H, Ehtemam-Haghighi S, Kent D, Dargusch MS. Recent developments and opportunities in additive manufacturing of titanium-based matrix composites: A review. International Journal of Machine Tools and Manufacture. 2018;**133**:85-102. DOI: 10.1016/j.ijmachtools.2018.06.003

[4] Perez-Soriano EM, Ariza E, Arevalo C, Montealegre-Melendez I, Kitzmantel M, Neubauer E. Processing by additive manufacturing based on plasma transferred arc of hastelloy in air and argon atmosphere. Metals. 2020;**10**(2):200. DOI: 10.3390/met10020200

[5] Sun W, Shan F, Zong N, Dong H, Jing T. A simulation and experiment study on phase transformations of Ti-6Al-4V in wire laser additive manufacturing. Materials and Design. 2021;**207**:109843. DOI: 10.1016/j.matdes.2021.109843

[6] Luo X, Yang C, Fu ZQ, Liu LH, Lu HZ, Ma HW, et al. Achieving ultrahigh-strength in beta-type titanium alloy by controlling the melt pool mode in selective laser melting. Materials Science and Engineering A. 2021;**823**:141731. DOI: 10.1016/j.msea.2021.141731

[7] Walker J, Middendorf JR, Lesko CCC, Gockel J. Multi-material laser powder bed fusion additive manufacturing in 3-dimensions. Manufacturing Letters. 2021. (In press). DOI: 10.1016/j.mfglet.2021.07.011

[8] Magana-Carranza R, Sutcliffe CJ, Patterson EA. The effect of processing parameters and material properties on residual forces induced in Laser Powder Bed Fusion (L-PBF). Additive Manufacturing. 2021;**46**:102192. DOI: 10.1016/j.addma.2021.102192

[9] Frazier WE. Metal additive manufacturing: A review. Journal of Materials Engineering and Performance. 2014;**23**(6):1917-1928. DOI: 10.1007/s11665-014-0958-z

[10] Sharples RV. The Plasma Transferred Arc (PTA) Weld. Surfacing Process. Cambridge, UK: Welding Institute; 1985

[11] Alberti EA, Bueno BMP, D'Oliveira ASCM. Additive manufacturing using plasma transferred arc. International Journal of Advanced Manufacturing Technology. 2016;**83**:1861-1871. DOI: 10.1007/s00170-015-7697-7

[12] Hoefer K, Mayr P. Additive manufacturing of titanium parts using 3D plasma metal deposition. Materials Science Forum. 2018;**941**:2137-2141. DOI: 10.4028/www.scientific.net/MSF.941.2137

[13] Shanping L, Hidetoshi F, Kiyoshi N. Effects of CO_2 shielding gas additions and welding speed on GTA weld shape. Journal of Materials Science. 2005;**40**(9):2481-2485. DOI: 10.1007/s10853-005-1979-7

[14] Jittavisuttiwong P, Poopat B. Effect of helium addition in argon shielding gas on metal transfer behavior in gas

metal arc welding of aluminium. Key Engineering Materials. 2013;**545**:219-224. DOI: 10.4028/www.scientific.net/KEM.545.219

[15] Vimal KEK, Naveen Srinivas M, Rajak S. Wire arc additive manufacturing of aluminium alloys: A review. Materials Today: Proceedings. 2021;**41**:1139-1145. DOI: 10.1016/j.matpr.2020.09.153

[16] da Silva LJ, Matos Scotti F, Bailoni D, Ruham F, Américo Scotti P. Effect of O_2 content in argon-based shielding gas on arc wandering in WAAM of aluminum thin walls. CIRP Journal of Manufacturing Science and Technology. 2021;**32**:338-345. DOI: 10.1016/j.cirpj.2021.01.018

[17] Arévalo C, Ariza Galván E, Perez-Soriano EM, Montealegre-Meléndez I. Effect of processing atmosphere and secondary operations on the mechanical properties of additive manufactured AISI 316L stainless steel by plasma metal deposition. Metals. 2020;**10**(9):1125. DOI: 10.3390/met10091125

[18] Ariza E, Montealegre-Meléndez I, Pérez-Soriano EM, Arévalo C, Neubauer E, Kitzmantel M. Processing of 17-4PH by Additive Manufacturing using A Plasma Metal Deposition (PMD) Technique. Maastricht, Netherlands: EuroPM2019; 2019

[19] Meuthen J, Bielik M, Ariza E, Neubauer E, Kitzmantel M, Plano S, et al. Concept for Mass Production of Automotive Parts Using Plasma Metal Deposition. Virtual: EuroPM2020; 2020

[20] Rodriguez J, Hoefer K, Haelsig A, Mayr P. Functionally graded SS 316L to Ni-based structures produced by 3D plasma metal deposition. Metals. 2019;**9**:620. DOI: 10.3390/met9060620

[21] Ariza E, Montealegre-Meléndez I, Pérez-Soriano EM, Arévalo C, Bielik M, Meuthen J, et al. Study on Processing Nickel Alloy Hastelloy C-22 by Additive Manufacturing Technique Plasma Metal Deposition (PMD). Virtual: EuroPM2020; 2020

[22] Ariza E, Neubauer E, Bielik M, Meuthen J, Bača L, Stelzer N, et al. Study of Nickel-Chromium Super Alloys processed with Plasma Metal Deposition (PMD) to enable Additive Manufacturing of Large Parts. Virtual: EuroPM2021; 2021

[23] Neubauer E, Ariza E, Meuthen J, Bielik M, Kitzmantel M, Bača L, et al. Analysis of the Anisotropy of Properties in Titanium Alloys made by Plasma Metal Deposition. Maastricht, Netherlands: EuroPM2019; 2019

[24] Ariza E, Bača L, Schnall M, Stelzer N, Neubauer E, Pambaguian L. Study on Processing Magnesium Alloy AZ91 by Plasma Metal Deposition (PMD) for Space Applications. Virtual: EuroMat2021; 2021

Chapter 11

Pure Copper: Advanced Additive Manufacturing

Lukas Stepien, Samira Gruber, Moritz Greifzu, Mirko Riede and Aljoscha Roch

Abstract

This book chapter elaborates on different additive manufacturing (AM) processes of copper and copper alloys. The scope is to give the reader a basic understanding of the state-of-the-art of copper additive manufacturing by different AM technologies, such as laser powder bed fusion (LPBF), laser metal deposition (LMD), binder jetting (BJ), and metal-fused filament fabrication (M-FFF). Furthermore, we want the reader to be able to use this knowledge to find and assess potential use cases. Recently, with the commercial availability of green laser sources, the difficulties for laser processing of pure copper were overcome, which gave AM technologies, such as LPBF and LMD new momentum and increased interest. AM technologies involving a subsequent sintering step. They are relatively new and gained interest due to fast build-up rates (BJ) or ease of operation (M-FFF). We will cover important material-related properties of copper and its implications for manufacturing and application (e.g. absorption, sinterability, conductivity, and its dependency on impurities). Further, we address applications for AM copper, present the state-of-the-art for above mentioned AM technologies and share our own recent research in this field.

Keywords: additive manufacturing, copper, electrical conductivity, applications

1. Introduction

Copper and copper alloys are one of the major groups of commercial metals. Pure copper is defined as having a minimum copper content of 99.3% [1].

While pure copper is used extensively for electrical components, such as cables and contacts, alloys like brass or bronze are used for thermal energy transfer applications, such as radiators and heat exchangers [2].

While the laser-based additive manufacturing of alloyed coppers, such as brass or bronze, was successfully done, approaches in processing pure copper with, at the time, available infrared laser sources were not satisfying in terms of electrical conductivity, density, and process stability. Electron beam-based AM technologies overcame this and reached densities close to 99.8% [3], however, the coarse powders combined with the high-thermal conductivity resulted in higher surface roughness and hindered de-powdering of fine channels. Sintered-based AM technologies recently reached densities above 95%, but as in metal injection molding (MIM), their mechanical properties are behind their laser or electron beam-melted counterparts.

With the availability of a powerful green laser source, some of the drawbacks in terms of the processing could be overcome resulting in highly dense and conductive

IntechOpen

parts. However technological aspects, such as a bigger laser spot diameter reduces the ability to produce for instance thin-walled or other intricate features.

2. Physical properties and their effects on manufacturing

Copper has unique properties that make it an outstanding engineering material, however, those properties can make the processing a particular challenge in the context of additive manufacturing and demands specific approaches.

2.1 Physical properties of pure copper

Copper possesses the second highest electrical and thermal conductivity of all metals. The high-thermal conductivity of copper is a particular challenge during welding processes whether it is during the direct laser metal deposition or powder bed laser processing. For powder bed, this results in higher surface roughness, because the heat zone (due to the heat spreading into the powder bed) is wider causing particles to partially sinter to the consolidated body.

Due to its crystalline structure (fcc), pure copper also has high ductility. This also remains after the processing of pure copper parts from powders. Internal stresses, typically a problem for additively manufactured materials, are very low. This is beneficial since process and geometry-induced distortions are usually not a big problem. Further, an stress-relief annealing, is in most cases, not necessary but may be useful for the homogenization of the microstructure. **Table 1** gives a brief overview of some physical properties of pure copper. While, based on the definition of pure copper, the absolute values often show deviations, however, the table should give an orientation.

The electrical conductivity of copper and its alloys is often given relative to a copper wire test sample (international annealed copper standard, IACS) which was established in 1914. For comparison, 100% IACS is defined as 58×10^6 S/m at 20 °C, while the absolute maximum electrical conductivity measured for pure copper (Cu-ETP-1 or Cu-OF-1) is 58.58×10^6 S/m at 20°C (referring to 101% IACS). Thus, some electrical conductivity values may also exceed 100% IACS [7]. For pure copper (99.999%) value is 103.06% IACS and for pure silver it is 106% IACS [2].

2.2 Absorption

The absorptivity of electromagnetic radiation into the material is wavelength-dependent. For all materials, the absorptivity generally increases with smaller wavelengths (**Figure 1**). For copper as a reflective material, there is a huge increase

Melting point	1083°C	[4]
Density	8.94 g/cm³ @ 20°C	[4]
Coef. thermal expansion	17.0 x 106 /C (20–100°C)	[4]
Thermal conductivity	401 W/mK @ 20°C	[4]
Electrical conductivity	59.6 MS/m @ 20°C	[5]
Ultimate tensile strength	210–390 MPa	[6]
Young's modulus	120 GPa	[6]

Table 1.
Physical properties of pure bulk copper.

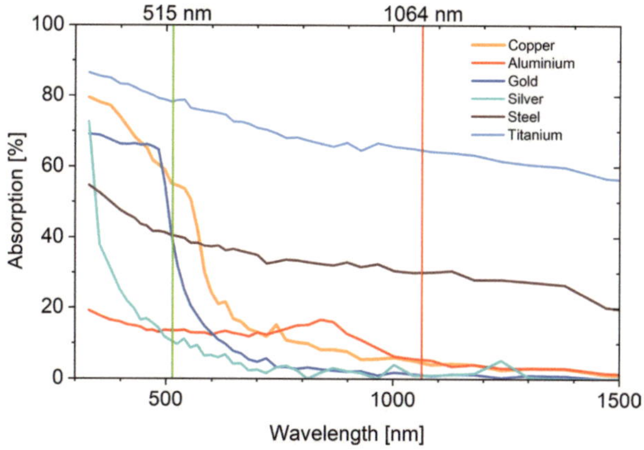

Figure 1.
Absorption of different solid metals. Data extracted from Spisz et al. [8].

in absorptivity at 515 nm (green wavelength) compared to 1064 nm (infrared wavelength). This can be used for laser-based AM processes to change the laser source to smaller wavelengths to increase the absorptivity, thus energy can be transferred more effectively resulting in higher efficiency.

To mitigate the low absorptivity in the infrared region, higher laser power can be used. Recently this approach become more attention due to the well-developed system technology, especially for big build sizes. However, the higher energy input into the powder bed can lead to smaller processing windows.

2.3 Sintering capability and impurities

While the absorptivity of copper does not affect the sintering capabilities of the copper powder, binder jetting and metal fused filament fabrication can be well compared to other powder metallurgical processes since a sintering step is clearly necessary to obtain functional metallic parts. For powder metallurgy of pure copper, the Copper Development Association Inc., an industrial board for copper, copper alloys, and their applications, mentions that "it is impractical to achieve a density of 8.94 g/cm^3 by pressing and sintering alone" [9]. To achieve high density, in classical powder metallurgy, non-spherical powders are used and pressure for compaction of 207–248 MPa is recommended. Pre-compaction at higher pressures of up to 730 MPa can further increase the sintered density of simple geometries up to 97.6% [10] but might be impractical for parts that are more complex. The sintering density of the parts is then a function of sintering time and temperature, as shown in **Figure 2**. To show a more recent example, hot pressing of copper for 4 minutes at 600, 700, and 800°C at 50 MPa resulted in density values between 97.9 and 99.1% [11]. Interestingly, also at the highest measured density, electrical conductivity was corresponding to 90.2% IACS. This example may illustrate, that even achieving high physical density is still no guarantee to achieve high electrical conductivity, too. Besides pressurized sintering, also sintering atmosphere or other modifications are mentioned to influence the sintering activity positively, as the use of reactive gases ore use of powders having a thin oxide layer.

Ott et al. investigated the heat conductivity of pressureless sintered Cu-powders and analyzed the influence of residual porosity, but also elemental impurities on that physical parameter and backed their analysis with simulated data. The conclusion of that study was, that impurities, especially Fe, cause a stronger depression of thermal conductivity than pores. According to that group, porosity of 2–5% causes

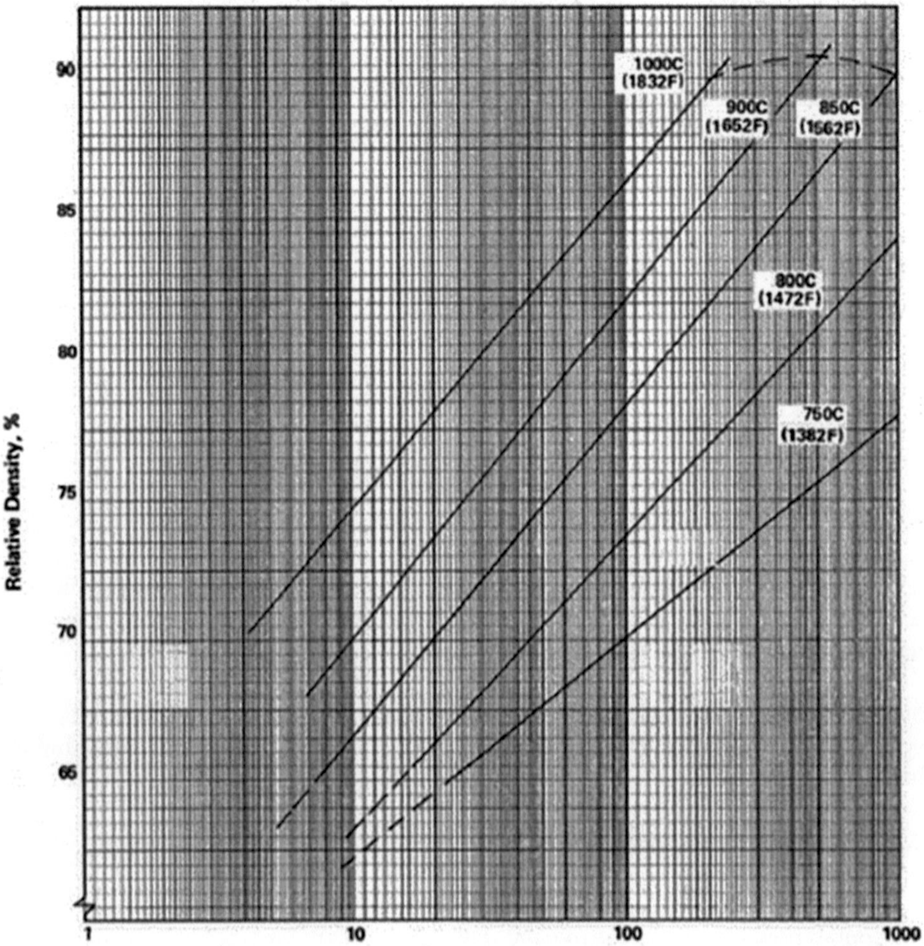

Figure 2.
Dependency of physical density from sintering temperature and time for copper powder compacts [9].

loss of 10 W/mK, while 200 mg/kg Fe cause ~40 W/mK [12]. Due to the connection between thermal and electrical transport, known as Wiedemann–Franz law, also the electrical conductivity is strongly affected by impurities (**Figure 3**).

During processing, oxygen from the ambient atmosphere or processing gas is the main contaminant. Fortunately, its effect on the conductivity is relatively small compared to other elements. However, using high-quality process gas (e.g. Argon with 99.999% purity) is highly recommended. Electrolytic-tough Pitch copper is allowed to have max. 400 ppm of oxygen. During LPBF processing we did not observe an additional rise in oxygen content for oxygen levels of 100 ppm in the processing gas during printing.

Ambient control with LMD is more challenging since normal shielding gas is often not enough to protect the part from oxidizing. Especially hot sections outside the working zone. Reasons are turbulences in the shielding gas stream down to the part. Technical solutions, such as a dedicated modular gas-shielding unit (e.g. COAXshield), showed good efficiency for Ti4Al4V but have to be verified for copper.

During the sintering of BJ and M-FFF parts, one can utilize hydrogen gas for the reduction of oxides and binder residue. However, during debinding carbon can potentially dissolve in copper causing a decrease in electrical conductivity.

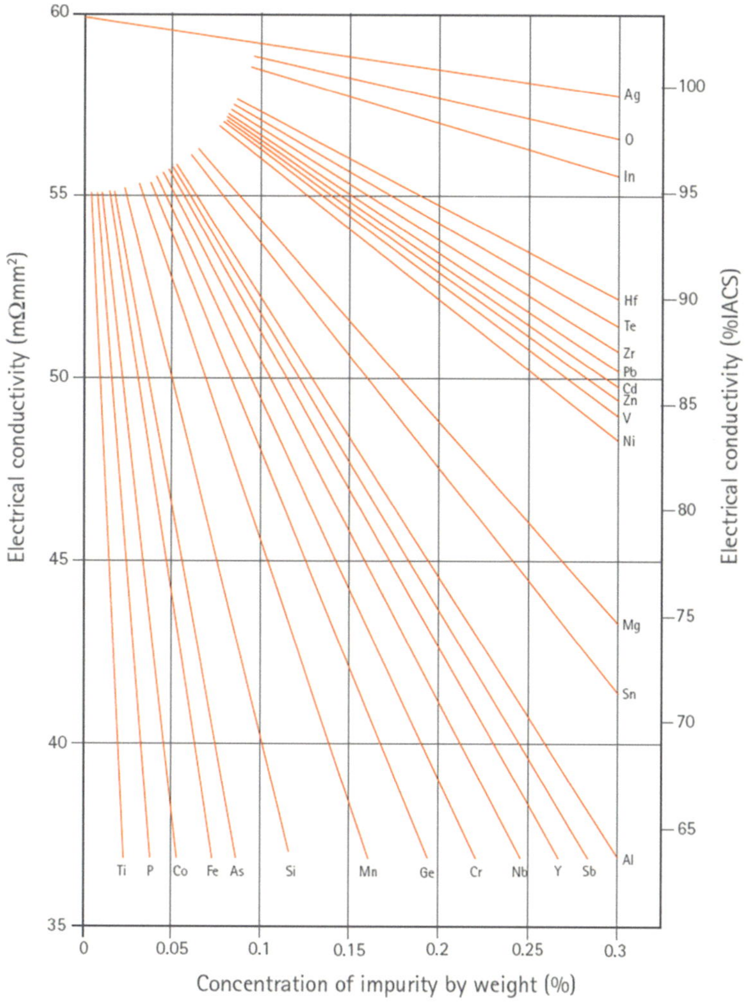

Figure 3.
Relation between electrical conductivity and concentration of impurities [13].

Jadhav et al. showed (here in the case of nanoparticle addition for LPBF) that small impurities of 0.055 wt.-% carbon in the printed part can also reduce the electrical conductivity to 22.7 $\times 10^6$ S/m (or 39.2% IACS) [14]. This explains the relatively lower electrical conductivity of binder-based sinter processes where complete binder burnout is often difficult to achieve.

3. State-of-the-art laser powder bed fusion

Laser powder bed fusion is an AM process with the following repeating process steps—metal powder particles are spread evenly onto a substrate with a recoating system, then a laser source selectively melts the metal powder with specified parameters according to a previously prepared computer file with scanning strategy and laser parameters, such as laser powder, scanning velocity, and distance of single scanning tracks. Then the substrate plate is lowered by a specific layer thickness, a new powder layer is spread, and the process is repeated until the part is finished. Commercially available systems range in build volume, maximum laser power, amount of used laser sources, and laser type. Since the absorption of pure copper is poor in the infrared wavelength and commonly, LPBF machines were equipped with infrared fiber lasers, the processing of pure copper with LPBF was challenging in the past [15–25]. The energy input into the material was insufficient for complete melting leaving a lack of fusion defects. The highest achievable density for pure copper parts when using a common 200 W infrared laser source was 83–88% [23, 24]. There have been two approaches in LPBF of pure copper to increase the density and subsequently the electrical conductivity—increase the infrared laser power to above 1 kW or switch to a green laser source. Colopi et al. and Ikeshoji et al. [19, 25] have used infrared laser powers of 1 kW and were able to increase the relative densities to 99.1 and 99.6%. However, melt-pool instabilities were observed due to the high difference in absorptivity in the solid and molten state of the pure copper which led to parts with low surface quality. Also, the high reflectivity can harm the optic system of the machine. TRUMPF has released an LPBF machine with an integrated green laser and could prove that high electrical conductivity can be achieved with such a system around 100% IACS [26]. With this machine, complex-shaped pure copper parts can be manufactured with high quality regarding density and electrical conductivity, and therefore, the technology is now ready to produce parts for various applications.

At Fraunhofer IWS such a TruPrint1000 Green Edition machine, equipped with a TruDisk1020 frequency-doubled laser emitting 515 nm wavelength, is available since mid-2020. The characteristics of the laser machine include a maximum laser power of 500 W, a spot diameter of 200 µm and a build volume of 100 mm diameter with 100 mm build height. Ongoing research concentrates on the following:

- process parameter development for pure copper and copper alloys to increase the build rate while maintaining the high part quality, such as density and electrical conductivity

- different post-processing techniques and their effects on surface quality and geometrical accuracy

- pure copper and copper alloy applications

The density of pure copper parts is above 99.5% and the electrical conductivity was proven to be above 100%IACS. The oxygen content in the final part is below 400 ppm.

As can be seen in **Figure 4**, the surface quality shows the high roughness of the pure copper parts. Therefore, the surface needs smoothening. With two benchmark geometries developed by Fraunhofer IWS (**Figure 5**) specific feature sizes and overhang angle

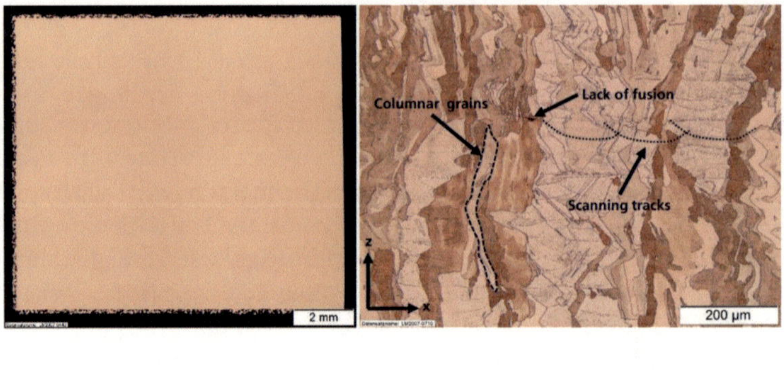

Figure 4.
Microsection of a density cube of pure copper (left), etched microstructure in the x-z axis (right) ©IWS.

roughness can be analyzed via 3D scan and tactile measurements. The effect of different post-processes, such as sandblasting, abrasive flow machining, or chemical processes, such as plasma or electropolishing can improve the surface quality. However, material removal can be irregular, and therefore the process itself and applied parameters must be adapted to each geometry and particular application (**Figure 5**).

Currently, possible applications investigated are components for the nuclear accelerator community, such as radiofrequency quadrupoles or nozzle geometries

Figure 5.
Benchmark for resolution and different features (left), benchmark for overhang angles (right) ©IWS.

Figure 6.
Example of pure copper inductor coil geometry ©IWS.

for laser metal deposition. Individualized inductor coils are also a field predestined for AM (**Figure 6**).

4. State-of-the-art laser metal deposition

Laser metal deposition (LMD) is an AM process that is assigned to the DED processes. Laser metal deposition is a well-established technology for coating and repair of metal components for more than a decade. Recently, it has been utilized for manufacturing metallic parts from micro to macro scale without any support structures. Compared to the well-known powder bed fusion process, LMD enhances manufacturing possibilities to overcome AM-specific challenges such as process inherent porosity, minor build rates, and limited part size. Moreover, the advantages aforementioned combined with conventional machining enable novel manufacturing approaches in various fields of applications.

For small and filigree additive manufactured components, LPBF is usually considered due to the freedom of design and short-lead times [27]. However, even this innovative technology has manufacturing constraints, such as the need for support structures or high build-up times. That affects cost efficiency and process stability. In contrast to powder bed processes or competing direct methods (e.g. WAAM and EBAM), additive manufacturing via powder LMD provides

- support-less manufacturing (cf. PBF),
- high productivity (cf. PBF),
- high flexibility due to local shielding (cf. PBF, EBAM),
- precise energy input—beneficial microstructure (cf. WAAM, EBAM),
- low porosity—HIP not needed (cf. PBF) and
- hybrid manufacturing in one machine (cf. PBF, EBAM)

That makes this technology suitable for the realization of high-performance component designs. Besides, a further advantage of LMD is that conventionally manufactured semi-finished parts can be used adding new features via LMD. This approach decreases manufacturing time and potentiates the advantages of hybrid AM processes. Hence, powder LMD has been established in several branches, e. g. aerospace, medical, or tooling industry for the production of components for jet engines, implants, or drilling tools [28]. To deposit material on a substrate, the powder material is blown into the process zone by a nozzle, partially preheated in the laser beam, and finally reabsorbed in the laser as illustrated in **Figure 7**.

During the manufacturing process, the bulk material is melted using a laser as a heat source and powder is transported via a carrier gas, like helium or argon [29], into the melting pool using a coaxial nozzle. The powder interacts there with the melting pool and gets absorbed to manufacture the desired part. To fully absorb the powder into the melting pool minimal energy is needed, which can be called line energy. The Marangoni effect causes a strong melt pool movement, which is driven by the surface tension of the melt and leads to a strong mixing of the filler (powder) and part of the substrate material [30]. That also results in potential pores being discharged, improved density, and increased building rates. The subsequent formation of a certain microstructure during solidification is mainly driven by the

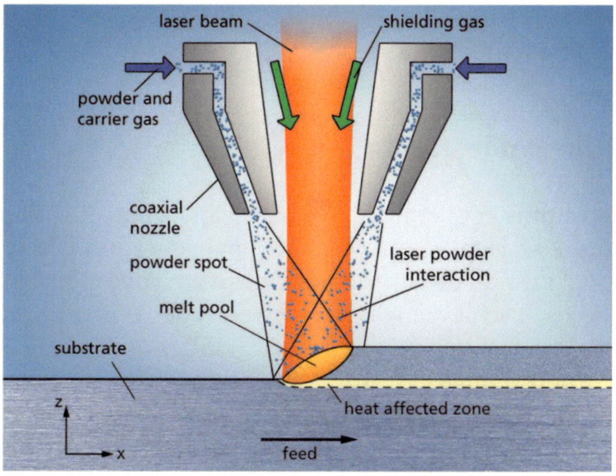

Figure 7.
Principle of laser metal deposition (LMD) using powder ©IWS.

material selection and the local and temporal gradient, which is affected by process parameters, material, and boundary conditions. When the powder is deposited, heat transfer through prior layers can result in an additional modification of the microstructure.

By tailoring energy input and distribution as well as powder particle size, a wide range of materials could be applied even on various substrate materials (e.g. Stellite on Inconel 718, Brass on Steel, Al2O3 on Al-Alloy) [31, 32]. However, the processing of pure copper using established infrared laser sources has been associated with major challenges. Low absorptivity ends up in a lack of fusion and high porosity [33]. High reflection can damage the laser source or may cause overheating of applied nozzles.

The use of green (495–570 nm) and blue (~445 nm) laser sources can increase the laser absorption of pure copper by a factor of 10 [34].

Specialized processing heads enable dense cooper parts manufactured on substrates, as well as complex prototypes [28, 35, 36].

Moreover, in contrast to powder bed-based additive manufacturing, LMD enables hybrid manufacturing (additive, subtractive) approaches and multi-material processes. Various powders could be applied, exchanged, and mixed *in situ* to achieve multi-material components with localized material properties. The latter recently was applied by IWS to significantly increase the performance of mold inserts by the local implementation of copper features and thus reduced cycle times [37] (**Figure 8**). In further developments, the essential intermediate and

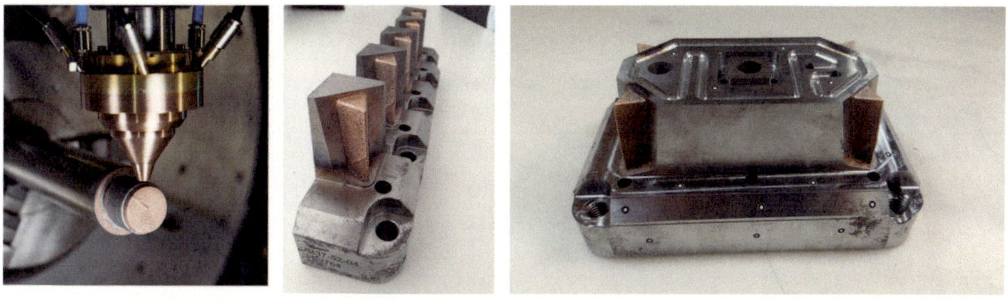

Figure 8.
Laser Metal Deposition with a green laser to build up multi-material mold inserts (pure Cu/steel 1.2764) ©IWS.

final machining could be fully incorporated in the LMD process chain, resulting in production tools close to industrial needs.

5. State-of-the-art binder jetting

Binder Jetting of pure copper has been intensively studied by Virginia Polytechnic Institute and State University. The main question of this research was how to increase the physical density of copper parts produced by binder jetting. Different approaches to achieve high density were taken under investigation of the influence of particle size of the feedstock (D50 = 15 μm or 75 μm) [38], including bimodal powder compositions [39] where a small fraction of very fine powder should fill up the spaces between the larger particles. Different sintering atmospheres (Ar and H_2) were also part of the analysis. Modified binders were investigated, comprising MOD (metal–organic decomposition) inks [40] and nanoparticle [41] enhanced binders. The expectation for the latter both approaches is that introduction of nanoparticles will decrease the temperature for the sintering process to start, but also introduce additional copper into the green body. Also, the influence of HIP post-treatment was investigated [42]. For all the previously described experiments sintering temperature was quite high 1075–1080°C compared to the theoretical melting point of copper 1084°C. Dwell time was varied between 2–10 h, most sintering regimes employed H_2 as reducing atmosphere.

The following **Table 2** sums up the results of that group.

Additionally, the same group published work using a copper feedstock that incorporates a foaming agent introduced by mechanical milling for modification of the porosity of printed parts [44].

It should be also mentioned that companies, active in the development of binder jetting machines, try to qualify materials to be processed on their equipment. Currently, DigitalMetal [45] and ExOne [46] have announced qualified processes with pure copper for applications, such as antennas, heat exchangers, and windings for electric drives.

To further investigate the influence of bimodal powder compositions on the electrical properties of binder-jetted parts, two powder feedstocks were selected, printed, and compared regarding the final part electrical conductivity at IWS. The powders were a monomodal and a bimodal composition, the latter consisted of 73% coarse and 27% fine powder. The powder size distribution of the feedstock is shown in **Table 3**.

Use of powder with D50 = 15 μm in comparison to D50 = 75 μm leads to 85.5 % instead of 63.2% of the theoretical density of copper, applying a 4 h @ 1080°C sintering regime	[38]
Using bi modal powders (30 μm + 5 μm with a mixing ratio of 17% + 73% respectively) results in a density of 92.3%. All bimodal compositions show significantly less shrinkage	[39]
By using HIP treatment of test samples from bimodal powders, the density could be further increased to 99.7%.	[42]
Using Metal-Organic-decomposition inks, the part density of the core section could be increased. The overall density however was lower as 73.3% in comparison to non-modified binder (80.8%).	[40]
By using nanoparticle loaded inks, the sintered part density is 86.1% compared to 80.9%, when using a neat binder,	[41]
Use of fine copper powders (~5 μm) with new recoating equipment.	[43]

Table 2.
Effect of different approaches by the Virginia polytechnic group on the relative density of BJ copper parts.

Powder	D_{10} [µm]	D_{50} [µm]	D_{90} [µm]
m4p PureCu.04	3	8	15
m4p PureCu1.0	18	26	38

Table 3.
D10, D50, and D90 of the fine (m4p PureCu.04) and coarse powder (m4p PureCu1.0).

Parts were printed on an ExOne binder jetter (model MFlex). After optimizing the parameters of roller speed, roller transverse speed, layer thickness, and binder saturation, a set of flat samples (25 × 25 × 1 mm^3) and cubes (10 × 10 × 10 mm^3) for measuring electrical conductivity, physical density, and dilatometry were printed. Although the focus was on the influence of the powder composition, also three different dwell times for sintering, and two different layer thicknesses during printing were compared. The density is analyzed by standard metallography, the electrical conductivity is measured by the eddy current test method (Sigmascope 350, Karl-Fischer), and the dilatometry was done with a DIL 402 Expedis Classic (Netzsch). As expected, longer dwell times lead to higher conductivity (**Figure 9**). The achieved maximum is found at 84.7% IACS for the bimodal powder and a layer thickness of 80 µm, while for the same configuration the monomodal sample led to 52.6%. For all sintering times, samples made of bimodal powder delivered better conductivity. The observation for the influence of layer thickness is that for monomodal powders, 50 µm leads to the same or slightly better results, while for the bimodal configuration the better values are found for 80 µm. Though, at 12 h that difference disappears.

The differences between the mono and bimodal powder distribution are apparent in **Figure 10**. The shrinkage of the mono (black) and bimodal (green) sample over the time of the applied temperature profile during sintering is shown. The plots do not contain any compensation for thermal expansion. Two main information can be extracted from the dilatometer experiment. First, the overall shrinkage for the bimodal powder is much lower (12.4%) than for the monomodal powder (17.3%). Second, the onset temperature for begin of shrinkage is ~37 K less for the bimodal powder at 987.5°C.

In **Figure 11**, two etched cross sections of bimodal samples, sintered for 2 h (left) and 12 h (right) are shown. After 2 h of sintering, the density is clearly still

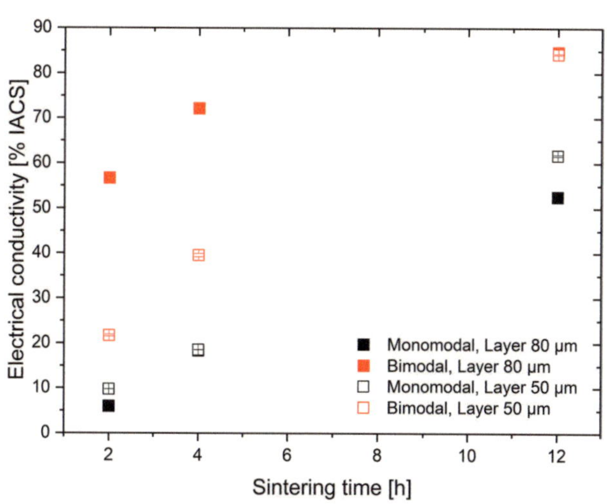

Figure 9.
Graph showing the relation between sintering time and electrical conductivity for mono- and bimodal feedstock and different layer thicknesses during the printing process.

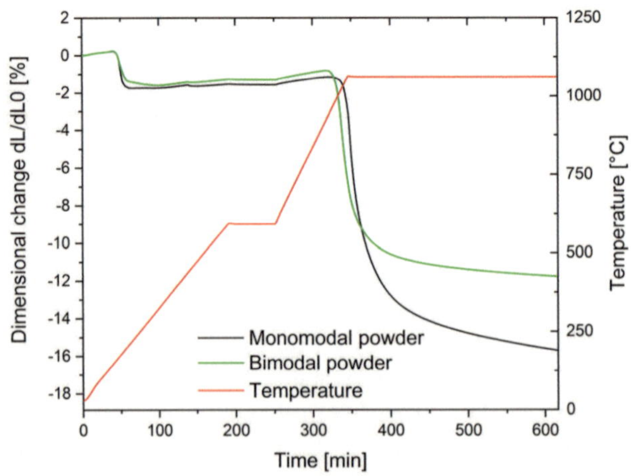

Figure 10.
Dilatometer plot, comparing shrinkage of mono and bimodal samples during the debinding and sinter profile.

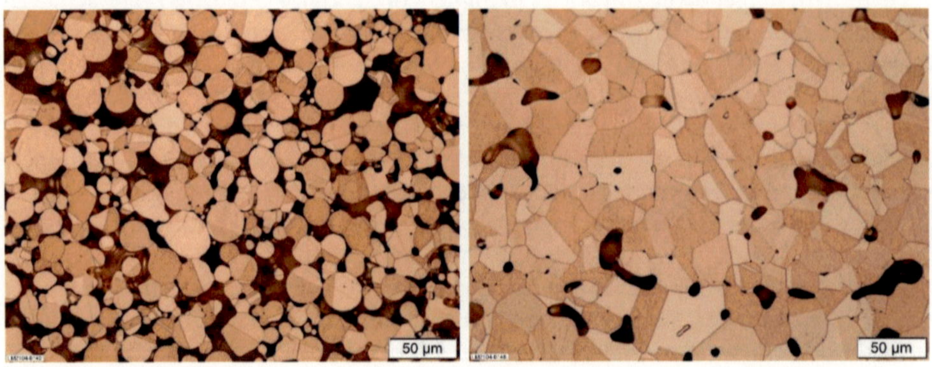

Figure 11.
Metallographic comparison of 2 h (left) and 12 h (right) sintering at 1080°C of bimodal samples.

low as it seems necking is just about to begin. After 12 h instead, a quite dense microstructure can be seen, nonetheless showing a lot and partially also quite large (> 50 μm) pores at the grain boundaries.

One of the main challenges in binder jetting obviously remains to achieve high sintered density since compaction of parts is not possible as in classic press and sinter processes. Bimodal powder compositions enhance green part density and stability, lead to higher sintered density, earlier sintering activity, and in the case of copper better electrical conductivity.

The sinter activity of shown samples is clearly low, as for comparison from **Figure 2** after 2 h about 90% relative density should be achievable in classic press and sinter. Using bimodal powder compositions seems to be one possible way to tackle that challenge even though 12 h sintering time is still very long. Possible reasons for the poor sintering activity might be insufficient powder bed compaction during the printing process, an incomplete debinding process, or sinter impeding surface oxides on the copper particles.

6. State-of-the-art metal-fused filament fabrication

Fused filament fabrication (FFF) belongs to the extrusion-based AM technologies. It was usually used for printing polymers, such as Acrylonitrile butadiene styrene

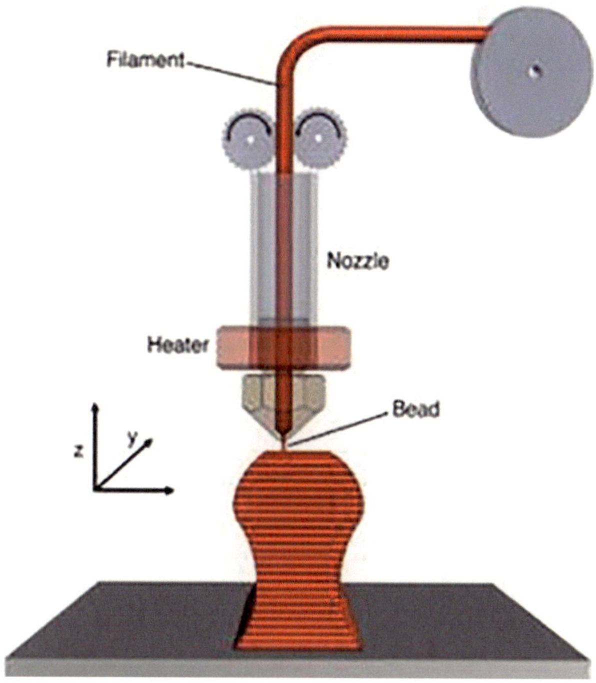

Figure 12.
Schematic representation of fused filament fabrication method [47].

(ABS) or Polylactide (PLA) [47], and became the most used AM technology worldwide due to its user-friendly handling [48, 49]. During the printing process, a filament is melted in a print head and extruded onto a build platform [50]. Layer after layer of molten filament is added to create a prototype or product. A sketch of the overall concept is shown in **Figure 12**.

Today, FFF is well established in many industries, such as the automotive sector [51, 52], in aviation (Airbus) [53], and the medical sector (printing biomedical implants, scaffolds, or other applications) [54]. The cost-efficiency of the FFF process suggested using FFF beyond polymers also for printing other materials.

Meanwhile, the upcoming metal FFF has demonstrated its capability in manufacturing sophisticated structures through a variety of materials [55–59]. Besides stainless steel (17-4PH) or titanium alloy (Ti6Al4V) [55, 60–63], Fe-parts for electrical engines or glass-ceramic scaffolds for medical application were printed [64]. Recently was published a multi-material approach by printing and sintering 17-4PH and ZrO2 together [65, 66].

During the process, a filament based on a polymer-binder, containing thermoplastic polymers [55, 56], infiltrated with metal powder, is fed into a print head where the binder is melted, and the material is extruded onto a building platform (**Figure 13**). After having printed, a so-called green-part layer-by-layer, a catalytic debinding step or solvent debinding step is required for removing a certain fraction of the binder. The solvent debinding step creates pores in the green part. These pores allow gases to escape during the thermal debinding of the remaining binder in a furnace. The polymer that remains after solvent debinding, stabilizes the structure as backbone until sintering of the particles takes place. The thermal debinding of the backbone by pyrolysis is crucial because escaping gases can cause deformations and cracks.

The part shrinks during sintering usually around 13–20% in x-, y- and z-direction, which needs to be predicted for near net shape fabrication.

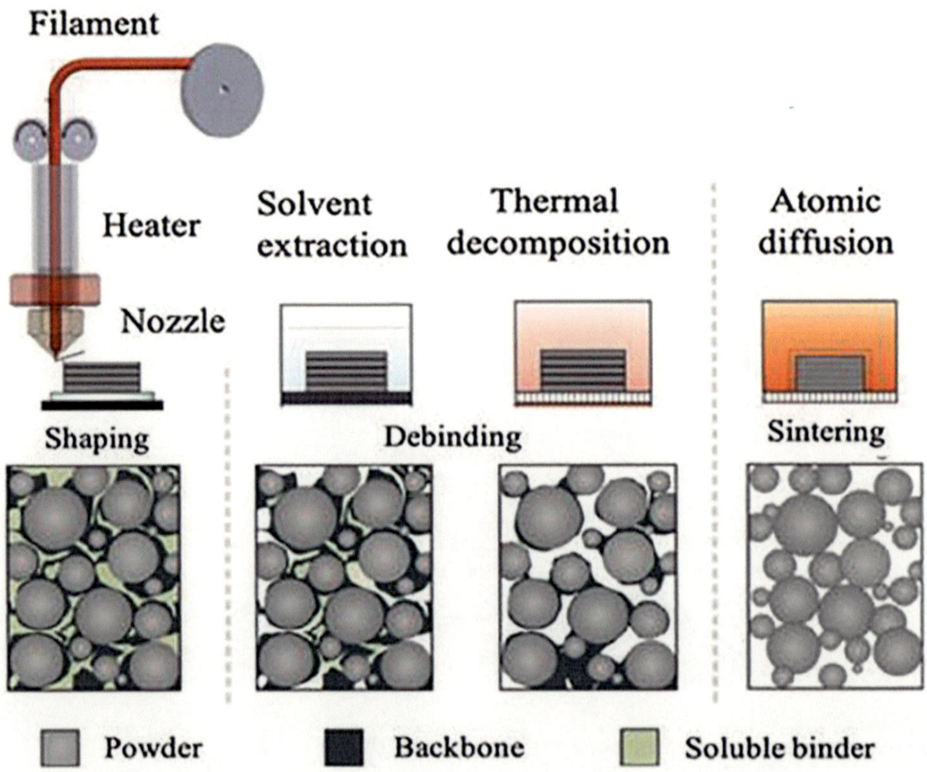

Figure 13.
FFF process, left to right: shaping the part by deposition of filament; two-step debinding process involving solvent extraction and thermal decomposition; finally sintering in a furnace, after [47, 67].

Significant advantages of FFF are as follows:

1. All kinds of powder materials and even nanoparticles can be utilized

2. Multi-material can be deposited by using different print heads

3. Microstructures related anisotropic mechanical behavior can be avoided due to homogeneously sintering [56]

4. Little investment costs and cost-efficient printing and sintering of metal and ceramic parts at atmospheric pressure

5. No powder particles are airborne, causing potential health problems for operators.

Additionally, high material throughput (1–10 g/min), material efficiency (no material waste), design freedom for printing even hollow structures, and the competitive material properties make FFF a highly competitive AM technology [56, 57, 65].

Nowadays companies, such as AM Extrusion GmbH [68] or BASF [69] offer an open filaments system for printing and sintering metal parts, such as copper, 316L, 17-4PH, or carbon steels, such as 440C, M2, or H13. Even filaments with unique materials can be prepared exclusively for customers.

Copper filaments by AM Extrusion GmbH (filled with 63 vol.% copper powder) can be printed with a modified BondTech extruder. Nozzle and print bed

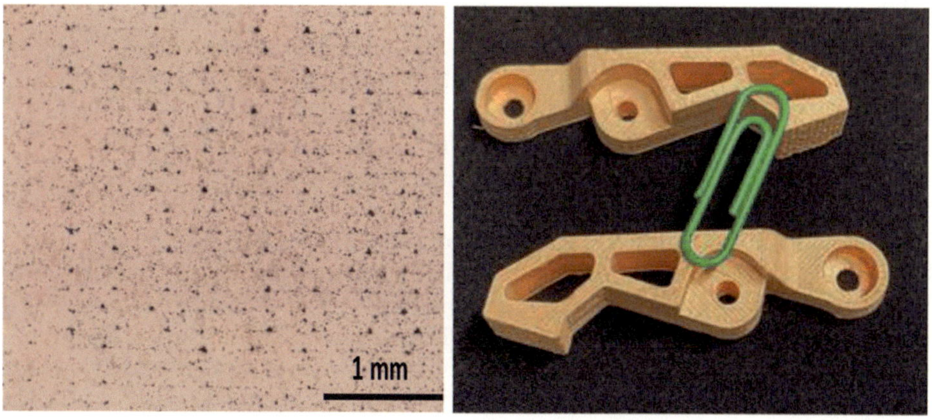

Figure 14.
(left) Properties of FFF printed copper measured by accredited test laboratory, (middle) cross section of sintered copper (@950°C, 90 min, H2), (right) sintered copper robot gripper [68].

temperatures are 120 and 70°C. The recommended nozzle is a 300 μm hardened steel nozzle. The standard layer height is 80–200 μm [68]. Using a 300 μm nozzle line, the width is 360 μm and print speed 1000–3000 mm/min.

After solvent debinding in acetone at 45°C and sintering at 950°C in H2, a relative density of 96% can be obtained [68]. The shrinkage during sintering is 13% in x-, y-, z-direction. A final part accuracy of < ± 80 μm can be obtained [68]. Material properties of FFF printed Cu and printed parts are shown in **Figure 14**.

Compared to powder-bed technologies, FFF is safe and user-friendly. During the FFF process, no powder can be airborne, which may cause health issues for employees. In general, FFF is capable of manufacturing medium-sized complex metal and ceramic structures in small serial production.

7. Conclusion

Additive manufacturing of copper is emerging and additive fabrication methods, such as laser powder bed fusion, laser metal deposition, binder jetting, fused filament fabrication, or electron beam melting become more refined.

Recently, it is possible to fabricate complex copper parts with an electrical conductivity of 100% IACS. In addition, the fabrication of hybrid material parts, including copper, is possible. Thus, additive manufacturing of pure copper keeps up and excels conventional manufacturing methods in terms of geometrical complexity.

Due to its unique properties, copper is primarily used for electrical or thermal applications. Already realized use cases are components for electric vehicles by LPBF [70], cooling sockets for milling tools by FFF [71], or a horn waveguide antenna [72].

Also increased research interest in found in the manufacturing of complex propulsion systems, such as aerospike thrusters, made from alloyed copper [73].

Further, printed heat sinks, heat pipes, and complex coils are already demonstrated.

Further improvements, especially impeccable material properties in combination with new fabrication approaches, are pursued. For instance, the modification of the copper powder feedstock with a coating of metal oxides or metal hydroxides (approx. 5–30% coverage) increases the absorptivity, especially when using standard infrared laser sources [74].

Another approach to utilize infrared lasers for the processing of pure copper is to use high laser power of 600–1000 W. Researchers from Politecnico di Milano achieved a density of ~ 97% using a 600 W laser on pure copper [75]. Yet this approach, in contrast to using green laser sources, has the advantage of using bigger build chambers. However, this advantage will disappear, since bigger LPBF setups with green laser sources are under development.

In addition, polymeric coatings of copper powder are under development for use in selective laser sintering machines. This process is advertised as cold metal fusion (or Metal SLS). Using this approach, lower laser powers are necessary to consolidate the powder, since only the polymer coating will be molten and sintered. Further, the commonly used infrared lasers can be used effectively. The printed part, however, needs to undergo a thermal sintering step though, comparable to binder jetting or FFF, to burn out the polymer and sinter the metal powder together [76].

Conflict of interest

All Authors declare that there is no conflict of interest.

Notes/thanks/other declarations

This research was conducted within the High-Performance Center »Smart Production and Materials« and partially funded by the Fraunhofer-Gesellschaft, the German Federal Ministry of Education and Research and the State of Saxony.

Author details

Lukas Stepien[1*], Samira Gruber[1], Moritz Greifzu[1,2], Mirko Riede[1] and Aljoscha Roch[3]

1 Fraunhofer Institute for Material an Beam Technology IWS, Dresden, Germany

2 Technical University Dresden, Institute of Materials Science, Dresden, Germany

3 AM Extrusion GmbH, Radebeul, Germany

*Address all correspondence to: lukas.stepien@iws.fraunhofer.de

References

[1] Li M, Zinkle SJ. Reference module in materials science and materials engineering, comprehensive nuclear materials, 4.20. Physical and Mechanical Properties of Copper and Copper Alloys. 2012;**4**:667-690. DOI: 10.1016/B978-0-08-056033-5.00122-1

[2] Davis JR. ASM speciality handbook: Copper and copper alloys. ASM International. 2001:3-9. DOI: 10.131/caca2001p003

[3] Sirris. Development of copper alloys with EBM [Internet]. 2021. Available from: https://www.sirris.be/development-copper-alloys-ebm [Accessed: November 23, 2021]

[4] Standards Handbook. Part 2, Alloy Data. New York: Copper Development Association Inc.; 1973

[5] Matula RA. Electrical resistivity of copper, gold, palladium, and silver. Journal of Physical and Chemical Reference Data. 1979;**8**(4):1147-1262. DOI: 10.1063/1.555614

[6] AZOM. An Introduction to Copper [Internet]. 2021. Available from: https://www.azom.com/properties.aspx?ArticleID=597 [Accessed: November 23, 2021]

[7] Deutsches Kupferinstitut. Elektrische Leiterwerkstoffe [Internet]. 2021. Available from: https://www.kupferinstitut.de/anwendungen/elektrotechnik-und-energie/elektrische-leiterwerkstoffe/ [Accessed: September 20, 2021]

[8] Spisz EW, Weigand AJ, Bowman RL, Jack JR. Solar absorptances and spectral reflectances of 12 metals for temperatures ranging from 300 to 500 K. NASA Technical Note. 1969:25

[9] Coble RI, Gupta TK. Intermediate stage sintering. In: Proceedings of Sintering and Related Phenomena. Gordon and Breach; 1967. pp. 423-444

[10] Dixit M, Srivastava RK. Effect of compaction pressure on microstructure, density and hardness of Copper prepared by Powder Metallurgy route. IOP Conference Series: Materials Science and Engineering. 2018;**377**: 012209. DOI: 10.1088/1757-899X/377/1/012209

[11] Islak S, Kir D, Buytoz S. Effect of sintering temperature on electrical and microstructure properties of hot pressed Cu-TiC. Since of Sintering. 2017;**46**: 5-21. DOI: 10.2298/SOS1401015I

[12] Ott J, Burghardt A, Britz D, Mücklich F. Influence of porosity and impurities on the thermal conductivity of pressure-less sintered Cu powder green bodies. Powder Metallurgy. 2021;**64**:85-96. DOI: 10.1080/00325899.2021.1871806

[13] Chapmann D. High Conductivity Copper for Electrical Engineering, Copper Development Association Publication No 122. European Copper Institute Publication No Cu0232. February 2016. Originally Published in May 1998

[14] Jadhav SD, Dadbakhsh S, Vleugels J, Hofkens J, van Puyvelde P, Yang S, et al. Influence of carbon nanoparticle addition (and impurities) on selective laser melting of pure copper. Materials. 2019;**12**:2469. DOI: 10.3390/ma12152469

[15] Yan X, Chang C, Dong D, Gao S, Ma W, Liu M, et al. Microstructure and mechanical properties of pure copper manufactured by selective laser melting. Materials Science and Engineering: A. 2020;**789**:139615. DOI: 10.1016/j.msea.2020.139615

[16] Silbernagel C, Gargalis L, Ashcroft I, Hague R, Galea M,

Dickens P. Electrical resistivity of pure copper processed by medium-powered laser powder bed fusion additive manufacturing for use in electromagnetic applications. Additive Manufacturing. 2019;**29**:100831. DOI: 10.1016/j.addma.2019.100831

[17] Kaden L, Matthäus G, Ullsperger T, Tünnermann A, Nolte S. Selective laser melting of copper using ultrashort laser pulses. Applied Physics A. 2017;**125**:596. DOI: 10.1007/s00339-017-1189-6

[18] Jadhav SD, Dadbakhsh S, Goossens L, Kruth JP, van Humbeeck J, Vanmeensel K. Influence of selective laser melting process parameters on texture evolution in pure copper. Journal of Materials Processing Technology. 2019;**270**:47-58. DOI: 10.1016/j.jmatprotec.2019.02.022

[19] Colopi M, Demir AG, Caprio L, Previtali B. Limits and solutions in processing pure Cu via selective laser melting using a high-power single-mode fiber laser. International Journal of Advanced Manufacturing Technology. 2019;**104**:2473-2486. DOI: 10.1007/s00170-019-04015-3

[20] Colopi M, Caprio L, Demir AG, Previtali B. Selective laser melting of pure Cu with a 1 kW single mode fiber laser. Procedia CIRP. 2018;**74**:59-63. DOI: 10.1016/j.procir.2018.08.030

[21] Becker B. Selektives Laserschmelzen von Kupfer und Kupferlegierungen [thesis]. Aachen: RWTH Aachen; 2014

[22] Lingqin X, Guang C, Luyu Z, Pan L. Explore the feasibility of fabricating pure copper parts with low-laser energy by selective laser melting. Materials Research Express. 2020;7:106509. DOI: 10.1088/2053-1591/abbd08

[23] Lykov PA, Safonov EV, Akhmedianov AM. Selective laser melting of copper. MSF. 2016;**843**: 284-288. DOI: 10.4028/www.scientific.net/MSF.843.284

[24] Trevisan F, Calignano F, Lorusso M, Lombardi M, Manfredi D, Fino P. Selective laser melting of chemical pure copper. Euro PM2017. 1-5 October 2017 Milan. Euro PM2017 Proceedings. p. 6

[25] Ikeshoji TT, Nakamura K, Yonehara M, Imai K, Kyogoku H. Selective laser melting of pure copper. JOM. 2018;**70**:396-400. DOI: 10.1007/s11837-017-2695-x

[26] Wagenblast P, Myrell A, Thielmann M, Scherbaum T, Coupek D. Additive manufacturing with green disk lasers. Proceedings of SPIE 11271. Laser 3D Manufacturing VII. 2020;**18**:8. DOI: 10.1117/12.2551150

[27] Correa J, Ferreira P. Analysis and design for rapid prototyping mechanism using hybrid flexural pivots. Procedia Manufacturing. 2015;**1**:779-791. DOI: 10.1016/j.promfg.2015.09.062

[28] Brueckner F, Riede M, Mueller M, Marquardt F, Willner R, Seidel A, et al. Enhanced manufacturing possibilities using multi-materials in laser metal deposition. Journal of Laser Applications. 2018;**30**:32308. DOI: 10.2351/1.5040639

[29] Azarniya A, Colear XG, Mirzaali MJ, Sovizi S, Bartolomeu F, Weglowski M, et al. Additive manufacturing of Ti–6Al–4V parts through laser metal deposition (LMD): Process, microstructure, and mechanical properties. Journal of Alloys and Compounds. 2019;**804**:163-191. DOI: 10.1016/j.jallcom.2019.04.255

[30] Mazumder J. Overview of melt dynamics in laser processing. Optical Engineering. 2019;**30**:1208-1219. DOI: 10.1117/12.55899

[31] Gasser A, Backes G, Kelbassa I, Weisheit A, Wissenbach K. Laser additive manufacturing. Laser Technik Journal. 2010;7:58-63. DOI: 10.1002/latj.201090029

[32] Bourell D, Kruth JP, Leu M, Levy G, Rosen D, Beese AM, et al. Materials for additive manufacturing. CIRP Annals. 2017;**66**:659-681. DOI: 10.1016/j.cirp.2017.05.009

[33] Popovich A, Sufiiarov V, Polozov I, Borisov E, Masaylo D, Orlov A. Microstructure and mechanical properties of additive manufactured copper alloy. Materials Letters. 2016;**179**:38-41. DOI: 10.1016/j.matlet.2016.05.064

[34] Siva Prasad H, Brueckner F, Volpp J, Kaplan AFH. Laser metal deposition of copper on diverse metals using green laser sources. International Journal of Advanced Manufacturing Technology. 2020;**107**:1559-1568. DOI: 10.1007/s00170-020-05117-z

[35] Hess A, Schuster R, Heider A, Weber R, Graf. Continuous wave laser welding of copper with combined beams at wavelengths of 1030 nm and of 515 nm. Physics Procedia. 2011;**12**:88-94. DOI: 10.1016/j.phpro.2011.03.012

[36] Asano K, Tsukamoto M, Funada Y, Sakon Y, Abe N, Sato Y, et al. Copper film formation on metal surfaces with 100 W blue direct diode laser system. Journal of Laser Application. 2018;**30**: 032602. DOI: 10.2351/1.5040635

[37] Polenz S, Kolbe C, Bittner F, López E, Brueckner F, Leyens C. Integration of pure copper to optimize heat dissipation in injection mould inserts using laser metal deposition. Journal of Laser Application. 2020;**33**: 012029. DOI: 10.2351/7.0000303

[38] Bai Y, Williams CB. An exploration of binder jetting of copper. Rapid Prototyping Journal. 2015;**21**:177-185. DOI: 10.1108/RPJ-12-2014-0180

[39] Bai Y, Wagner G, Williams CB. Effect of particle size distribution on powder packing and sintering in binder jetting additive manufacturing of metals. Journal of Manufacturing Science and Engineering. 2017;**139**: 081019. DOI: 10.1115/1.4036640

[40] Bai Y, Williams CB. Binder jetting additive manufacturing with a particle-free metal ink as a binder precursor. Materials and Design. 2018;**147**:146-156. DOI: 10.1016/j.matdes.2018.03.027

[41] Bai Y, Williams CB. The effect of inkjetted nanoparticles on metal part properties in binder jetting additive manufacturing. Nanotechnology. 2018;**29**:395706. DOI: 10.1088/1361-6528/aad0bb

[42] Kumar A, Bai Y, Eklund A, William CB. Effects of hot isostatic pressing on copper parts fabricated via binder jetting. Procedia Manufacturing. 2017;**10**:935-944. DOI: 10.1016/j.promfg.2017.07.084

[43] Miyanaji H, Rahman KM, Da M, William CB. Effect of fine powder particles on quality of binder jetting parts. Additive Manufacturing. 2020;**36**:101587. DOI: 10.1016/j.addma.2020.101587

[44] Miyanaji H, Ma D, Atwater MA, Darling KA, Hammond VH, Williams CB. Binder jetting additive manufacturing of copper foam structures. Additive Manufacturing. 2020;**32**:100960. DOI: 10.1016/j.addma.2019.100960

[45] Digital Metal, Available from: https://digitalmetal.tech/digital-metal-first-in-the-world-to-offer-pure-copper-as-printing-material-to-its-market-leading-high-precision-metal-binder-jetting-system/ [Accessed: February 25, 2021]

[46] ExOne. Available from: https://www.exone.com/en-US/Resources/News/ExOne_Maxxwell_3D_Print_Copper_Coil_Winding [Accessed: September 2, 2021]

[47] Osswald TA, Puentes J, Kattinger J. Fused filament fabrication melting model. Additive Manufacturing. 2018;**22**:51-59. DOI: 10.1016/j.addma.2018.04.030

[48] Hwang S, Reyes EI, Moon K, Rumpf RC, Kim NS. Thermo-mechanical characterization of metal/polymer composite filaments and printing parameter study for fused deposition modeling in the 3D printing process. Journal of Electronic Materials. 2015;**44**:771-777. DOI: 10.1007/s11664-014-3425-6

[49] Masood SH, Song WQ. Development of new metal/polymer materials for rapid tooling using fused deposition modelling. Materials and Design. 2004;**25**:587-594. DOI: 10.1016/j.matdes.2004.02.009

[50] Singh R, Davim JP, editors. Additive Manufacturing: Applications and Innovations. 1st ed. Boca Raton, FL: CRC Press/Taylor & Francis Group; 2018. p. 280. ISBN 9780367780944

[51] Business Wire. McLaren Formula 1 Racing Deploys Stratasys Additive Manufacturing to Improve 2017 Car Performance [Internet]. https://www.businesswire.com/news/home/20170407005040/en/McLaren-Formula-1-Racing-Deploys-Stratasys-Additive [Accessed: August 23, 2018]

[52] Forsmark, JH. Additive Manufacturing at Ford Motor Company: Past, Present, and Future. 2 June 2016. Retrieved from Smart Manufacturing Seminar Series: Available from: http://www.sme.org/uploadedFiles/Smart_Manufacturing_Education_Series/Forsmark.pdf

[53] Wohlers TT, Caffrey T, Campbell RI. Wohlers Report 2016: 3D Printing and Additive Manufacturing State of the Industry: Annual Worldwide Progress Report. Colorado, USA: Wohlers Associates; 2016. ISBN 978-0-9913332-2-6

[54] Singh S, Ramakrishna S, Singh R. Material issues in additive manufacturing: A review. Journal of Manufacturing Processes. 2017;**25**:185-200. DOI: 10.1016/j.jmapro.2016.11.006

[55] Zhang Y, Bai S, Riede M, Garratt E, Roch A. A comprehensive study on fused filament fabrication of Ti-6Al-4V structures. Additive Manufacturing. 2020;**34**:101256. DOI: 10.1016/j.addma.2020.101256

[56] Zhang Y, Poli L, Garratt E, Foster S, Roch A. Utilizing fused filament fabrication for printing iron cores for electrical devices. 3D Printing and Additive Manufacuring. 2020;7:279-287. DOI: 10.1089/3dp.2020.0136

[57] Liu B, Wang Y, Lin Z, Zhang T. Creating metal parts by fused deposition modeling and sintering. Materials Letters. 2020;**263127252**. DOI: 10.1016/j.matlet.2019.127252

[58] Allahverdi M, Danforth SC, Jafari M, Safari A. Processing of advanced electroceramic components by fused deposition technique. Journal of the European Ceramic Society. 2001;**21**:1485-1490. DOI: 10.1016/S0955-2219(01)00047-4

[59] Yang H, Yang S, Chi X, Evans JRG. Fine ceramic lattices prepared by extrusion freeforming. Journal of Biomedical Materials Research Part B: Applied Biomaterials. 2006;**79B**:116-121. DOI: 10.1002/jbm.b.30520

[60] Agarwala MK, Bandyopadhyay A, van Weeren R, Langrana NA, Safari A, Danforth SC, et al. Fused deposition of ceramics (FDC) for structural silicon nitride components. In: 1996 International Solid Freeform Fabrication Symposium. Austin, Texas: University of Texas; 1996. pp. 335-344; http://utw10945.utweb.utexas.edu/1996TOC

[61] Iyer S, McIntosh J, Bandyopadhyay A, Langrana N, Safari A, Danforth SC, et al.

Microstructural characterization and mechanical properties of Si3N4 formed by fused deposition of ceramics. International Journal of Applied Ceramic Technology. 2008;**5**:127-137. DOI: 10.1111/j.1744-7402.2008.02193.x

[62] Venkataraman N, Rangarajan S, Matthewson MJ, Harper B, Safari A, Danforth SC, et al. Feedstock material property—Process relationships in fused deposition of ceramics (FDC). Rapid Prototyping Journal. 2000;**6**:244-253. DOI: 10.1108/13552540010373344

[63] Lombardi JL, Hoffman RA, Waters JA, Dragan P. Issues associated with EFF & FDM ceramic filled feedstock formulation. In: 1997 International Solid Freeform Fabrication Symposium. Austin: University of Texas; 1997. http://hdl.handle.net/2152/71415

[64] Marsh AC, Zhang Y, Poli L, Hammer N, Roch A, Crimp M, et al. 3D Printed bioactive and antibacterial silicate glass-ceramic scaffold by fused filament fabrication. Materials Science and Engineering: C. 2021;**118**:111516. DOI: 10.1016/j.msec.2020.111516

[65] Abel J, Scheithauer U, Janics T, Hampel S, Cano S, Müller-Köhn A, et al. Fused filament fabrication (FFF) of metal-ceramic components. Journal of Visualized Experiments. 2019;**143**: 57693. DOI: 10.3791/57693

[66] Scheithauer U, Bergner A, Schwarzer E, Richter HJ, Moritz T. Studies on thermoplastic 3D printing of steel–zirconia composites. Journal of Materials Research. 2014;**29**:1931-1940. DOI: 10.1557/jmr.2014.209

[67] Gonzalez-Gutierrez J, Cano S, Schuschnigg S, Kukla C, Sapkota J, Holzer C. Additive manufacturing of metallic and ceramic components by the material extrusion of highly-filled polymers. A Review and Future Perspectives. Materials. 2018;**11**:36. DOI: 10.3390/ma11050840

[68] AM Extrusion. Metal and Ceramic 3D Printing [Internet]. 2021. Available from: https://am-extrusion.com [Accessed: October 10, 2021]

[69] BASF. Material Portfolio [Internet]. 2021. Available from: https://forward-am.com/?utm_term=ultrafuse%20316l&utm_campaign=mf%20%7C%20EN%20%7C%20DACH%20%7C%20Search%20%7C%20Products&utm_source=adwords&utm_medium=ppc&hsa_acc=3431802958&hsa_cam=11406127154&hsa_grp=111450800786&hsa_ad=476572011030&hsa_src=g&hsa_tgt=kwd-837123926108&hsa_kw=ultrafuse%20316l&hsa_mt=p&hsa_net=adwords&hsa_ver=3&gclid=EAIaIQobChMIuNHP1LTA8wIVkgsGAB1bdwOtEAAYASAAEgJm4fD_BwE [Accessed: October 10, 2021]

[70] Simpson N, Jung J, Helm A, Mellor P. Additive manufacturing of a conformal hybrid-strand concentrated winding topology for minimal AC loss in electrical machines. In: IEEE Energy Conversion Congress and Exposition (ECCE). Vancouver; 2021. pp. 3844-3851. DOI: 10.1109/ECCE47101.2021.9595059

[71] Prototypen Zentrum. Kupfer additiv gefertigt!? Wo? Natürlich bei PTZ! [Internet]. Available form: https://www.ptz-prototypen.de/de/unternehmen/aktuelles/235-kupfer-additiv-gefertigt.html [Accessed: September 15, 2021]

[72] 3D Printing Industry. Digital Metal Launches New Pure Copper Powder for 3D Printing [Internet]. 2021. Available from: https://3dprintingindustry.com/news/digital-metal-launches-new-pure-copper-powder-for-3d-printing-185259/ [Accessed: October 10, 2021]

[73] Space News. Pangea Aerospace Tests Aerospike Engine [Internet]. 2021.

Available from: https://spacenews.com/pangea-aerospace-tests-aerospike-engine/ [Accessed: January 4, 2022]

[74] Infinite Flex GmbH, Powder for Laser Sintering, and Use [Patent]. WO002020254108A1. Publication 2020

[75] 3DPrint. Researchers Use Single Mode 1 kW Fiber Laser to 3D Print Pure Copper Powder [Internet]. 2018. Available from: https://3dprint.com/224845/3d-printing-pure-copper-powder/ [Accessed: August 20, 2021]

[76] Headmade Materials. Materials [Internet]. 2021. Available from: https://www.headmade-materials.de/en/materials [Accessed: November 13, 2021]

Chapter 12

Nanometric 3D Printing of Functional Materials by Atomic Layer Deposition

David Muñoz-Rojas, Matthieu Weber, Christophe Vallée, Chiara Crivello, Abderrahime Sekkat, Fidel Toldra-Reig and Mikhael Bechelany

Abstract

Atomic layer deposition (ALD) is a chemical vapour deposition (CVD) method that allows the layer-by-layer growth of functional materials by exposing a surface to different precursors in an alternative fashion. Thus, thanks to gas-solid reactions that are substrate-limited and self-terminating, precise control over thickness below the nanometer level can be achieved. While ALD was originally developed to deposit uniform coatings over large areas and on high-aspect-ratio features, in recent years the possibility to perform ALD in a selective fashion has gained much attention, in what is known as area-selective deposition (ASD). ASD is indeed a novel 3D printing approach allowing the deposition of functional materials (for example metals to oxides, nitrides or sulfides) with nanometric resolution in Z. The chapter will present an introduction to ALD, which will be followed by the description of the different approaches currently being developed for the ASD of functional materials (including initial approaches such as surface pre-patterning or activation, and newer concepts based on spatial CVD/ALD). The chapter will also include a brief overview of recent works involving the use of ALD to tune the properties of 3D printed parts.

Keywords: spatial atomic layer deposition, spatial chemical vapor deposition, area-selective deposition, atomic layer infiltration, surface nanoengineering, 3D printing of functional materials

1. Introduction

Vapor-phase techniques are powerful approaches for the deposition of functional thin films of different materials, including metals and compounds such as oxides, nitrides, and even organic materials and composites onto a substrate [1–3]. There are two types of vapor deposition methods, namely physical and chemical. Physical vapor deposition (PVD) methods involve a change of state (i.e., evaporation and recondensation) of a source, and include, among other, sputtering, pulsed laser deposition or different evaporation approaches [4]. While PVD methods yield materials of high quality with tunable properties, they are performed in high vacuum and often high temperatures, using sophisticated equipment. Finally, the

IntechOpen

low vacuum process results in a line-of-sight type of coating (i.e., only taking place on the directly exposed surfaces.

The possibility to pattern and 3D print materials at different scales has a tremendous impact on many technologies and applications. Over the years. Different 3D printing approaches have been developed allowing such patterning. This include, to name a few, aerosol jet printing (also known as Maskless Mesoscale Materials Deposition or M_3D) [5], ink jet and screen printing [6], laser chemical vapor deposition (LCVD) [7], laser-induced forward transfer (LIFT) [8, 9] or micro stereo lithography and multiphoton lithography [10]. The interested readers are encouraged to the cited references for more details on these methods. In this chapter, we focus on 3D printing approaches based on ALD. A brief introduction to CVD and ALD is thus presented next.

Chemical vapor deposition (CVD) approaches on the other hand rely on chemical reactions between different precursors on and over a surface. In conventional CVD, the precursors are injected in the reactor at the same time and the reaction is activated by heat (hot substrate) or by other energy sources, such as plasma. A scheme representing the reaction chamber is shown in **Figure 1a** [11]. This technique allows the deposition of high-quality films [12], and is largely used by the industry. Nonetheless, CVD is governed by the diffusion of the different gas precursors, and therefore, the deposition of extremely thin films with a thickness control at the sub-nanometer level [13], and the uniform coating of large areas or high-aspect-ratio/porous features is extremely difficult [14].

Such limitations prompted the development of an alternative method, namely, atomic layer deposition (ALD). ALD is indeed a CVD method but it is characterized by having the substrate exposed to the different precursors one at a time, and not simultaneously as in CVD. Thus, in typical ALD processes, a precursor is first injected in a deposition chamber where the substrate is located. The precursor can then react with active sites on the surface (i.e., undergoing a chemisorption) until the latter is saturated. A purge step is then applied to eliminate excess precursor and reaction byproducts. Then a second reactant is injected that will react with

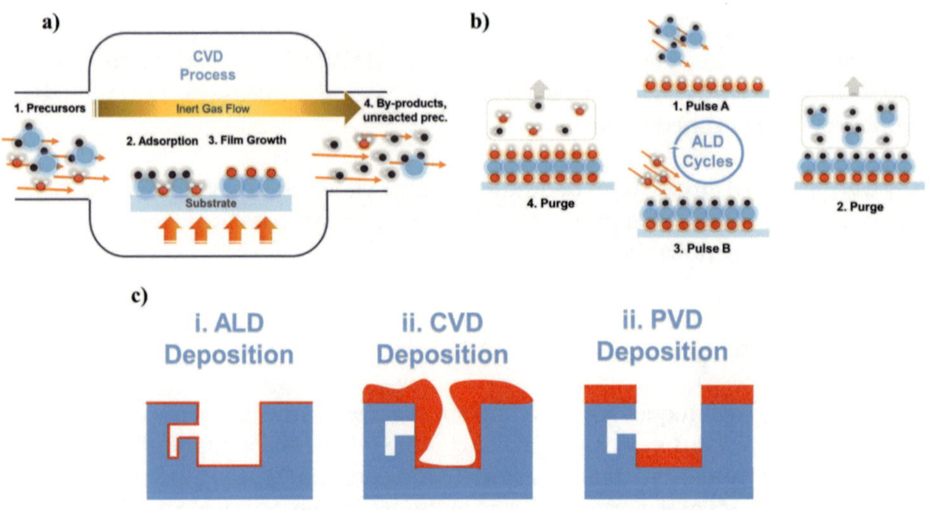

Figure 1.
(a) CVD mechanism where the precursor is adsorbed on the surface at relatively high-temperature followed by the film growth and a release of volatile byproducts, (b) ALD process: Schematic of one ALD cycle of monolayer growth. The first step consists in exposing the substrate to the precursor followed by a purge step to remove all the byproducts an excess precursor, then another step with a co-reactant agent and the final step in which the byproducts an excess precursors are purged again, (c) illustration of edge coverage for ALD, CVD, and PVD.

the preciously adsorbed layer. After the reaction is completed, again a purge step is necessary to eliminate excess reactant and reaction byproducts. Such an ALD cycle is shown in **Figure 1b**. As a result of this sequential exposure to the different reactants, the ALD process is surface-selective and self-terminating, which in turn offers unique control over film thickness at the angstrom level (i.e., a given growth per cycle, GPC, being obtained for each process as a function of the reactor geometry and precursors used) and allows the conformal coating of porous, complex and high-aspect-ratio substrates. The films are also compact and free of pinholes and can be obtained at low temperatures (even room temperature) due to the high reactivity of ALD precursors. The reader is referred to reviews and books dedicated to ALD for more information [1–3]. **Figure 1c** shows a sketch of the different types of coating obtained over the high-aspect-ratio features when using the different techniques discussed.

Over the years, the number of materials that can be deposited by ALD has grown enormously, including pure elements (e.g., metals), nitrides, sulfides, oxides, fluorides, etc. (see the atomic limits site, with includes an ALD materials database that is permanently being updated [15]). While at the origin the main motors of the ALD development were the deposition of homogeneous coatings over large areas or high aspect-ratio features, in the last years, there have been innovative developments in the ALD field that allow the localized and topological deposition of functional materials. This opens the door to its application as a new nano-to-macro 3D printing technology based on gas precursors. These recent developments, namely, area-selective deposition (ASD) and different spatial approaches, are presented in Sections 2 and 3, respectively. Finally, the unique assets of the ALD technique are ideal to tune the properties of pieces fabricated by conventional 3D printing approaches. Section 4 presents a brief overview of recent results on this line. The chapter finishes with some conclusive remarks.

2. Area-selective deposition (ASD)

For more than 50 years, the shrinking of microelectronic devices has involved successive steps of deposition, lithography and etching. Indeed, unlike building a house, it is not possible to directly draw the walls or pillars of a chip on a 300 mm substrate. It is therefore necessary to first cover the whole substrate with a thin layer, before removing part of it by the steps of lithography (to draw the object) and etching (to remove what should not remain on the surface). This is called a top-down approach. The reduction of the dimensions of microelectronic devices in the last 10 years to nanometric scales has greatly complicated these steps and increased their cost. Indeed, for many years, the wavelengths used to draw were greater than the desired line thickness. It was therefore necessary to make lithography more complex by integrating etching/deposition steps to achieve the desired dimensions, such as multi-patterning (Self Aligned Double or Quadrupole Patterning—SADP and SAQP). Thus, these steps allowing to obtain locally nanometric materials on the substrate are now complex, time-consuming and expensive. They must also be done with nanometric placement precision, which is already a real challenge.

The alternative solution to this increasingly complex approach is to deposit the material directly and selectively on the desired surface without having to resort to lithography steps. This so-called selective growth on a surface is a bottom-up approach and is known as area-selective deposition (ASD) [16, 17]. In an ideal ASD process, a thin film should be uniformly deposited in the desired growth region while no deposition should be observed in the desired no-growth region. This requires the use of a surface selective deposition process, with controlled growth at the atomic scale, and thus ALD is the one that seems to be the most adapted. Indeed, a growing number of researchers working on the ALD process are now trying to establish strategies from this process to have a material deposited selectively on a surface. The three main strategies are: (i) to use an inherent selectivity of the precursor/substrate couple [18, 19]; (ii) to block the growth on the no-growth area by a pre-deposition treatment [20, 21]; (iii) to promote the growth on the growth area by a pre-deposition treatment [22]. Whatever the strategy, we observe growth on all surfaces after a certain number of cycles, or at best, a little defectivity with nuclei on the no-growth area, i.e., the selectivity fades out during the successive ALD cycles. It was then proposed to regularly add the surface treatment step (passivation step) in the ALD cycles changing a cycle from a (treatment + AB) process to an (ABC) cycle with the treatment reinjected regularly [23]. Another proposed solution is to use super-cycles with the injection of etching steps every x ALD cycles, this is called ASD by super-cycles of deposition-etching [24–28]. In the end, with these different ASD strategies, passivation or super-cycles, [29] the possibility to deposit thin films of more than 10 nm on a previously selected surface has been successfully demonstrated. However, the additional steps (passivation, etching) increase the processing time. It takes more than 24 h for passivation by a self-assembled monolayer (SAM) using chemical baths. The insertion of etching steps in an ALD process also increases the processing time by a factor of 2–3 compared to a conventional ALD process. Finally, the insertion of passivation or etching chemistries can induce contamination of the deposited layer as well as process drifts (with for example a modification of the chemistry of the ALD reactor walls during the etching step). Nevertheless, ASD has the potential to deposit functional

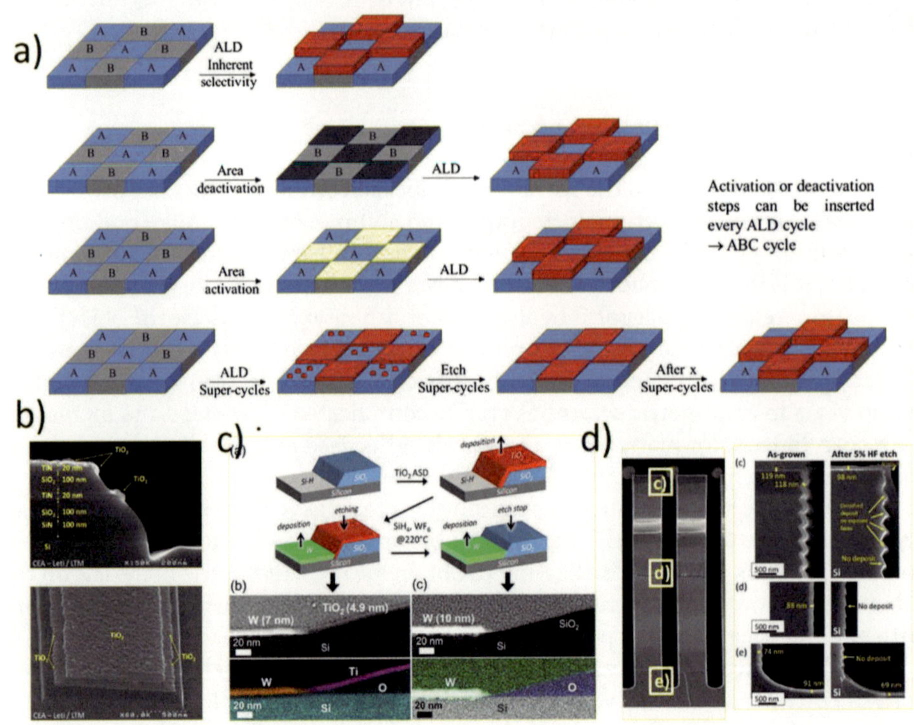

Figure 2.
(a) Illustration of four different strategies for an area selective deposition where a is the growth area and B the no-growth area; (b–d) examples of ASD using deposition and etch: (b) TiO_2 on TiN vs. Si/SiO_2 (reprinted with permission) [25], (c) patterning at the microscale of TiO_2 and W films (reprinted with permission) [30], and (d) topographical selective deposition of Ta_2O_5 on Si (reprinted with permission) [31].

materials with complex 3D shapes and nanometric resolution, well beyond the possibilities offered by standard 3D printing methods. This can be done by different approaches, as shown in **Figure 2**.

Although, ASD offers a huge potential for the 3D printing of functional materials at resolutions orders of magnitude below what can be achieved with conventional 3D printing approaches, the different steps it implies (i.e., surface prepattering, regeneration of the selectivity) make them harder to work with. It would thus be desirable to develop an ALD approach that could allow the direct deposition of patterned materials. This can indeed be achieved, as detailed in the next section, by using different *spatial* approaches. The question that arises now is whether this approach can be used to develop ASD processes on larger surfaces and at dimensions that are no longer nanometric but micrometric or millimetric. The natural answer is yes. Indeed, an optimized ALD process is uniform across m-scale objects. Using a spatial ALD technology, we can then imagine selectively printing on a large surface with ALD: 3D printing. This is described in the next section.

3. 3D printing of functional materials based on spatial CVD/ALD approaches

As explained in the introduction, the unique assets of ALD are the result of having a surface-limited, self-terminated reaction between gas reactants and the surface of a substrate. To limit the reaction to the surface, the ALD is based in alternate exposures of the precursors to the substrate. Traditionally, this has been done by sequential injection of the precursors in a deposition chamber followed by purging steps, thus in a temporal approach, as detailed in **Figure 1b** above and in the scheme below (**Figure 3a**). An alternative approach consists in having a continuous injection of the different reactants but in different locations of the reactor, keeping them separated by a region of inert gas. Then, by alternatively exposing the substrate to the different regions, the ALD cycle is reproduced (**Figure 3b**). This approach is known as Spatial ALD (SALD) [33–36]. The first advantage of processing in the spatial mode is that the process can become much faster (up to two orders of magnitude) since no purging step is required.

The SALD concept is very versatile and can indeed be applied in different ways [33, 37]. SALD can even be performed at atmospheric pressure (i.e., no vacuum processing) and even in the open air (i.e., no deposition chamber), and this is sometimes referred to as Atmospheric-Pressure SALD (AP-SALD). This is the case of the close-proximity approach based on a manifold injection head, originally presented by Kodak [38]. In this particular approach, the different reactants are carried to the injection head where they are distributed along alternate parallel channels (**Figure 3c**) [39]. By proper design of the head, the different flows can be kept separated provided the substrate is at close proximity of the head (i.e., 50–200 μm). Then by scanning the substrate back and forth under the head the ALD cycles are achieved. It is worth noting that since the size and area of the deposition depend on the head size and substrate scan distance, this SALD approach can already be seen as an ASD approach at the cm scale.

Close-proximity SALD approaches based on injection heads have several extra appealing advantages. The first one is that deposition can be also performed in spatial CVD (SCVD) mode. Then, crosstalk between the different reactants above the surface of the substrate is allowed. In this case, the deposition rate can be faster, but care must be taken since the properties of the materials deposited could change [32]. The impact of the change in the film properties when passing from the SALD to SCVD mode has to be evaluated depending on the intended application, but

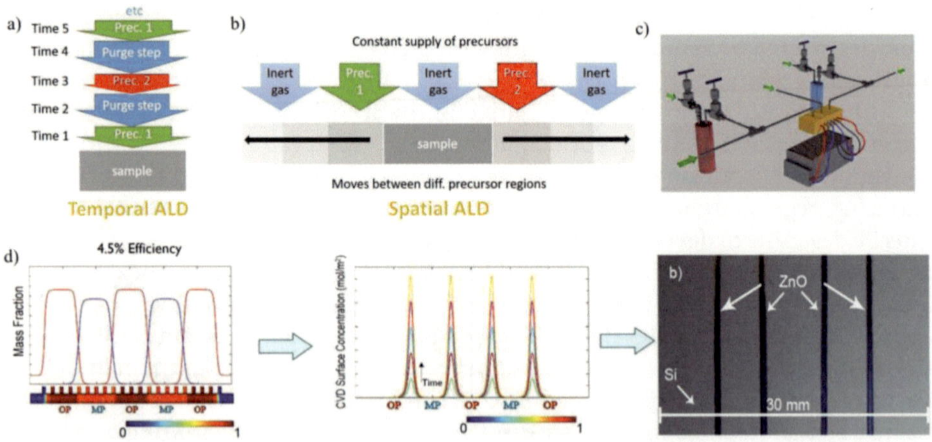

Figure 3.
(a) Schematic representation of the classical temporal ALD approach with the different characteristic steps of an ALD cycle: (1) injection of the first precursor, (2) purging step, (3) injection of the second precursor, (4) purging step, separated in time. (b) Schematic representation of the spatial ALD approach, where the precursors are injected continuously in the reaction chamber in different locations separated by an inert gas and the sample is exposed to the different regions to reproduce the ALD cycle. (c) Scheme of the close-proximity AP-SALD approach based on a manifold injection head: the precursors are carried out from the containers of the head where they are distributed in parallel alternative channels. (d) COMSOL simulation of the mass fraction of each precursor present in different areas of the substrate (left). In these cases the evacuation of the precursors is not efficient and thus cross-talk is observed, yielding a CVD reaction on the zones where the precursors meet (see COMSOL simulation in the center). If a deposition is made in static mode (i.e., without moving the substrate), 4 lines of oxide can be obtained, as shown in the optical image (right) where 4 lines of ZnO have been deposited on a Si wafer in this way (adapted from Ref. [32]).

several works have demonstrated that the SCVD can be used to deposit components for functional devices [40]. In addition, the possibility of having SCVD opens the door to a new ASD approach. Indeed, the CVD reaction can be located in different areas above the substrate. **Figure 3d** presents a computational fluid dynamics (CFD) simulation that shows the areas over the substrate where the different reactants meet and thus react when the deposition is performed in certain SCVD conditions. Then, by performing a static deposition (i.e., without the substrate scan that is needed to perform the spatial ALD cycles) growth of the films can be localized to the regions where the reactants meet (see the four ZnO lines obtained by this approach in **Figure 3d**). This constitutes a new alternative approach of ASD at a higher scale and much faster deposition rate than the traditional ASD approaches based on ALD that have been described in the previous section [32].

The second advantage of using a close-proximity SALD approach based on an injection head is that the system can be customized by simply modifying the injection head. While this is so, the modification and fabrication of the head can result very difficult, if not impossible, thus limiting the potential of the approach (see **Figure 4a** where the scheme of a standard SALD head is shown. It comprises several parts that need to be fabricated separately and then soldered, and the distribution of the different gas flows to the head is quite complex involving many pipes). To overcome this limitation, D. Muñoz-Rojas' group at the Laboratoire des Matériaux et du Génie Physique (LMGP, Grenoble, France) has introduced the utilization of 3D printing for the fabrication of customized SALD injection heads [41]. This allows having more freedom to design the head and, for example, the gas distribution can be incorporated in the body of the head (**Figure 4b** and **c**) [41–43]. Plastic heads can be printed for depositions taking place at low temperatures while metal 3D printing is also possible for higher temperatures [44]. Thanks to 3D printing, the design of the heads can be easily customized. This is very convenient to easily

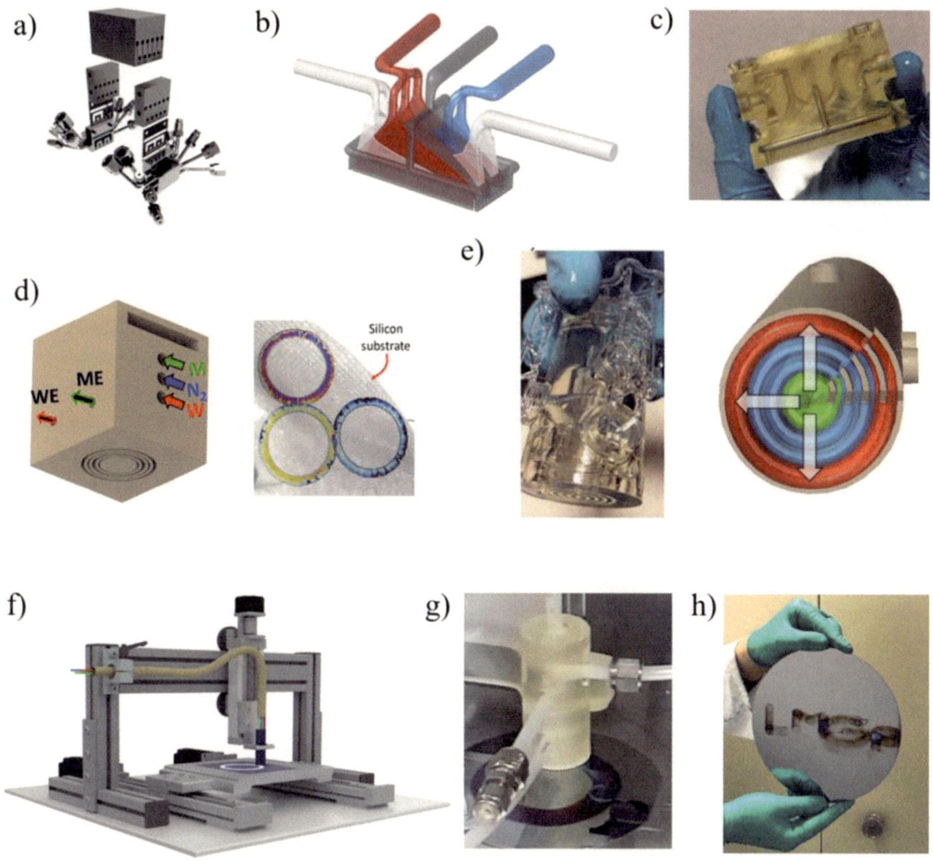

Figure 4.
(a) Scheme of a close-proximity SALD head made of several parts and fabricated by conventional approaches. (b) 3D scheme of a head design integrating the gas distribution for the different gases inside its body: metallic precursor in green, co-reactant in red, inert gas in blue and exhaust in black. (c) Head printed with clear resin where the distribution channels can be observed. (d) 3D scheme of a head designed for circular shape deposition in static SCVD mode. ZnO circles with different thicknesses are shown. (e) Picture of a printed SALD pen (left), bottom view of the concentric gas outlets in the SALD pen approach allowing deposition in any direction (right). (f) Scheme of the SALD pen installed in a 3D table. (g) Scheme of a SALD pen implemented in the XYZ table and drawing ZnO in a circular pattern. (h) LMGP initials on a Si wafer drawn with the 3D printed SALD pen (adapted with permission from Ref. [41]).

modify the area of deposition, and also to have free-form patterns when performing SCVD with custom heads (**Figure 4d**) [41].

The possibility to deposit free-form patterns without having to modify the head for each design would also be appealing. This can indeed be done if instead of using parallel channels, the head is designed so that concentric channels are used. In this way, no matter which direction the head moves, the substrate will be exposed to the different reactants, thus leading to ALD film growth (**Figure 4e**). Such a head can again be readily implemented by 3D printing. D. Muñoz-Rojas' group demonstrated that such a SALD pen can be printed and used to deposit free-form patterns when installed in an XYZ table, in this case with a resolution going down to several mm (**Figure 4f-h**) [41]. This represents a new 3D printing approach that is based on gas precursors and that offers nanometric resolution in Z. Here again, the resolution of the obtained patterns in X-Y depends on the head design and the possibility to scale it down. Indeed, the latter work by Midani et al. presented a similar concept in which sub-millimiter resolution was achieved by inserting a capillary in the central metal precursor channel of the SALD pen [45].

Certainly, the advances in the different 3D printing technologies will allow de fabrication of SALD heads with smaller channels, which will extend the possibilities of SALD for depositing patterns of functional materials down to the micrometer scale in X-Y.

4. ALD for conventional additive manufacturing: recent works involving the use of ALD to tune the surface and properties of 3D printed parts

Additive manufacturing (AM), also known as 3D printing, is recognized as a revolutionary technology, which has primarily been used in the field of engineering to create customized prototypes [46–48]. 3D printing has now become a subject of great interest and is extensively applied in many areas, such as prototyping, medicine [49] or aerospace [50], since it allows new products with complex geometries and microarchitecture (multiple pore shape and size) to be imagined, designed and fabricated. However, the material from which the designed products are made is still limited by the 3D-printing material itself. Even if the number of available materials that can be printed is expanding [48], most of the manufactured objects are made of polymer or stainless steel. Thus, a post-treatment may be required to control the nature and chemistry of the product surface and offer it its desired functionality. As illustrated in **Figure 5** and discussed below, ALD is a highly appealing technique to expand the potential of 3D printing through coating or infiltration of the printed parts.

As seen previously, ALD can be used as an innovative and novel 3D printing route, to prepare customized and complex 3D structures at the nano-to-cm scale. In addition, this technology can also be used to precisely tune the surfaces of 3D printed objects that were manufactured using more "conventional" additive manufacturing approaches such as fused deposition modeling, inkjet printing, stereolithography, selective laser sintering (SLS), powder bed fusion or even bioprinting [46–48]. ALD allows the preparation of thin films with a sub-nanometer thickness control, high uniformity and excellent conformality even on high aspect ratios substrates, a unique capability, as discussed in the first section of this chapter [3, 53–58]. As ALD allows the conformal coating of complex substrates with nanolayers made of an expanding

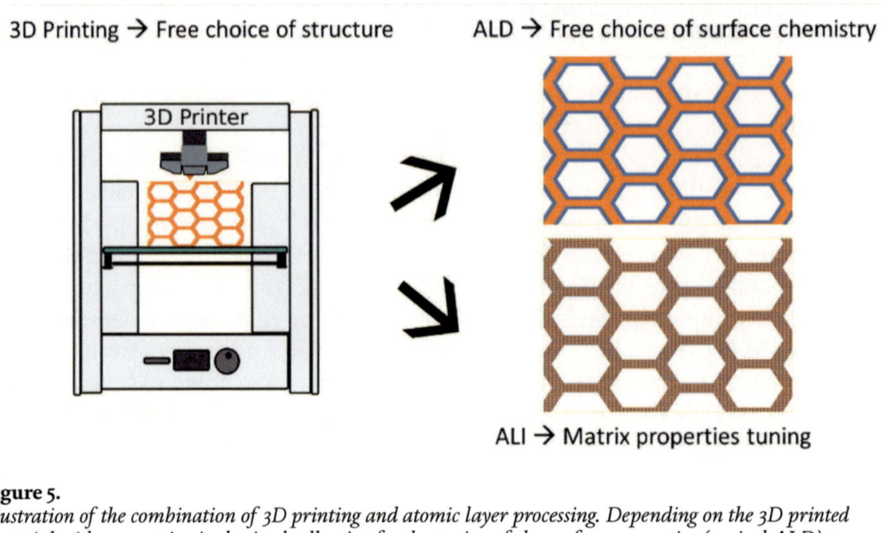

Figure 5.
Illustration of the combination of 3D printing and atomic layer processing. Depending on the 3D printed material, either a coating is obtained, allowing for the tuning of the surface properties (typical ALD); or inorganic components are introduced to the subsurface of the 3D objects (atomic layer infiltration, ALI [51, 52]), permitting the tuning of the matrix.

number of materials [2, 15], such as oxides [59, 60], metals [61], nitrides [62] and sulfides [63], the combination of this route with 3D printing can be beneficial to a myriad of applications.

A large number of 3D printed objects are made of polymers, the current mainstream materials being ABS (acrylonitrile butadiene styrene) and PLA (polylactic acid). However, when performing ALD on 3D printed objects based on such polymeric materials, some considerations must be taken. The first obvious consideration is related to the ALD process temperature, which has to be lower than the polymer melting point. 3D printing materials such as ABS and PLA will already be deformed when the temperature is higher than 200°C. The ALD processes must therefore be compatible with rather low temperatures. Furthermore, as depicted by the review of Parsons and co-workers [64], the ALD precursors often infiltrate and react with polymeric substrates, which can alter the eventual 3D printed products. As shown by numerous studies, the risk of reaction between the polymer and the precursors increases with temperature and long exposures times. In addition, the presence of functional groups in the polymer chains also increases the potential infiltration of the ALD precursors [51, 52, 59, 64–68]. Thus, even if most of the ALD processes are compatible with the coating of 3D printed objects, these considerations must be taken into account and the processes have to be tuned accordingly to coat certain 3D printed materials. When the processes developed are compatible, the unique capability of ALD to coat complex objects with such control over the layer deposited, makes this route particularly relevant and attractive. This innovative combinatorial approach has been used for different and various applications, such as aerospace, photoelectrocatalysis, filtration, biomedicine, or solid-state batteries.

Kestila et al. combined polymeric additive manufacturing and an ALD-coating to produce satellite propulsion components with improved structural integrity and thermal resistance [69]. The components were made of two different polymers, namely acrylonitrile butadiene styrene (ABS) and polyamide, and were coated with alumina by ALD. The Al_2O_3 layer allowed to enhance the structural integrity for the polymeric restrictors and progressively smoothed out the PA surface improving the argon flow through the restrictor, which might be due to increased surface smoothness [69]. Heikinen et al. have recently shown that ALD of alumina on porous 3D printed ABS plastics permits to considerably lower their vacuum degassing. Nyman et al. have also confirmed the low outgassing of ABS, but also polyether ether ketone (PEEK), polycarbonate (PC), and nanodiamond-doped polylactide (ND-PLA) 3D printed materials with an Al_2O_3 ALD coating [70]. Thus, the combination of plastic 3D printing with ALD opens prospects for the fabrication of laboratory vacuum tools, and is also suited for spacecraft tools and in-space manufacturing applications [70, 71]. Moll et al. also coupled powder bed additive manufacturing with CVD and ALD of nitrides, to prepare 3D Ti-6Al-4 V structures highly resistant to high-temperature oxidizing environments. Coupling CVD and ALD on the 3D printed objects permitted to obtain thick coating and roughness reduction by CVD, and filling of narrow defects and reactivity mitigation by ALD [72].

Browne et al. employed additive manufacturing and ALD for photoelectrocatalysis, by depositing TiO_2 onto 3D-printed electrodes. These electrodes were initially printed in inert stainless steel, and gained their catalytic functionality thanks to the ALD coating. The conformality allowed by ALD successfully permitted these 3D-printed electrodes to be used as photoanodes for water oxidation. The results presented have shown that the 3D-printed stainless steel electrode coated with ALD of TiO_2 were considerably more active towards the water oxidation, and that the catalytic activity was enhanced by increasing the number of ALD cycles applied [73]. The team of Pumera et al. recently applied ALD to 3D-printed nanocarbon/polylactic acid electrodes to coat them with metal dichalcogenide MoS_2 nanolayers [74]. The

MoS_2 coated electrodes were then successfully applied for photoelectrocatalytic hydrogen evolution reaction (HER). Recently the group of M. Bechelany from the Institut Européen des Membranes (IEM, Montpellier, France) has developed in collaboration with the University of Zaragoza the functionalization of 3D printed ABS filters with MOF (Metal-Organic Framework) for toxic gas removal [75]. The fabrication approach at low temperature includes ALD of Zinc oxide on the ABS 3D printed filter followed by the hydrothermal conversion of ZnO to ZIF-8, Zeolitic Imidazolate Framework. The obtained filters show a good adsorption performance for dimethyl methylphosphonate, thus demonstrating their potential for toxic gas capture applications. Such types of 3D printed filters with an active MOF layer could have a wide range of applications in environmental fields such as adsorption systems for removing toxic gases or water pollutants.

In the biomedical field, the combinatorial approach has been applied to prepare silver-coated titanium orthopedic implants. [76] Using the selective laser melting (SLM) 3D-printing technique, titanium orthopedic implants have been fabricated with intricate geometries. The surface chemistry of the prepared implants has then been modified by coating them with a silver nanolayer by ALD. The inhibition of bacterial colonization obtained thanks to the silver coating resulted in the drastic reduction of the pathogenic biofilm. This result, combined with the increase of the vascularization and the osseointegration observed, opens a new path to this combinatorial approach for clinical orthopedic applications [76]. The "pure marriage" between 3D printing and ALD has also been exploited by Xue et al., who tailored the surface of 3D printed plastic earplugs using plasma-assisted ALD [75]. By combining 3D printing, plasma-assisted ALD and hydrothermal process, they loaded a layer of ZnO nanoarrays on the surface of the earplugs and thus improved the antibacterial properties of the earplugs, which enhanced the safety of the ear devices. In addition, they have shown that the sound insulation performances were higher than those of traditional earplugs. Finally, the field of solid-state batteries benefited as well from the combination of 3D printing and ALD. For example, thanks to an innovative 3D-printing ink formulation, a cell-based on a 3D-printed stacked array of LLZ ($Li_7La_3Zr_2O_{12}$, a solid lithium conductor) and lithium electrodes was fabricated, and ALD of alumina has been performed at the surface of the LLZ to allow the wetting of lithium [77, 78]. The ability to 3D-print solid electrolytes enables the manufacturing of unique ordered structures, and ALD permits their efficient functionalization, improving the overall efficiency of the battery device.

These few selected studies demonstrate the great potential of combining additive manufacturing and ALD. The combinatorial approach allows the fast prototyping of functional products with the additional precise control over their surface chemistry. As depicted in the presented examples, the benefits of combining 3D printing and ALD nanocoatings can be applied to many complex surfaces, and the lack of materials that can be 3D printed is at least partially solved by the use of ALD coatings. Thus, this novel approach allows synthesizing precisely integrated and customized architectures with tailored surface performance, and/or eventually the bulk properties of the materials thanks to ALI, paving the way towards innovative and functional products, and opening prospects for many potential applications.

5. Conclusion

Although ALD was initially developed to exploit the possibility it offered to obtain continuous, pin-hole-free thin films even over large areas, in the last years there have been different approaches to perform ALD in a localized fashion, giving rise to the ASD field. As it has been discussed, these methods are based on different approaches allowing either a high spatial resolution in XY (at the nanometer) or simpler more direct approaches that provide direct patterning at the millimeter and micrometer level in XY. In any case, and given these approaches are based on the ALD method, the control in Z is nanometric. The possibility to have spatial control over the ALD process can be exploited as a new gas-based technique for the 3D printing of functional materials at different scales, providing a unique approach to the fabrication of functional materials with complex shapes. Beyond using ALD as a 3D printing technique in itself, the possibility it offers to coat (even infiltrate) complex shapes in a highly controlled way and with a large amount of different materials is ideal to nanoengineer the properties of pieces obtained by standard 3D printing approaches, thus expanding the range of applications that can be achieved. ALD should thus experience an important penetration in the 3D printing field in the coming years.

Acknowledgements

D.M.-R. acknowledges support from the European Union's Horizon 2020 FETOPEN-1-2016-2017 research and innovation program under Grant Agreement 801464, and through the Marie Curie Actions (FP7/2007-2013, Grant Agreement No. 63111). The Agence Nationale de la Recherche (ANR, France) is also acknowledged for funding via the programs ANR-16-CE05-0021 (DESPATCH) and ANR-20-CE09-0008 (ALD4MEM). The French National Research Agency (in the framework of the "Investissements d'avenir" program (No. ANR-15-IDEX-02) through the project Eco-SESA) is acknowledged for a PhD Grant.

Author details

David Muñoz-Rojas[1*], Matthieu Weber[1], Christophe Vallée[2], Chiara Crivello[1], Abderrahime Sekkat[1], Fidel Toldra-Reig[1] and Mikhael Bechelany[3]

1 Univ. Grenoble Alpes, CNRS, Grenoble INP, LMGP, Grenoble, France

2 SUNY POLY, CNSE, Albany, New York, USA

3 Institut Européen des Membranes, IEM, UMR-5635, ENSCM, CNRS, Univ Montpellier, Montpellier Cedex 5, France

*Address all correspondence to: david.munoz-rojas@grenoble-inp.fr

References

[1] Leskelä M, Mattinen M, Ritala M. Review article: Atomic layer deposition of optoelectronic materials. Journal of Vacuum Science and Technology B. 2019;**37**:030801

[2] Johnson RW, Hultqvist A, Bent SF. A brief review of atomic layer deposition: From fundamentals to applications. Materials Today. 2014;**17**:236-246

[3] George SM. Atomic layer deposition: An overview. Chemical Reviews. 2010;**110**:111-131

[4] Mattox DM. Physical vapor deposition (PVD) processes. Metal Finishing. 2002;**100**:394-408

[5] Paulsen JA, Renn M, Christenson K, Plourde R. Printing conformal electronics on 3D structures with aerosol jet technology. In: 2012 Futur. Instrum. Int. Work. Proc. NY: IEEE; 2012. pp. 1-4

[6] Salmerón JF, Molina-Lopez F, Briand D, Ruan JJ, Rivadeneyra A, Carvajal MA, et al. Properties and printability of inkjet and screen-printed silver patterns for RFID antennas. Journal of Electronic Materials. 2014;**43**:604-617

[7] Fotovvati B, Dehghanghadikolaei A, Namdari N. Laser-assisted coating techniques and surface modifications: A short review. Particulate Science and Technology. 2021;**39**:738-747

[8] Serra P, Piqué A. Laser-induced forward transfer: Fundamentals and applications. Advanced Materials Technologies. 2019;**4**:1-33

[9] Molina R, Ertuğrul M, Larrea NR, Rico V, Yubero F, González-Elipe AR, et al. Laser-induced scanning transfer deposition of silver electrodes on glass surfaces: A green and scalable technology. Applied Surface Science. 2021;**556**:149673. DOI: 10.1016/j.apsusc.2021.149673

[10] Schmidleithner C, Kalaskar DM. Stereolithography. In: Cvetković D, editor. 3D Printing. London, UK: IntechOpen; 2018. pp. 1-22

[11] Jones AC, Hitchman ML. Chemical Vapour Deposition-Precursors, Processes and Applications. Cambridge, UK: Royal Society of Chemistry; 2009. p. 40

[12] Kurek A, Gordon PG, Karle S, Devi A, Barry ST. Recent advances using guanidinate ligands for chemical vapour deposition (CVD) and atomic layer deposition (ALD) applications. Australian Journal of Chemistry. 2014;**67**:989-996

[13] Knoops HCM, Potts SE, Bol AA, Kessels WMM. Atomic layer deposition. In: Handb. Cryst. Growth. Amsterdam, Netherlands: Elsevier; 2015. pp. 1101-1134

[14] Muñoz-Rojas D, Nguyen VH, Masse de la Huerta C, Aghazadehchors S, Jiménez C, Bellet D. Spatial Atomic Layer Deposition (SALD), an emerging tool for energy materials. Application to new-generation photovoltaic devices and transparent conductive materials. Comptes Rendus Physique. 2017;**1**:391-400

[15] Kessels WMME Atomiclimits ALD Database. Available from: https://www.atomiclimits.com/alddatabase/ [Accessed: April 16, 2021]

[16] Parsons GN, Clark RD. Area-selective deposition: Fundamentals, applications, and future outlook. Chemistry of Materials. 2020; **32**:4920-4953

[17] Mackus AJM, Schneider JR, MacIsaac C, Baker JG, Bent SF. Synthesis of doped, ternary, and quaternary materials by atomic layer deposition: A

review. Chemistry of Materials. 2019;**31**:1142-1183

[18] Oh I-K, Sandoval TE, Liu T-L, Richey NE, Bent SF. Role of precursor choice on area-selective atomic layer deposition. Chemistry of Materials. 2021;**33**:3926-3935

[19] Maeda E, Nabatame T, Hirose M, Inoue M, Ohi A, Ikeda N, et al. Correlation between SiO_2 growth rate and difference in electronegativity of metal–oxide underlayers for plasma enhanced atomic layer deposition using tris(dimethylamino)silane precursor. Journal of Vacuum Science and Technology A. 2020;**38**:032409

[20] Yarbrough J, Shearer AB, Bent SF. Next generation nanopatterning using small molecule inhibitors for area-selective atomic layer deposition. Journal of Vacuum Science and Technology A. 2021;**39**:021002

[21] Liu T-L, Bent SF. Area-selective atomic layer deposition on chemically similar materials: Achieving selectivity on oxide/oxide patterns. Chemistry of Materials. 2021;**33**:513-523

[22] Singh JA, Thissen NFW, Kim W-H, Johnson H, Kessels WMM, Bol AA, et al. Area-selective atomic layer deposition of metal oxides on noble metals through catalytic oxygen activation. Chemistry of Materials. 2018;**30**:663-670

[23] Merkx MJM, Jongen RGJ, Mameli A, Lemaire PC, Sharma K, Hausmann DM, et al. Insight into the removal and reapplication of small inhibitor molecules during area-selective atomic layer deposition of SiO_2. Journal of Vacuum Science and Technology A. 2021;**39**:012402

[24] Vallat R, Gassilloud R, Eychenne B, Vallée C. Selective deposition of Ta_2O_5 by adding plasma etching super-cycles in plasma enhanced atomic layer deposition steps. Journal of Vacuum Science and Technology A: Vacuum, Surfaces and Films. 2017;**35**:01B104

[25] Vallat R, Gassilloud R, Salicio O, El Hajjam K, Molas G, Pelissier B, et al. Area selective deposition of TiO_2 by intercalation of plasma etching cycles in PEALD process: A bottom up approach for the simplification of 3D integration scheme. Journal of Vacuum Science and Technology A. 2019;**37**:020918

[26] Lee J, Lee J, Oh H, Kim C, Kim J, Kim DH, et al. Inherently area-selective atomic layer deposition of SiO_2 thin films to confer oxide versus nitride selectivity. Advanced Functional Materials. 2021;**31**:2102556

[27] Vos MFJ, Chopra SN, Ekerdt JG, Agarwal S, Kessels WMM(E), Mackus AJM. Atomic layer deposition and selective etching of ruthenium for area-selective deposition: Temperature dependence and supercycle design. Journal of Vacuum Science and Technology A. 2021;**39**:032412

[28] Song SK, Saare H, Parsons GN. Integrated isothermal atomic layer deposition/atomic layer etching supercycles for area-selective deposition of TiO_2. Chemistry of Materials. 2019;**31**:4793-4804

[29] Mameli A, Roozeboom F, Poodt P. Area-selective atmospheric-pressure spatial ALD of SiO_2 Using Interleaved Back-Etch steps yielding selectivity >10 nanometer. In: ECS Meet Abstr. Fairfax, Virginia: ECS; 2019. DOI: 10.1149/MA2019-02/24/1138

[30] Song SK, Kim JS, Margavio HRM, Parsons GN. Multimaterial self-aligned nanopatterning by simultaneous adjacent thin film deposition and etching. ACS Nano. 2021;**15**:12276-12285

[31] Yeghoyan T, Pesce V, Jaffal M, Lefevre G, Gassilloud R, Posseme N, et al. Low temperature topographically

selective deposition by plasma enhanced atomic layer deposition with ion bombardment assistance. Journal of Vacuum Science and Technology A: Vacuum, Surfaces and Films. 2021;**39**:032416

[32] Masse de la Huerta C, Nguyen VH, Dedulle J, Bellet D, Jim C, Muñoz-Rojas D. Influence of the geometric parameters on the deposition mode in spatial atomic layer deposition: A novel approach to area-selective deposition. Coatings. 2018;**9**:5

[33] Poodt P, Cameron DC, Dickey E, George SM, Kuznetsov V, Parsons GN, et al. Spatial atomic layer deposition: A route towards further industrialization of atomic layer deposition. Journal of Vacuum Science and Technology A: Vacuum, Surfaces and Films. 2012; **30**:010802

[34] Muñoz-Rojas D, MacManus-Driscoll J. Spatial atmospheric atomic layer deposition: A new laboratory and industrial tool for low-cost photovoltaics. Materials Horizons. 2014;**1**:314-320

[35] Hoye RLZ, Muñoz-Rojas D, Nelson SF, Illiberi A, Poodt P, Roozeboom F, et al. Research update: Atmospheric pressure spatial atomic layer deposition of ZnO thin films: Reactors, doping, and devices. APL Materials. 2015;**3**:040701

[36] Muñoz-Rojas D, Maindron T, Esteve A, Fabien Piallat J, Decamps KJM. Speeding up the unique assets of atomic layer deposition. Materials Today Chemistry. 2018;**12**;96-120

[37] Muñoz-Rojas D, Nguyen VH, Masse de la Huerta C, Jiménez C, Bellet D. Spatial atomic layer deposition. Chemical Vapor Deposition Nanotechnology. 2019:1-25. DOI: 10.5772/intechopen.82439

[38] Levy DH, Nelson SF, Freeman D. Oxide electronics by spatial atomic layer deposition. Journal of Display Technology. 2009;**5**:484-494

[39] Muñoz-Rojas D, Sun H, Iza DC, Weickert J, Chen L, Wang H, et al. High-speed atmospheric atomic layer deposition of ultra thin amorphous TiO_2 blocking layers at 100 °C for inverted bulk heterojunction solar cells. Progress in Photovoltaics: Research and Applications. 2013;**21**:393-400

[40] Musselman KP, Muñoz-Rojas D, Hoye RLZ, et al. Rapid open-air deposition of uniform, nanoscale, functional coatings on nanorod arrays. Nanoscale Horizons. 2017;**2**:110-117

[41] Huerta CAM, Nguyen VH, Sekkat A, Crivello C, Toldra-Reig F, Veiga PB, et al. Gas-phase 3D printing of functional materials. Advanced Materials Technologies. 2020;**5**:2000657

[42] Nguyen VH, Sekkat A, Masse De La Huerta CA, et al. Atmospheric plasma-enhanced spatial chemical vapor deposition of SiO_2 using trivinylmethoxysilane and oxygen plasma. Chemistry of Materials. 2020;**32**:5153-5161

[43] Alshehri AH, Loke JY, Nguyen VH, et al. Nanoscale film thickness gradients printed in open air by spatially varying chemical vapor deposition. Advanced Functional Materials. 2021;**31**:2103271

[44] Sekkat A, Nguyen VH, Masse de La Huerta CA, Rapenne L, Bellet D, Kaminski-Cachopo A, et al. Open-air printing of Cu_2O thin films with high hole mobility for semitransparent solar harvesters. Communications Materials. 2021;**2**:78

[45] Midani L, Ben-Yahia W, Salles V, Marichy C. Nanofabrication via maskless localized atomic layer deposition of patterned nanoscale metal oxide films. ACS Applied Nano Materials. 2021;**4**:11980-11988. DOI: 10.1021/acsanm.1c02550

[46] Wong KV, Hernandez A. A review of additive manufacturing. International Scholarly Research Notices. 2012;**2012**. Article ID 208760. DOI: 10.5402/2012/208760

[47] Herzog D, Seyda V, Wycisk E, Emmelmann C. Additive manufacturing of metals. Acta Materialia. 2016; **117**:371-392

[48] Ngo TD, Kashani A, Imbalzano G, Nguyen KTQ, Hui D. Additive manufacturing (3D printing): A review of materials, methods, applications and challenges. Composites. Part B, Engineering. 2018;**143**:172-196

[49] Dawood A, Marti BM, Sauret-Jackson V, Darwood A. 3D printing in dentistry. British Dental Journal. 2015;**219**:521-529

[50] Fiske M, Edmunson J, Fikes J, Johnston M, Case M. 3D printing in space: A new paradigm for America's military. Military Engineer. 2018; **110**:74-77

[51] Petit RR, Li J, Van de Voorde B, Van Vlierberghe S, Smet PF, Detavernier C. Atomic layer deposition on polymer thin films: On the role of precursor infiltration and reactivity. ACS Applied Materials & Interfaces. 2021;**13**:46151-46163

[52] Lee S-M, Pippel E, Gösele U, Dresbach C, Qin Y, Chandran CV, et al. Greatly increased toughness of infiltrated spider silk. Science (80-). 2009;**324**:488-492

[53] Puurunen RL. Surface chemistry of atomic layer deposition: A case study for the trimethylaluminum/water process. Journal of Applied Physics. 2005; **97**:121301

[54] Ritala M, Leskela M. Atomic layer deposition. In: Handb. thin Film Mater. Cambridge, Massachusetts, United States: Academic Press; 2001. pp. 103-159

[55] Leskelä M, Ritala M. Atomic layer deposition (ALD): From precursors to thin film structures. Thin Solid Films. 2002;**409**:138-146

[56] Weber M, Julbe A, Ayral A, Miele P, Bechelany M. Atomic layer deposition for membranes: Basics, challenges, and opportunities. Chemistry of Materials. 2018;**30**:7368-7390

[57] Dendooven J, Goris B, Devloo-Casier K, Levrau E, Biermans E, Baklanov MR, et al. Tuning the pore size of ink-bottle mesopores by atomic layer deposition. Chemistry of Materials. 2012;**24**:1992-1994

[58] Knez M, Nielsch K, Niinistö L. Synthesis and surface engineering of complex nanostructures by atomic layer deposition. Advanced Materials. 2007;**19**:3425-3438

[59] Chawla V, Ruoho M, Weber M, Chaaya AA, Taylor AA, Charmette C, et al. Fracture mechanics and oxygen gas barrier properties of Al_2O_3/ZnO nanolaminates on PET deposited by atomic layer deposition. Nanomaterials. 2019;**9**:88

[60] Ott AW, Klaus JW, Johnson JM, George SM, McCarley KC, Way JD. Modification of porous alumina membranes using Al_2O_3 atomic layer controlled deposition. Chemistry of Materials. 1997;**9**:707-714

[61] Maina JW, Merenda A, Weber M, Pringle JM, Bechelany M, Hyde L, et al. Atomic layer deposition of transition metal films and nanostructures for electronic and catalytic applications. Critical Reviews in Solid State and Materials Sciences. 2020;**46**:1-22

[62] Weber M, Coy E, Iatsunskyi I, Yate L, Miele P, Bechelany M. Mechanical properties of boron nitride thin films prepared by atomic layer deposition. CrystEngComm. 2017;**19**:6089-6094

[63] Dasgupta NP, Meng X, Elam JW, Martinson ABF. Atomic layer deposition of metal sulfide materials. Accounts of Chemical Research. 2015;**48**:341-348

[64] Parsons GN, Atanasov SE, Dandley EC, et al. Mechanisms and reactions during atomic layer deposition on polymers. Coordination Chemistry Reviews. 2013;**257**:3323-3331

[65] Pinna N, Knez M, editors. Atomic layer deposition of nanostructured materials. John Wiley & Sons; 2012. 472 p

[66] Spagnola JC, Gong B, Arvidson SA, Jur JS, Khan SA, Parsons GN. Surface and sub-surface reactions during low temperature aluminium oxide atomic layer deposition on fiber-forming polymers. Journal of Materials Chemistry. 2010;**20**:4213

[67] Gong B, Parsons GN. Quantitative in situ infrared analysis of reactions between trimethylaluminum and polymers during Al_2O_3 atomic layer deposition. Journal of Materials Chemistry. 2012;**22**:15672-15682

[68] Leng CZ, Losego MD. Vapor phase infiltration (VPI) for transforming polymers into organic–inorganic hybrid materials: A critical review of current progress and future challenges. Materials Horizons. 2017;**4**:747-771

[69] Kestilä A, Nordling K, Miikkulainen V, Kaipio M, Tikka T, Salmi M, et al. Towards space-grade 3D-printed, ALD-coated small satellite propulsion components for fluidics. Additive Manufacturing. 2018;**22**:31-37

[70] Nyman L, Kestilä A, Porri P, Pudas M, Salmi M, Silander R, et al. Constructing spacecraft components using additive manufacturing and atomic layer deposition: First steps for integrated electric circuitry. Journal of Aerospace Engineering. 2021; **34**:4021049

[71] Heikkinen ITS, Marin G, Bihari N, Ekstrum C, Mayville PJ, Fei Y, et al. Atomic layer deposited aluminum oxide mitigates outgassing from fused filament fabrication–based 3-D printed components. Surface and Coatings Technology. 2020;**386**:125459

[72] Moll A, Blandin J-J, Dendievel R, Gicquel E, Pons M, Jimenez C, et al. Coupling powder bed additive manufacturing and vapor phase deposition methods for elaboration of coated 3D Ti-6Al-4V architectures with enhanced surface properties. Surface and Coatings Technology. 2021;**415**:127130

[73] Browne MP, Plutnar J, Pourrahimi AM, Sofer Z, Pumera M. Atomic layer deposition as a general method turns any 3D-printed electrode into a desired catalyst: Case study in photoelectrochemisty. Advanced Energy Materials. 2019;**9**:1900994

[74] Ng S, Zazpe R, Rodriguez-Pereira J, Michalička J, Macak JM, Pumera M. Atomic layer deposition of photoelectrocatalytic material on 3D-printed nanocarbon structures. Journal of Materials Chemistry A. 2021;**9**:11405-11414

[75] Pellejero I, Almazán F, Lafuente M, Urbiztondo MA, Drobek M, Bechelany M, et al. Functionalization of 3D printed ABS filters with MOF for toxic gas removal. Journal of Industrial and Engineering Chemistry. 2020;**89**:194-203

[76] Devlin-Mullin A, Todd NM, Golrokhi Z, Geng H, Konerding MA, Ternan NG, et al. Atomic layer deposition of a silver nanolayer on advanced titanium orthopedic implants inhibits bacterial colonization and supports vascularized de novo bone ingrowth. Advanced Healthcare Materials. 2017;**6**:1700033

[77] Han X, Gong Y, Fu KK, He X, Hitz GT, Dai J, et al. Negating interfacial

impedance in garnet-based solid-state Li metal batteries. Nature Materials. 2017;**16**:572-579

[78] McOwen DW, Xu S, Gong Y, Wen Y, Godbey GL, Gritton JE, et al. 3D-printing electrolytes for solid-state batteries. Advanced Materials. 2018;**30**:1707132

Chapter 13

Perspective Chapter: Direct Energy Deposition of Cu-Fe System Functionally Graded Materials – Miscibility Aspects, Cracking Sources, and Methods of Assisted Manufacturing

Konstantin Makarenko, Oleg Dubinin and Igor V. Shishkovsky

Abstract

Direct energy deposition is a reliable additive manufacturing method of producing components with highly sophisticated geometry from a single material or combination of different materials with high manufacturing freedom and efficiency. The assembly operations are not required after the direct energy deposition: such complex parts as a rocket combustion chamber, a nuclear reactor element, a heat exchanger, and so on, could be fabricated layer-by-layer during one technological step. Promising applications are associated with Cu-Fe system laser deposited functionally graded components, which allow combining good oxidation resistivity, antifrictionality, thermal, and electrical conductivity of copper with mechanical strength, processability, and corrosion resistance of stainless steel. The main issue, which appears in the case of laser deposition of such materials, is internal stresses caused by significant inequality of physical properties of copper/bronze and steel, their limited miscibility, forming of brittle phases at the interface, and complexity of variation of mechanical and physical properties of the resulted alloy. The mentioned factors could cause various cracking in resulted parts. Specific techniques such as ultrasonic assistance, implementation of the external magnetic field, and post-treatment (hot isostatic pressing, machining), could be suggested to improve the quality of laser deposited Cu-Fe system functionally graded materials.

Keywords: direct energy deposition, functionally graded materials, Cu-Fe, Cu-Fe-Al, capillary convection, diffusion, cracking, ultrasonic-assisted DED, magnetically-assisted DED, hot isostatic pressing, machining

1. Introduction

Metal laser additive manufacturing (AM) includes a complex of precise rapid fabrication methods (selective laser sintering – SLS, also known as Direct Metal Laser Sintering™ – DMLS; selective laser melting – SLM, also known as laser powder bed fusion – LPBF, and LaserCUSING™; direct energy deposition – DED, also known as laser directed energy deposition – LDED; laser metal deposition – LMD, direct metal deposition – DMD, Laser Engineered Net Shaping – LENS™, direct laser deposition – DLD, laser solid forming – LSF, laser metal deposition shaping – LMDS, and 3D laser cladding), which allows the creation of complex parts without assembly operations along with the low material waste. The AM also provides a possibility of in situ control and adjustment of process parameters for the better quality of the resulted parts using methods of laser-induced breakdown spectroscopy [1], acoustic emission sensors [2], machine learning [3], which allows the prediction of resulted phase composition and mechanical behavior of the parts, co-axial spatially integrated pyrometry [4], direct metal tooling [5], etc. The AM allows the fabrication of products not only from a single material but from a combination of two and more materials, which may have significantly different properties (such as alloys of Cu and Ni [6], Ni and Fe [7], Ni and Ti [8], Ti and Fe [9, 10], Cu and Fe [11, 12], Al and Fe [13], Al and Ti [14]). The resulted parts with the gradient of properties within their volume, created from various materials, are specified as functionally graded materials (FGMs) or compositionally graded materials [15–18]. The significant interest is attributed to the manufacturing of materials of the Cu-Fe system, which combine thermal expansion properties, electrical and thermal conductivity of bronze with high rigidity, mechanical strength (yield stress, ultimate tensile strength (UTS), flexure strength, creep resistivity), and corrosion resistance of stainless steel [19]. Moreover, the Cu-Fe alloys, especially multi-layered, are characterized by significant values of magnetoresistance [20–25]. The applicability of different AM technologies for Cu-Fe FGMs fabrication was justified by the results of the following research:

a. For the DED:

- X. Zhang et al. [26] successfully produced SS 304 L – commercially pure (CP) copper FGM with ~370% average thermal diffusivity and ~ 100% heat conductivity improvement in comparison with CP SS, through the intersections of a nickel-based alloy using the intermediate section technique [18];

- Authors of study [27] fabricated Cu-Fe FGM via the direct joining [18] with the resulted morphology of partially elongated columnar dendrites, and observed a microstructure refinement (up to 50 μm grain size) due to rapid solidification rate, but identified poor yield stress (123 MPa) and UTS (250 MPa), caused by the issues of direct bonding of such dissimilar materials;

- H.S. Prasad et al. [28] deposited a 99.9% CP copper on various metal substrates (aluminum, steel, and titanium) using a high-absorbable green (515 nm) disk laser source instead of common industrial infrared emitters. The longitudinal cracking in the case of steel-copper FGM was observed, and it was noted that the substrate preheating can be applied for the wettability of dissimilar materials improvement, which affects the resulted bonding parameters.

b. For other AM technologies:

- Y. Bai et al. [29] prepared the SS 316 L – C52400 metal composite via the SLM with two different interfaces (transition from SS 316 L to C52400 and vice versa) without any observable brittle intermetallics except CuNi;

- K.S. Osipovich et al. [30] fabricated the bimetallic samples from electrolytic tough pitch copper C11000 and SS 304 wire materials by the electron beam wire-feed AM technology with appropriate metallurgical bonding between SS and Cu with free of the defects transition zone.

The key difficulties with AM of Cu-Fe system FGMs are the stepping junction of Young's modulus of elasticity and coefficient of linear thermal expansion during the transition from Cu-based to Fe-based part; mismatch of the lattice parameters; poor mutual miscibility of steel and copper/bronze (especially during rapid solidification); embrittlement due to intermetallic phases forming [5]. The current study is devoted to the discussion of factors that have an influence on miscibility and intermixing in the liquid phase of Cu-Fe system alloys, possible sources of cracking of DED-fabricated Cu-Fe FGMs, and the assisted manufacturing techniques, which could be used for the improvement of the resulted parts quality.

2. Technological steps during the DED of FGMs, and miscibility aspects of the Cu-Fe system

The common process of the DED of FGM could be divided into several stages briefly described below.

a. Preheating of the substrate (optional; decreases the temperature difference between substrate and material, reduces stresses and strains in the resulted part, prevents warping and the separation from the substrate [31]);

b. Laser heating of the substrate along with heating and melting of the first particles of powder, which falls on the substrate and particularly absorbs the energy from the laser beam;

c. While the laser head moves in the X-Y coordinate plane, the new powder particles become heated and melted; simultaneously previous areas rapidly cool down and solidify;

d. When the first layer of the part is finished, the laser turns off, and the powder stream stops (it may take a small time delay between the shutdown of the laser and the moment when the last powder particles are thrown away from the powder nozzle by the feeding gas; thus, these particles can anyway sinter with the substrate or previous layer and cause satellite defects such as balled-up protrusion [31]. These defects could be eliminated by machining, or prevented by the proper scan path planning, appropriate terminations, and suitable powder delivery [31–33]);

e. Cooling time between layers: a process is paused for several seconds to let the previous layer(−s) cool down. On the one hand, it prevents overheating and re-melting of the previous layer(−s) (especially if they were fabricated from another material characterized by higher heat conductivity, higher laser radiation absorption, and lower melting point) and provides a more appropriate thermal history (therefore, in some cases, the cooling time could be artificially increased to provide more intense cooling), but on the other hand, it could increase the further undesired thermal strains and stresses due to growth of the temperature gradient, and cause cracking. Besides that, it is significantly simpler to embed the new material inside the previously fabricated layers if they are preheated;.

f. The next layer is deposited on the previous, and the steps №№a)-e) are repeated; the difference from the first layer is that the heat is spread not only in the substrate but inside the previous layer(−s) too. Depth of the laser influence z, [m] in the case of laser-pulsed DED could be approximately estimated by the following Equation [34]:

$$z \simeq 2 \cdot \sqrt{\frac{a \cdot \tau_p}{\pi}} - \frac{T \cdot \kappa}{q}, \tag{1}$$

whereτ_p, [s] is an average duration of the laser pulse; T, [K] is the temperature of the point with a coordinate z; q, [W/m^2] is a laser power density; a, [m^2/s] and κ, [W/(m·K)] are a thermal diffusivity and thermal conductivity of the material respectively (in the case of two or more component materials it is permissible, for a first approximation, to apply the rule-of-mixture equation to estimate the average a and κ of the complex system);

g. The fabrication of FGM by the transition from one material to another requires gradual modification of the powder chemical composition starting from the definite layer №n (smooth transitioning method), or beginning of the layer №n from completely new material (direct joining) [18]. In both these cases, a joint melt pool emerges, where both materials and the admixtures are distributed by two main mechanisms: capillary convection (Marangoni effect) and diffusion through the moving phase interface. It also should be mentioned here that the fluid motion in the melt pool could be also caused by the pressure of vapor, which evaporates from a front wall of the shallow vapor cavity [35], but this mechanism, associated with boiling, is common for several other laser treatment technologies such as laser cutting and laser alloying, but not relevant for the DED, where laser power is less than that is needed for boiling: $P < P_{boil}$. Therefore, two main mechanisms of intermixing, actual for the DED technology – a Marangoni effect and diffusion, are briefly specified below.

2.1 Marangoni effect

The capillary convection is associated with the movement of liquid near the phase interface caused by the dependence of surface tension on two factors – temperature and admixture concentration [35]: in particular, these parameters are different near the crystallization front and far away from it. Let us consider a horizontal layer of liquid with $\zeta < z < -h$ height ($\zeta \equiv 0$ if the liquid is not disturbed), which is heated from above by the source of concentrated energy like a laser beam, which is absorbed with the flow density Q. A temperature gradient dT_0/dz is constant. About 30 years ago, E.B. Levchenko and A.L. Chernyakov investigated, in this case, a propagation of heat- and mass-transporting capillary waves on the surface of nonuniformly heated liquid [36], and demonstrated that, using the linearized Navier–Stokes and heat conduction equations, supposing that liquid is incompressible, from the system of equations:

$$\begin{cases} \Delta\varphi = 0; \\ \frac{\partial \vec{A}}{\partial t} = \nu \cdot \Delta\vec{A}; \\ div\vec{A} \equiv 0; \\ \frac{\partial T_1}{\partial t} + v_z \cdot \frac{dT_0}{dz} = \chi \cdot \Delta T_1; \\ v = \nabla\varphi + rotA, \end{cases} \tag{2}$$

with boundary conditions at $z = 0$:

$$\begin{cases} \rho \cdot \frac{\partial^2\varphi}{\partial t^2} - \alpha \cdot \left(\frac{\partial^2 v_z}{\partial x^2} + \frac{\partial^2 v_z}{\partial y^2}\right) + \rho \cdot g \cdot v_z + 2 \cdot \eta \cdot \frac{\partial^2 v_z}{\partial t \partial z} = 0; \\ \frac{d\alpha}{dT} \cdot \left(\frac{\partial^2 T}{\partial t \partial x} + \frac{dT_0}{dz} \cdot v_z\right) = \eta \cdot \frac{\partial}{\partial t}\left(\frac{\partial v_x}{\partial z} + \frac{\partial v_z}{\partial x}\right); \\ \kappa \cdot \frac{dT}{dz} = Q, \end{cases} \tag{3}$$

and boundary conditions at $z = -h$: temperature $T = const$, velocity $v = 0$,

where φ and $\vec{A}$ are the scalar and vector potentials of velocity respectively, ρ, c,κ, and $\eta = \rho{\cdot}\nu$ are the density, specific heat, thermal conductivity, and viscosity of the liquid phase respectively, it could be shown that two wave types exist in the considered liquid:

- capillary-gravity waves:

$$\omega_0 = \sqrt{g \cdot k + \frac{\sigma \cdot k^3}{\rho}} \tag{4}$$

- thermocapillary waves (sonic-type):

$$\omega = k \cdot c, \tag{5}$$

where g is gravitational acceleration, k is a wave number,$c^2 \propto dT/dz$, σ is a surface tension coefficient. These two mechanisms (thermocapillary and capillary-gravity waves) are responsible for elemental intermixing at the interface area of FGM due to convection and have a significant influence on the resulting mechanical bonding strength.

The first factor of capillary convection – a dependence between surface tension and concentration – plays an important role in stimulating intermixing in binary systems such as Cu-Fe. If the expression for the molar Gibbs free energy of the multi-component liquid mixture is written in a form:

$$g(T,P) = x_1 \cdot \left(g_1^0(T,P) + R \cdot T \cdot ln\, x_1\right) + x_2 \cdot \left(g_2^0(T,P) + R \cdot T \cdot ln\, x_2\right) + g^{ex}, \tag{6}$$

where g^{ex} is an excess molar Gibbs energy of the liquid phase (could be defined using a Wilson equation), x_1, x_2 and $g_1^0(T,P)$, $g_2^0(T,P)$ are the mole fraction and molar Gibbs free energy at the temperature T and pressure P for components 1 and 2 respectively, and, assuming the equality of cross-interaction energy to the arithmetic mean of pure components 1 and 2, then the concentration-dependent surface tension of the liquid mixture σ_m (J/m^2) could be expressed as the following [37]:

$$\sigma_m = x_1 \cdot \sigma_1 + x_2 \cdot \sigma_2 - \left(\frac{x_1 \cdot x_2}{x_1 + x_2 \cdot a} \cdot b + \frac{x_1 \cdot x_2}{x_2 + x_1 \cdot c} \cdot d \right), \tag{7}$$

where σ_1 and σ_2, [J/m^2] are surface tension coefficients of the mixture components 1 and 2; a = Λ_{12}, b = R·T·$(\partial\Lambda_{12}/\partial A)$, c = Λ_{21}, and d = R·T·$(\partial\Lambda_{21}/\partial A)$ – adjustable parameters, different for various binary systems (A – surface area, [m^2], Λ_{12}, Λ_{21} – Wilson equation parameters). The Eq. (7) could be used for estimation of the surface tension for binary liquid mixtures and alloys but could be extended (in the general case) to the more complicated systems of three and more components, such as Cu-Fe-Al, Cu-Fe-Sn, Cu-Fe-Cr, etc., which are more relevant to the real FGMs.

Due to the second factor of the capillary convection – a temperature-dependent character of surface tension – the capillary convection has as much importance, so concentrated and narrow the energy source is [35]. The temperature depends on the coordinate along the part's surface, the absolute value of the surface tension in a liquid melt pool relies on the temperature of the surface and commonly decreases while the temperature grows. This gradient of the temperature (and the surface tension) is a source of a resultant force, which acts from the center of the laser heat spot to the periphery. If the surface is melted, this force initiates a movement of the liquid, which leads to an increase of intermixing between base material and admixtures – in the case of single-component alloy; between two base materials – in the case of Cu-Fe system alloys.

2.2 Diffusion

The diffusion through the moving phase interface (crystallization front) in the steady (quasi-stationary) process (while the admixture concentration in the solidified area is distributed regularly alongside normal to crystallization front) is described by the following Equation [19, 38]:

$$\frac{dC_l}{dt} = v_{cryst} \cdot \frac{dC_l}{dx}, \tag{8}$$

where C_l is an admixture distribution in the liquid phase near the crystallization front, v_{cryst} is a crystallization rate, t – time. If the process is unsteady (admixture concentration beside the crystallization front in the liquid phase is time-dependent), so the diffusion process in moving coordinate system $0x'$, associated with the crystallization front, will be described by the integral Equation [38]:

$$\frac{\partial C_l}{\partial t} = D \cdot \frac{\partial^2 C_l}{\partial (x')^2} + v_{cryst} \cdot \frac{\partial C_l}{\partial x'}. \tag{9}$$

This differential equation could be solved without significant difficulties in four different cases [38]:

1. Initial saturation during crystallization of the first portions of metal;

2. Quasistationary state;

3. Final saturation period, which happens when the last portions of metal crystallize;

4. Unsteady process – transient concentration variation, which appears when the crystallization rate changes.

The last case is the most suitable for the FGM fabrication via the DED, characterized by extremely non-equilibrium conditions. Therefore, the solution of the Eq. (9) in the fourth case via Laplas' integral transformation, is performed in consideration of three conditions [38]:

$$\begin{cases} C_l = C_0 \text{ } if \text{ } x' = \infty \text{for each } t \geq 0; \\ C_l = C_0 \cdot \left(1 + \frac{1-k}{k} \cdot exp\left(-\frac{v_{cryst}}{D} \cdot x'\right)\right) \text{for} \forall x' \geq 0 \text{ at initial time } t = 0; \\ D \cdot \frac{C_l}{\partial x'} + v_{cryst} \cdot (C_l - C_s) = D \cdot \frac{C_l}{\partial x'} + v_{cryst} \cdot (1-k) \cdot C_l = 0 \text{ if } x' = 0, \end{cases} \tag{10}$$

which describes the unsteady diffusion process occurring along with the immediate change of the crystallization rate from the value v_{cryst} to the new value v'_{cryst}, and the transition from the distribution attributed to v_{cryst} to the distribution associated with v'_{cryst}, could be expressed by the following equation:

$$\frac{C_s(x_1)}{C_0} = 1 - \frac{1}{2} \cdot erfc\left(\frac{\sqrt{\frac{v_{cryst}}{D}} \cdot x_1}{2}\right) + (1-k) \cdot \left(\frac{\frac{v_{cryst}}{2 \cdot v'_{cryst}}}{k - \frac{v_{cryst}}{v'_{cryst}}}\right)$$
$$\cdot exp\left(-\frac{v_{cryst}}{v'_{cryst}} \cdot \left(1 - \frac{v_{cryst}}{v'_{cryst}}\right) \cdot \frac{v'_{cryst}}{D} \cdot x_1\right)$$
$$\cdot erfc\left(\left(\frac{v_{cryst}}{v'_{cryst}} - \frac{1}{2}\right) \cdot \sqrt{\frac{v'_{cryst}}{D}} \cdot x_1\right) +$$
$$+ \left(\frac{2 \cdot k - 1}{2}\right) \cdot \left(\frac{1 - \frac{v_{cryst}}{v'_{cryst}}}{k - \frac{v_{cryst}}{v'_{cryst}}}\right) \cdot exp\left(-k \cdot (1-k) \cdot \left(\frac{v'_{cryst}}{D}\right) \cdot x_1\right) \cdot erfc\left(\left(k - \frac{1}{2}\right) \cdot \sqrt{\frac{v'_{cryst}}{D} \cdot x_1}\right), \tag{11}$$

where x_1 is a distance from the point, where the crystallization rate abruptly changed from v_{cryst} to v'_{cryst}, to the considered section, D is a diffusion coefficient, C_s is an admixture distribution in the liquid phase near the crystallization front, k is an admixtures distribution constant, C_0 is an initial admixture concentration, and erfc is a complement error function.

The diffusion processes commonly have a positive influence on the mechanical properties of as-deposited parts due to intensification of the intermixing, but the increase of a crystallization rate (what is common for laser technologies such as DED) leads to a decrease in the liquid alloy lifetime and lowers a diffusion intermixing in the liquid phase. Moreover, in specific Cu-I binary systems (where I = Fe, Co, Nb, Cr, and so on [39]), a liquid phase separation exists that prevents intermixing and makes difficulties for the desired alloy development. A discussion of this phenomenon along with parameters of temperature-and-concentration-dependent behavior of the material should be conducted using a Cu-Fe binary system phase diagram, which is demonstrated in **Figure 1** [20, 39–42].

As it was mentioned above, a Cu-Fe system is characterized by the existence of a metastable liquid miscibility gap [20, 39, 41–43] (binodal, or coexistence curve), which shows a specific transient temperature for each concentration of Cu and Fe: if the liquid is undercooled below this temperature (what is convenient for rapid solidification induced by laser treatment), a liquid separation of immiscible Cu and

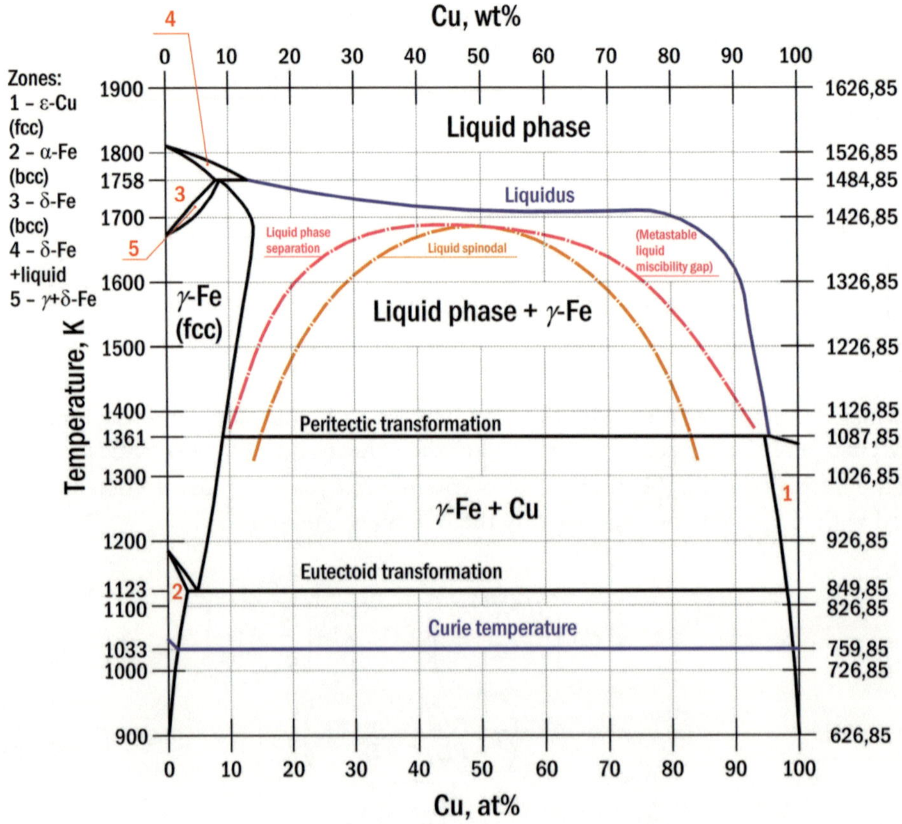

Figure 1.
A Cu-Fe system phase diagram [20, 39–42].

Fe appears [39], followed by further coagulation and dendritic crystallization after cooling. The position of this gap is shown by the pink line in **Figure 1**. A spinodal curve [42] (orange line in **Figure 1**) denotes the stable states, where such separation of undercooled liquid phases is not observed anymore (these states are located below this curve). The spinodal curve could be analytically described as a geometrical locus of points, where the Gybbs free energy second partial derivative with respect to concentration equals zero. The intersection point of these two curves (binodal and spinodal) is known as a critical point. Under that, a black horizontal line at the level of 1361 K shows the position of peritectic transformation (forming of an equilibrium solid solution consisting of a solid-state ε-Cu matrix around the γ-Fe dendrites, from the γ-Fe primary phase and surrounding liquid). The next line, associated with eutectoid transformation, corresponds to the temperature of 1123 K and shows a reversible disassociation of equilibrium solid solution into the two stable phases: α-Fe and ε-Cu. The Curie temperature of the Cu-Fe system equals to approximately 1033 K [39] (1043 K in the case of pure Fe [42]) (blue line in **Figure 1**). The Fe-rich areas of the phase diagram (located near its left side) are characterized mostly by face-centered cubic (FCC) crystal lattice; Cu-rich (right side) – by body-centered cubic (BCC). The phases observed at different temperatures and various Cu-Fe ratios are shown on the left top side of the diagram (and titled "zones") or are specified at the coordinate plane directly.

As it was said before, the existence of a metastable miscibility gap, caused by huge positive enthalpy between Cu and Fe, has a negative influence on the mechanical properties and applicability of Cu-Fe system FGMs [20]. The possible way of solution

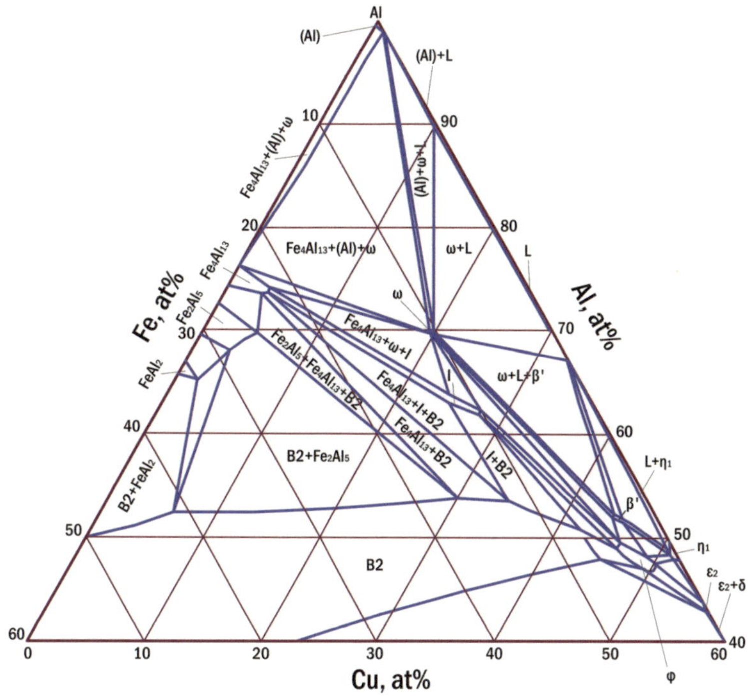

Figure 2.
A 600°C experimental isothermal section of the Cu-Fe-Al system ternary phase diagram [44–46].

to this problem is a consideration of a ternary system Cu-Fe-X, where X is a third chemical element, which reduces the incompatibility of Cu and Fe by forming sustainable equilibrium phases, including the case of a high-speed solidification. Aluminum could be suggested as the possible example of this element according to the Cu-Fe-Al system ternary phase diagram shown in **Figure 2** [44–46]. The Cu-Fe-Al system has three established ternary phases: tetragonal $\omega FeCu_2Al_7$ (ω in **Figure 2**), stable icosahedral quasicrystalline $\tau_i FeCu_2Al_6$ forming between 750 and 800°**C** (I in **Figure 2**), and $\varphi FeCu_{10}Al_{10}$ (φ in **Figure 2**), which forms in a solid-state and exists at 600°**C** [44–46]. In practice, such ternary system could be realized by using a SS – aluminum bronze FGM, which was discussed, for example, in studies [5, 19]. Other notations in **Figure 2**: B2 – ordered BCC phase, L – liquid phase, ε_2 – binary compound $\varepsilon_2 Cu_3Al$, δ – binary compound δ in the Al-Cu system, η_1 – binary compound $\eta_1 CuAl$, (Al) – solid solution based on disordered FCC Al phase, β' – phase-separated from the solution of FeAl in the ternary system [46].

3. Types of cracking in laser deposited FGMs and their sources

During the laser treatment processes, associated with the melting of metal (laser welding, laser cladding, DED, SLM, SLS), different types of cracking processes may appear. They are intensive in the case of the DED of Cu-Fe FGMs (especially at their interface areas [5]) because of the significant difference between physical properties, chemical compositions of base materials, and their limited miscibility, which was discussed above. Common cracking types include hot (solidification) cracking, liquation cracking, ductility dip cracking, cold cracking, and rewarming cracking (including post-weld heat treatment cracking). Below all these types of cracking and their sources are briefly described.

3.1 Hot cracking

Hot cracking, also known as solidification cracking, is a brittle intercrystallite (intergranular) failure, which appears along the boundaries of grains during the material crystallization [47, 48]. Elasticoplastic strains taking place during solidification cause hot cracking if their values exceed the strain capacity of the material [34]. One of the most significant parameters at this stage is a strain rate:

$$\alpha = \frac{\partial \varepsilon}{\partial T}. \tag{12}$$

The thermophysical properties of base metals of the FGM, their rigidity, and the operation conditions determine a strain rate in the high-temperature range significantly. There is a specific temperature interval, where the plasticity and strength of both components of the FGM, or of one of them, are low; it is called a brittle temperature range (BTR) [34]. This interval is characterized by the decrease of plasticity and is the most probable for cracking. Overall, three factors play a leading role in forming of hot cracks:

1. Elasticoplastic strain rate;

2. BTR range;

3. Minimal plasticity of the material within BTR.

3.2 Liquation cracking

The liquation cracking is also a kind of intergranular failure, which occurs during solidification in the partially melted ("mushy") zones of the material because of grain boundary liquation [47–49]. This kind of cracking is the most common for Al-, Ni-, and Fe-based systems [49]. High thermal contraction of bronze and the presence of intermetallics provokes the increase of this type of cracking in the case of Cu-Fe system FGMs [47, 49]. Several works state that a high-energy input of the AM could also be a reason for the liquation cracking [50, 51].

3.3 Ductility dip cracking

The ductility dip cracking, associated with a local ductility loss, occurs between the different grains only in a solid-state of material [47, 49, 52, 53] in an elevated temperature range of $(0.5 \cdot T_{melt}) \lesssim \mathrm{T} \lesssim T_{sol}$, where T_{sol} is a recrystallization (solidus) temperature (according to the other data – in a temperature range of about $0.4 \cdot T_{melt}$–$0.7 \cdot T_{melt}$ [50]). This type of cracking is actual for FCC alloys including Fe-, Ni- and Cr-based, but the systems that contain Cu also may suffer the ductility dip cracking: for instance, the Al-9Cu-6Ce (wt%)-based AM-fabricated alloys demonstrated the ductility dip at elevated temperatures ($\leq 400°C$), which was attributed to a concomitant dip in a strain-rate sensitivity of deformation [54]. It should be noted that as distinguished from the liquation cracking, the ductility dip cracking does not exhibit a liquation [49].

3.4 Cold cracking

Cracking that appears in the material during its cooling at $\mathrm{T} \lesssim 473$ K or within several days after printing, is called cold cracking. Cold cracking has a character of a slow failure. A long-time influence of the internal residual stresses causes elasticoplastic strain on the borders between different grains. Boundaries of grains have less stress resistivity compared with grain bodies because most part of the crystal lattice distortions is concentrated at the boundaries of grains [34]. Therefore, the most common areas of cold cracks appearance are the boundaries of the grains. Besides, further movement of a crack may include boundaries, such as the bodies of the grains. The high solidification rate, which is common for laser 3D printing, may provide intensification of many phase transformations in material, such as $\gamma \Longrightarrow \alpha$ transformation in stainless steel, and decrease the cold cracking resistivity. The grain refinement could contrarily increase this parameter (possible ways that lead to this result are discussed further in §4). Overall, the three most common sources of cold cracking could be specified:

1. Source №1 is a forming of the hard and brittle phases in the material during its solidification. If some regions gain low plasticity, high hardness, and increased specific volume (in other words, it may be said that these regions suffer full or partial hardening), their interface areas become saturated by internal stresses.

2. Source №2 is the existence of hydrogen (in several specific cases). Admixture of the hydrogen also may result in a cold cracking in the case of special kinds of steels or titanium alloys. The solubility of hydrogen in these materials strongly increases while the temperature grows. Therefore, the liquid metal during the DED may include a lot of hydrogen, which can be taken from the environmental gases. It is one of the reasons for the necessity of the shielding gas during the DED.

3. Source №3 is a combined source. The majority of cold cracks in the real cases (specifically of steels and titanium alloys) are provided by both sources simultaneously – the appearance of the low-plasticity phases along with filling with the hydrogen.

In the case of DED of Cu-Fe system FGMs, the cold cracks may be observed in three different regions of the FGM:

- in the steel area due to all three sources described before (hardening phases, which appear due to transformation of austenite to martensite, inclusion of hydrogen, or combined source);

- in the bronze area due to precipitation of hardening phases common for Cu-based alloys (β'-phase – a structured solid solution based on electron compound of Cu and Zn with BCC lattice – is more usual for brasses; Sn-enriched phases, such as $Cu_{31}Sn_8$; Pb-based phases); and the mixture of hydrogen or oxygen (the last one is the most undesired chemical element in pure copper or bronze) [34].

- in the border area if it consists of:

    - another material (in case of intermediate section method) – the source depends on the chemical composition of this material;

    - mixture of both materials (gradient path method) due to forming of the new brittle phases based on steel and bronze elements;

- alternating layers of steel and bronze (alternating layers technique [5]) – the same as in the previous case.

If the border area is a narrow interface between steel and bronze (direct joining method), the most common types of cracking are hot cracks and liquation cracks.

3.5 Rewarming cracking

The rewarming cracking, which occurs due to cyclically repeated heating of the previously solidified layers is also a widespread source of cracking in the laser deposited FGMs, including Cu-Fe system alloys. The low laser radiation absorption coefficient of copper in the infrared area of the spectrum (which is common for the most part of commercially available industrial fiber lasers, such as erbium-dopped and ytterbium-dopped) could be the trigger for the fast-appeared rewarming cracking: the necessity of the high laser power implementation during the deposition of Cu-based layers leads to significant overheating of the volume of the part that lays beneath; the depth of such influence could be estimated using the Eq. (1). The slow-appeared rewarming cracking could appear during the heat post-treatment conducted after printing (post-weld heat treatment cracking) [50], such as high tempering, which is carried out under a DED-fabricated part to decrease its residual stresses. The rewarming cracking is characterized by its own BTR, which is lower than the BTR of the hot cracking and is commonly presented by the intercrystalline failure in the coarse-grained area of the part.

3.6 Summary

In the case of DED of Cu-Fe system FGMs, the most widespread types of cracking are hot, liquation, and rewarming cracking. Low-temperature cracking (ductility dip and cold) appears rather more uncommonly. The methods, which could be suggested to struggle with the cracking in these materials are:

- an increase of intermixing in the interface zones, which could be achieved by the implementation of several specific techniques such as ultrasonic-assisted manufacturing, which is described below in §4;

- grain refinement using such techniques as ultrasonic vibration assistance, magnetization of the melt pool by the external field, heat treatment (e.g. annealing) by the external laser source during the printing;

- decrease of microstructural defects such as porosity, shrinkage cavities, unmelted particles, and foreign inclusions, which could appear as stress concentrators (several methods of these parameters controlling are also discussed below in §4);

- exploitation of another laser sources instead of commercial fiber lasers (such as green disk solid-state lasers) with higher radiation absorption coefficient for Cu-based areas of FGM to decrease the excessive heat input into the material;

- realization of new experimental techniques such as hybrid laser-arc directed energy deposition;

- application of the intermediate section method to suppress a formation of brittle intermetallics, which could be also provoked by the increase of intermixing as its negative side effect;

- formation of stable phases during solidification by supplementation of additional constituents such as aluminum (see §2.2) and nickel (e.g. D22 alloy);

- improvement of the thermal history of the alloy by changing its fabrication process parameters;

- implementation of the alternating layers technique, which could be also suggested as an effective method of cracking reduction in Cu-Fe laser deposited FGMs.

4. Technological features for improvement of the DED-fabricated parts quality

In §§ 2–3 above, the aspects of miscibility and common sources of cracking in laser deposited FGMs, especially in Cu-Fe system, were described. Both these phenomena are linked to the mechanical strength of the resulted parts: an increased intermixing along with grain refinement commonly leads to a decrease in cracking, and vice versa: liquid phase separation and coarse-grained structure are factors of mechanical strength lowering, especially if they are followed by crystal lattice defects. Several specific techniques could be applied for increase of miscibility and decrease of cracking. The techniques are aimed to provide an increase in the quality of a) as-built DED-fabricated parts in situ; b) parts after the DED and post-treatment. The examples of valuable assisted manufacturing techniques are discussed specifically below.

4.1 In situ improvement of the DED-fabricated parts quality

4.1.1 Ultrasonic-assisted DED process

Our scheme of the ultrasonic-assisted DED is presented in **Figure 3**. The source of the ultrasonic frequency (~21 kHz) current signal (master oscillator in the scheme) is used for the generation of the mechanical wave inside the magnetostrictor (magnetostrictive transducer – a device that converts an ultrasonic frequency alternating current energy to the mechanical energy of ultrasonic frequency vibration) located in the stainless steel cooling reservoir. The wave is conducted to the waveguide, which is placed under the surface of the substrate with rigid fixation. In **Figure 4**, you can see our practical implementation of this scheme (left picture) and the resulted microstructure (right picture). The ultrasonic frequency generator UZG-2 M (is not shown in the picture) with 2 kW ultimate output power, 1.8–1.9 kW magnetostrictor PMS2–20 (pos. 2, placed in the Ø150x335 mm cylindrical stainless steel cooling vessel) with 21.5–22 kHz frequency range, and the titanium alloy waveguide (pos. 4) with rectangular base surface 120x18 mm were manufactured by OOO "Ultra-Rezonans" (Yekaterinburg, the Russian Federation). Water at room temperature was applied as a cooling liquid (2 L/min flow). The DED process was performed using InssTek MX-1000 3D-printer (InssTek, Daejeon, Republic of Korea) (pos. 1) in a direct tooling mode [5, 19] realized by two cameras (pos. 3). The microstructure image was obtained using an optical microscope Altami MET 1C (OOO Altami, Saint Petersburg, the Russian Federation). Pressing,

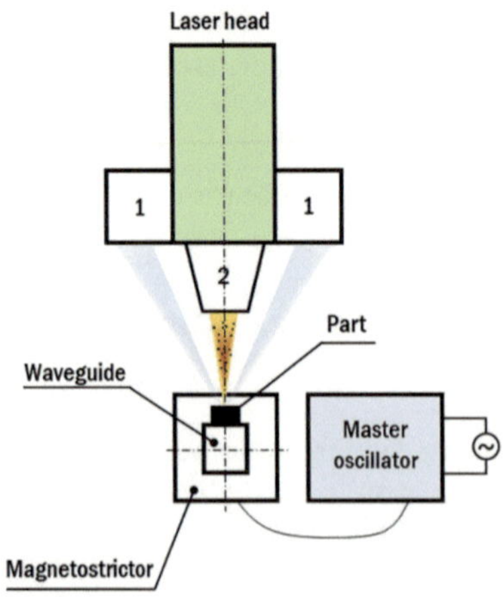

Figure 3.
The ultrasonic-assisted DED process scheme.

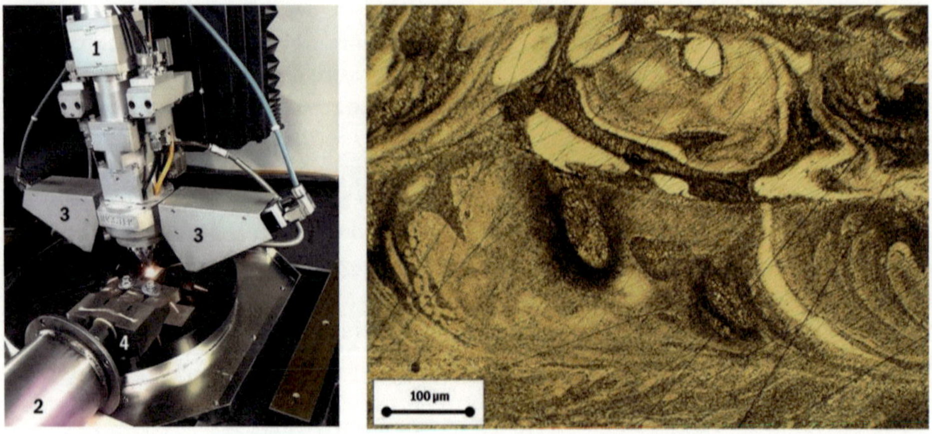

Figure 4.
The ultrasonic-assisted DED process. Left picture, positions: 1 – The laser head; 2 – The cooling vessel of the magnetostrictive transducer (see the description above); 3 – Cameras; 4 – Waveguide. Right picture: Example of the fine-grained intermixed microstructure of ultrasonic-assisted DED-fabricated part of 50% aluminum bronze with 9.5% Al and 1.0% Fe content (chemical composition similar to UNS C61800) and 50% SS 316 L. light areas: Islands of stainless steel, dark areas: The copper-based regions.

grinding, and polishing was conducted via TechPress 2™ and MetPrep 3™/PH-3™ (Allied High Tech Products, Inc., CA, United States) with 6 μm minimal grain size. The specimen was preliminarily etched for 9 s by 50 ml HCl + 50 ml C_2H_5OH + 2.5 g $CuCl_2$ [55–62].

The ultrasonic-assisted DED process is suggested to achieve the following advantages:

1. Refinement of the grain structure: X.H. Wang et al. [55] demonstrated the refinement of Fe-based composite coating after ultrasonic-assisted LMD, the disappearance of the columnar dendrites with 200 W ultrasonic power, and the appearance of the equiaxed dendrites with 400 W; C.J. Torado et al. [56] also observed the grain refinement of 3D-printed Inconel 625 using the ultrasonic assistance, decrease of the epitaxial growth and improvement of homogeneity.

2. Increase of microhardness and wear resistance: X.H. Wang et al. [55] showed that ultrasonic assistance increases the wear resistance of coating up to 2.4

times in comparison with a coating without ultrasonic assistance in the context of Fe-based composite LMD-fabricated coating deposited on the 5CrNiMo substrate. D. Zhang et al. [57] observed the increase of microhardness from ~380–450 HV1.0 to ~435 ... 515 HV1.0 at all levels of laser output power due to grain refinement and reduction of the porosity. Method of strengthening of a part surface, an increase of average impact energy, hardness, a decrease of a wear mass loss, and redistribution of the reinforcement particles leading to their uniform dispersion, provided by the ultrasonic assistance, was also shown by the authors of the patent [58].

3. Increase of Young's modulus: D. Zhang et al. [57] showed that the ultrasonic vibration improves Young's modulus of the DED-fabricated parts from ~45–55 GPa to ~50–65 GPa.

4. Increase of the tensile strength: C.J. Torado et al. [56] pointed at the increase of tensile properties mostly due to the β-grain refinement: UTS with ultrasonic assistance was equal to ~1160 MPa (yield strength ~1100 MPa) while the results without ultrasonic assistance were: UTS ~1020 MPa, yield strength ~990 MPa (about 12% improvement of both parameters) of 3D-printed Ti-6Al-4V alloy. Y. Zhang et al. [59] achieved 1.4–1.6 times increase of tensile strength of Al 4047 parts due to grain refinement using the ultrasonic-assisted DED instead of traditional casting technology and observed the microstructure consisted of the columnar Al dendrites with equiaxed Si particles at boundaries of the layers along with equiaxed Al crystals surrounded by fine Si phases in the middle zone of the alloy.

5. Reduction of the eutectic spacing: S. Yan et al. [60] reported the decrease of this parameter in the carbon fiber toughening nanoscale Al_2O_3-ZrO_2 laser deposited eutectic, fabricated with ultrasonic assistance. The resulted value reached 50 ± 5 nm.

6. Improvement of the fracture toughness: S. Yan et al. [60] reported about 2.5–4 times increase in the carbon fiber toughening nanoscale Al_2O_3-ZrO_2 eutectic due to the grain refinement and the whisker toughening of the carbon fiber.

7. Removal of cracking: the authors of the patent [61] described the laser deposition of Al-12Si eutectic alloy and found out that cracking in the deposited structure, which was seen after the common DED, wasn't observed when the ultrasonic-assisted DED was implemented, and the microstructure of the sedimentary layer was changed.

Because of a serious lack of experimental studies related to the topic of ultrasonic-assisted DED of Cu-Fe system materials, it is struggling to predict indisputably that all the mentioned changes, described above, will be seen in the materials of this system too. Nevertheless, our first tensile tests of the binary Cu(50%)-Fe(50%) alloy (see **Figure 4** and its description) were conducted in accordance with ASTM E8/E8M-16a at 2.7 mm/min rate using INSTRON 5969 dual column machine showed that the average ultimate tensile strength of common DED-fabricated parts equals 848.3 MPa, while this result in case of the ultrasonic-assisted DED reaches 952.7 MPa, what is 1.12 times higher. Cu-based parts created from the tin bronze powder without any Fe-based constituents also demonstrate the responsiveness to the agitation of a melt pool by the ultrasonic frequency waves during the DED: A. Gorunov [62] observed the intermixing between the tin bronze clads and material of a substrate along with substrate cracks bridging and claimed that it is possible to variate size, shape, and intermixing rate of the deposited material by changing the ultrasonic-related parameters of the process.

It is known that there is a dependence between mechanical performance and build orientation of parts fabricated via the traditional DED process. For instance, K. Zhang et al. [63] showed the difference between fracture morphology and anisotropic mechanical performance in specimens stretched parallel and perpendicular to the build direction; P. Guo et al. [64] reported that the higher elongation at failure was seen at 0° build direction rather than at 90°, what was an unexpected result because in the second case the direction of the external load was parallel to the dendritic grains; E. Azinpour et al. [65] observed the lower UTS and yield stress values of parts fabricated at 0° building direction in respect to the direction of load, in comparison with that of 90°. The build orientation of the part during the ultrasonic-assisted DED plays even a more important role in further mechanical characteristics, microstructure, and morphology parameters because of the significance of the mutual disposition between the ultrasonic wave front and the axes of the solidified grains of the material. Besides, the absorption of mechanical ultrasonic wave energy during its propagation within the volume of the part points to influence of a part's size and geometry on the ultrasonic-affected changes in its microstructural and mechanical properties. It is expected to observe a lower ultrasonic influence at the highest layers of a tall part if it is built in a vertical direction. Contrarily, if the part is a built-in horizontal direction, the ultrasonic influence will be more uniform and is expected to have a more regular distribution. The evaluation of ultrasonic attenuation in polycrystalline metals occurring mostly because of scattering from grains was conducted by T. Stepinski and P. Wu [66] for the pure copper specimens. A frequency-dependent attenuation coefficient $\zeta(f)$ was determined by the spectral shift method (based on the measurement of a signal reflected from the front and backs surfaces of the metal plate) using the following equation:

$$\zeta(f) = \left(\frac{2 \cdot \pi}{B}\right)^2 \cdot \frac{f_i - f_0}{2 \cdot D} \cdot f, \tag{13}$$

where B is the bandwidth of the input signal, f_i and f_0 are the central frequencies of the input and output signal respectively, and 2·D is a full path length of an ultrasonic signal (D – a thickness of the plate). The resulted measured attenuation amounted from $0.3684^{+0.0398}_{-0.0469}$ to $0.4613^{+0.0489}_{-0.0645}$ for 5.35 MHz central frequency in dependence on the specimen thickness (from 36 to 41.5 mm) and nominal grain size (from 125 to 175 to 250–350 μm). These results show that at the high frequencies the intensity of an ultrasonic wave in Cu polycrystal decreases by approximately 1.09–1.11 times per each mm propagated. Therefore, the Cu specimen even of a 40 mm height will suffer more than 30 times decrease in the ultrasonic wave intensity at its top surface at this frequency (5.35 MHz).

Besides, during our experiments, we observed that parts fabricated with ultrasonic assistance in a vertical build direction showed the defects increasing from the bottom to the top side (see the left picture in **Figure 5**). Nevertheless, the similar parts, fabricated at the same treatment regimes in a horizontal orientation, had a regular shape and normal roughness (**Figure 5**, middle picture). The same was shown by parts, deposited in a vertical direction without ultrasonic assistance (**Figure 5**, right picture). The increase in a surface roughness of the first part could be caused by a spatter due to significant vibration, or by the difference in the cooling rate of the lowest and the highest layers [31]. This difference comes from the fact that layers, larger affected by ultrasonic vibration, undergo more intense stirring provided by cavitation and acoustic flow effects [67] in a melt pool. The

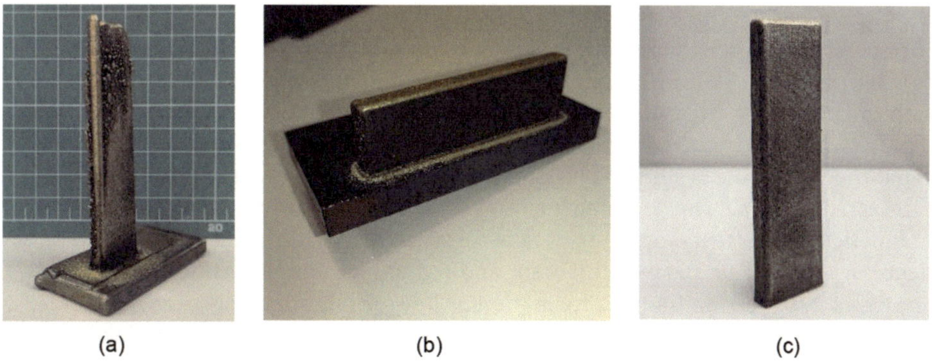

Figure 5.
Parts fabricated: a) with ultrasonic assistance (built in a vertical direction) (left picture); b) with ultrasonic assistance (built in a horizontal direction) (middle picture); c) without ultrasonic assistance (built in a vertical direction) (right picture).

cavitation effect is associated with nonlinearly expanding, contracting, oscillating, shrinking, and collapsing cavitation bubbles in liquid under alternating negative and positive pressure. The collapses of the bubbles produce instantaneous high temperature and pressure [68] in the surrounding area followed by the generation of high-speed liquid microjets and new bubbles nuclei keeping the ultrasonic cavitation process and promoting the liquid flow in the melt pool. The acoustic flow effect is a result of a sound pressure gradient caused by attenuation of the ultrasonic wave during its propagation in the melt. The sound flow slows down by reaching the bottom of the cavity, spreads upward along its side wall, and forms the circulation. The induced acoustic flow effect effectively promotes the flow in the molten pool, amplifying the effect of convection and diffusion.

The changes in the part's shape in its top section could appear because of a disproportional distribution of the powder due to vibration of the mounting layers and substrate. These defects were partially observed not only in parts fabricated via the ultrasonic-assisted DED, but the presence of ultrasonic vibration increased them. Columnar-shaped vertical tall deposit is seen only on the left side of the specimen because of the specificity of the track pattern: the path of the nozzle within each layer is finished near this area, therefore the last fallen powder particles are sintered to the hot surface even if the laser is already turned off. The mentioned defects could be reduced or eliminated by changing a part build direction, as it is shown in **Figure 5**, or a decrease of an ultrasonic generator output power.

4.1.2 Magnetization of the melt pool during the DED process

Magnetization of the melt pool by the external magnetic field during the AM processes, such as DED and SLM, could provide a slight change of phase distribution and intensification of iron oxidation [69] due to dominant thermoelectric magnetohydrodynamic convection. Other microstructural parameters of AM-fabricated parts could be also affected by the external magnetic field: J. Wang et al. [70] reported on the influence of the external magnetic field on the microstructure of laser deposited materials and observed a transformation from continuous to discrete morphology that was demonstrated for DED-fabricated SS 316 L. A similar transformation is also expected to be observed in Cu-Fe system SS 316 L-based FGMs, especially in the regions of the pure SS and mixture of SS and Cu/bronze. Then, D. Du, A. Dong et al. [71] discussed the increase of dendrite spacing and epitaxial nucleation of the laser deposited Inconel 718 alloy with the growth of the magnetic field flux density. The external magnetic field decreases the Marangoni convection, so the forming of the equiaxed grains is suspended, and the growth of the columnar grain is increased, which is an undesirable property in most part of further practical applications, but this disadvantage is compensated by the formation of new equiaxed grains because of increase of the nucleation sites ahead of the dendrite front. Such an increase could also partially suppress the effects of liquid separation in Cu-Fe system FGMs and provide a slight increase in quality of as-fabricated parts. D. Du, J.C. Haley et al. [72] studied the influence of the external magnetic field on the properties of SLM-processed AlSi10Mg alloy, and observed a decrease of porosity, grain refinement and reduction of columnar grains, what approves a suggestion that the formation of new cellular dendrites prevails under the decrease of the amount of the equiaxed grain due to decrement of Marangoni convection, what was mentioned above. Grain refinement and increase of the amount of the equiaxed grain provide higher mechanical strength of the parts fabricated with the external magnetic field implementation, which was proved by the results of the mentioned study [72] (an increase of the UTS from 300 to 330 MPa to 390–410 MPa), but, nevertheless, wasn't demonstrated by A.M. Filimonov et al. [73] even with higher magnetic induction (0.2 T versus 0.12 T). It could point at the fact that this technique is not characterized by universality and could be applied for grain refinement and increase of the tensile strength in the case of specific groups of materials. The lack of experimental data does not allow us to evidently expect these phenomena in the case of Cu-Fe FGMs; therefore, this problem needs a more detailed experimental investigation.

Besides the increase of the mechanical strength, the quality of as-deposited parts could be enhanced via the magnetically-assisted DED by the improvement of the powder catchment efficiency leading to the better compression of fabricated tracks, a decrease of their skewness and dilution [74, 75]. If the ferromagnetic Fe-based powder is applied, these results could be also expected for Cu-Fe FGMs in the Fe-based area and regions with the presence of both Fe and Cu.

The practical interest of a magnetically-assisted DED approach in the case of Cu-Fe system is significantly associated with the formation of ferromagnetic phases in the material even it is fabricated from the non-magnetic initial components. The appearance of ferromagnetic BCC phase leading to 49 emu/g specific magnetizations of Cu-Fe binary alloy in presence of Al was shown by O. Dubinin et al. [76]. Similar magnetic properties were repeatedly approved by the authors of the current article during the investigation of the Cu-Fe binary alloy fabricated by the ultrasonic-assisted DED, which was discussed above in §4.1.1: the specific magnetization of this alloy amounted 58 emu/g (measured using the vibrating-coil magnetometer LakeShore 7410). Therefore, due to forming of ferromagnetic phases in Cu-Fe alloys, the external magnetic field is expected to have an influence on their parameters during the fabrication even there is no possibility to improve the catchment efficiency and variate the powder bed characteristics directly because of the utilization of the non-magnetic powders only.

The main disadvantage of the external magnetic field-assisted DED is expressed in the fact that, on the one hand, a low-power magnetic field could not provide significant microstructural changes [73, 69], and, on the other hand, the high-power fields have a negative influence on the components of a technological installation.

4.2 Post-processing treatment of the DED-fabricated parts

4.2.1 Hot isostatic pressing (HIP)

The part during a HIP is processed by the isostatic pressure of the confining inert gas in the high-temperature conditions [77]. The HIP post-processing is widely used for porosity closing and homogenization of various metals (first of all, for Ti-based alloys) in casting, powder metallurgy, and AM [50], improvement of fatigue performance [77]. The volume fraction of the defects can be significantly reduced after the HIP, but nevertheless, several defects such as unmelted particles, porosity, and defects of the surface can be left after this post-treatment [78]. The phase structure could be changed and ductile phases such as α-martensite could be transformed into the ductile α–β phase [78] increasing the ductility of the processed part. For instance, P. Li et al. [79] reported the increase in cycle fatigue performance of LPBF-fabricated Ti-6Al-4V due to neutralization of the crack initiating defects under the part's surface.

The HIP could be also implemented for post-treatment of Cu-based AM-fabricated parts for enhancement of their physical properties [80] and quality. For instance, M. Agarwala et al. [81] applied the HIP for SLS-fabricated bronze-nickel alloy and showed that this technique allows achievement of the almost full-density parts with suppressed Kirkendall porosity, which is, due to the difference between a diffusion rate of Cu and Fe atoms, an important factor in a Cu-Fe system too (a self-diffusion coefficient of high-purity γ-Fe equals $0.18 \cdot \exp.(-64{,}500/(RT))$ cm^2/s [82], of Cu – $0.468 \cdot \exp.(-47{,}140/(RT))$ cm^2/s over the range 685–1062°*C* [83]).

However, the HIP of large-sized products, such as real space industry parts, requires huge technological installations, which is a limiting factor of this technique. The related industrial solutions include such examples as invention [84], technological installation [85], and large HIP systems [86].

4.2.2 Machining

Machining (including milling, turning, vibratory grinding, abrasive, chemical and electrical polishing, ultrasonic nanocrystal surface modification) of the DED-fabricated parts is a common process, which improves their surfaces' overall quality [87]. These techniques provide a decrease of the roughness of the laser deposited part surface and eliminate the satellite defects (such as protrusions), various edge defects (such as unsuitable levels of edges and their shape), and the morphological defects (such as waviness, blobs, and zits) [31]. Particularly for the Cu-based system, a magnetically driven abrasive polishing was implemented by I. Karakurt et al. [88] to improve the quality of electron beam melting-fabricated copper samples. The results showed a decrease in surface roughness from 35 μm to 4 μm. The Fe-based DED-manufactured material – A131 steel with a moderate percentage of copper (≤ 0.35 wt%) – was treated by the milling post-processing by Y. Bai et al. [89]: a significant surface roughness modification from >20 μm to <1 μm was gained, and the slight changing of microhardness was also observed. Nevertheless, the cracking, which is typical for the DED of Cu-Fe FGMs, as it was largely described above, is not expected to be eliminated by common machining techniques, which are mostly targeted to increase the overall surface quality and lowering of roughness.

5. Conclusion

In the current chapter, three interdependent aspects, important for the DED of Cu-Fe system FGMs, were observed: specificities of intermixing and miscibility; factors, leading to cracking; assisted manufacturing techniques, aimed at enhancement of the quality of the parts. It was pointed out that the typical FGM DED process could be divided into 7 stages, including the preliminary phase and the phase of the conjoining of the base metals A and B. The interfacial area between the base metals A and B is characterized by the joint melt pool, where intermixing is dependent on two physical processes: capillary convection (including capillary-gravity waves and thermocapillary waves) and diffusion. The limited miscibility, precipitation of the hardening phases, excessive strain rates, porosity, unmelted particles, and other defects may cause cracking in Cu-Fe FGMs, especially at their interfacial area, which could be divided into the hot, liquation, rewarming, cold, and ductility dip cracking (three first types are prevalent). The limited miscibility between Cu and Fe could be improved by supplementation of a third component e.g. Al that provides forming of binary and ternary stable phases in the Al-Cu, Al-Fe, and Al-Cu-Fe systems, such as $\omega FeCu_2Al_7$, $\tau_i FeCu_2Al_6$, and $\varphi FeCu_{10}Al_{10}$. Using other laser sources (such as green disk solid-state lasers) with higher radiation absorption coefficient for Cu-based areas of FGM instead of commercial fiber lasers, it is possible to decrease the excessive heat input into the material. The specific technological approaches could be implemented for the DED-manufactured parts quality improvement, including the elimination of cracking. These approaches could be divided into two broad groups: techniques of the in situ influence, and the methods of post-processing. The ultrasonic-assisted DED and the DED in the external magnetic field were discussed specifically among the first group, the HIP, and the machining – among the second one. The ultrasonic assistance provides better intermixing, grain refinement, an increase in tensile strength, elasticity modulus, microhardness, fracture toughness, and wear resistance. The magnetically-assisted DED could also increase tensile strength, decrease porosity, refine the grain structure, improve the ferromagnetic powder catchment efficiency and compression of the tracks. The HIP provides homogenization and porosity closing (including the Kirkendall porosity), increasing the ductility and fatigue performance. The machining improves the overall surface quality of the parts, decreases roughness, eliminates satellite, edge, and morphological defects, e.g. waviness, blobs, and zits.

Acknowledgements

Oleg N. Dubinin is grateful to the Ministry of Science and Higher Education of the Russian Federation as part of the World-class Research Center Program: Advanced Digital Technologies (contract No. 075-15-2020-903 of November 16, 2020).

Author's contributions

Konstantin I. Makarenko: writing orignal manuscript, conducting experiments; Oleg N. Dubinin: conducting experiments; Igor V. Shishkovsky: providing guidance and revision.

Conflict of interest

The authors declare no conflict of interest.

Appendices and Nomenclature

AM	additive manufacturing;

BCC	body-centered cubic;
BTR	brittle temperature range;
CP	commercially pure;
DED	direct energy deposition;
DLD	direct laser deposition;
DMD	direct metal deposition;
DMLS™	Direct Metal Laser Sintering™;
FCC	face-centered cubic;
FGM	functionally graded material;
HIP	hot isostatic pressing;
LDED	laser directed energy deposition;
LENS™	Laser Engineered Net Shaping™;
LMD	laser metal deposition;
LPBF	laser powder bed fusion;
SLM	selective laser melting;
SLS	selective laser sintering;
SS	stainless steel.

Author details

Konstantin Makarenko*, Oleg Dubinin and Igor V. Shishkovsky
Skolkovo Institute of Science and Technology, Moscow, Russia

*Address all correspondence to: konstantin.makarenko@skoltech.ru

References

[1] Lednev VN, Sdvizhenskii PA, Asyutin RD, Tretyakov RS, Grishin MY, Stavertiy AY, et al. In situ elemental analysis and failures detection during additive manufacturing process utilizing laser induced breakdown spectroscopy. Optics Express. 2019;**27**(4):4612-4628. DOI: 10.1364/OE.27.004612

[2] Tang Z, Liu W, Wang Y, Saleheen KM, Liu Z, Peng S, et al. A review on in situ monitoring technology for directed energy deposition of metals. The International Journal of Advanced Manufacturing Technology. 2020;**108**:3437-3463. DOI: 10.1007/s00170-020-05569-3

[3] Nasiri S, Khosravani MR. Machine learning in predicting mechanical behaviour of additively manufactured parts. Journal of Materials Research and Technology. 2021;**14**:1137-1153. DOI: 10.1016/j.jmrt.2021.07.004

[4] Grasso M, Remani A, Dickins A, Colosimo BM, Leach RK. In-situ measurement and monitoring methods for metal powder bed fusion: An updated review. Measurement Science and Technology. 2021;**32**:112001. DOI: 10.1088/1361-6501/ac0b6b

[5] Makarenko K, Dubinin O, Shornikov P, Shishkovsky I. Specific aspects of the transitional layer forming in the aluminium bronze – Stainless steel functionally graded structures after laser metal deposition. Procedia CIRP. 2020;**94**:346-351

[6] Onuike B, Heer B, Bandyopadhyay A. Additive manufacturing of Inconel 718-copper alloy bimetallic structure using laser engineered net shaping (LENS™). Additive Manufacturing. 2018;**21**: 133-140. DOI: 10.1016/j.addma. 2018.02.007

[7] Banait SM, Paul CP, Jinoop AN, Kumar H, Pawade RS, Bindra KS. Experimental investigation on laser directed energy deposition of functionally graded layers of Ni-Cr-B-Si and SS316L. Optics and Laser Technology. 2020;**121**:105787. DOI: j.optlastec.2019.105787

[8] Shishkovsky I, Kakovkina N, Scherbakov V. Fabrication of heat-resisting nickel composite gradient structures with TiC nano additive during powder bed fusion process. Procedia CIRP. 2018;**74**:68-71. DOI: j.procir.2018.08.032

[9] Li W, Karnati S, Kriewall C, Liou F, Newkirk J, Taminger KMB, et al. Fabrication and characterisation of a functionally graded material from Ti-6Al-4V to SS316 by laser metal deposition. Additive Manufacturing. 2017;**14**:95-104. DOI: 10.1016/j. addma.2016.12.006

[10] Reichardt A, Dillon RP, Borgonia JP, Shapiro AA, McEnerney BW, Momose T, et al. Development and characterization of Ti-6Al-4V to 304L stainless steel gradient components fabricated with laser deposition additive manufacturing. Materials & Design. 2016;**104**:404-413. DOI: 10.1016/ j.matdes.2016.05.016

[11] Tan C, Zhou K, Ma W, Min L. Interfacial characteristic and mechanical performance of maraging steel-copper functional bimetal produced by selective laser melting based hybrid manufacture. Materials & Design. 2018; **155**:77-85. DOI: 10.1016/j. matdes.2018.05.064

[12] Tan C, Chew Y, Bi G, Wang D, Ma W, Yang Y, et al. Additive manufacturing of steel–copper functionally graded material with ultrahigh bonding strength. Journal of Materials Science and Technology. 2021; **72**:217-222. DOI: 10.1016/j.jmst.2020. 07.044

[13] Pęska M, Karczewski K, Rzeszotarska M, Polański M. Direct synthesis of Fe-Al alloys from elemental powders using laser engineered net shaping. Materials. 2020;**13**(3):531. DOI: 10.3390/ma13030531

[14] Shishkovsky I, Missemer F, Smurov I. Direct Metal Deposition of Functional Graded Structures in Ti- Al System. 2012;**39**:382-391. DOI: 10.1016/j.phpro.2012.10.052

[15] Ghanavati R, Naffakh-Moosavy H. Additive manufacturing of functionally graded metallic materials: A review of experimental and numerical studies. Journal of Materials Research and Technology. 2021;**13**: 1628-1664. DOI: 10.1016/j.jmrt.2021.05.022

[16] Saleh B, Jiang J, Fathi R, Al-hababi T, Xu Q, Wang L, et al. 30 years of functionally graded materials: An overview of manufacturing methods, applications and future challenges. Composites Part B Engineering. 2020; **201**:108376. DOI: 10.1016/j.compositesb.2020.108376

[17] Heer B, Bandyopadhyay A. Compositionally graded magnetic-nonmagnetic bimetallic structure using laser engineered net shaping. Materials Letters. 2018;**216**:16-19. DOI: 10.1016/j.matlet.2017.12.129

[18] Yan L, Chen Y, Liou F. Additive manufacturing of functionally graded metallic materials: A review of experimental and numerical studies. Journal of Materials Research and Technology. 2021;**13**:1628-1664. DOI: 10.1016/j.jmrt.2021.05.022

[19] Makarenko K, Dubinin O, Shishkovsky I. Analytical evaluation of the dendritic structure parameters and crystallization rate of laser-deposited Cu-Fe functionally graded materials. Materials. 2020;**13**:5665. DOI: 10.3390/ma13245665

[20] Sun X, Hao W, Geng G, Ma T, Li Y. Solidification microstructure evolution of undercooled Cu-15 wt.% Fe alloy melt. Advances in Materials Science and Engineering. 2018;**2018**:6304518. DOI: 10.1155/2018/6304518

[21] Cao MM, Zhou ZM, Tang LW, et al. Research progress of high strength and high conductivity Cu-Fe alloy. Materials Guide: Nano and New Material Album. 2011;**25**:2. in Chinese

[22] Imada R, Fujiwara Y, Tsunashima S. Structure and giant magnetoresistance effect of Fe/Cu and FeCoNi/Cu multilayers. Journal of Applied Physics. 1997;**81**:8. DOI: 10.1063/1.364462

[23] Monchesky TL, Heinrich B, Urban R, Myrtle K, Klaua M, Kirschner J. Magnetoresistance and magnetic properties of Fe/Cu/Fe/GaAs(100). Physical Review B. 1999;**60**:10242. DOI: 10.1103/PhysRevB.60.10242

[24] Monchesky TL, Urban R, Heinrich B, Klaua M, Kirschner J. Giant magnetoresistance of Fe/Cu/Fe(001) trilayers grown directly on GaAs(001). Journal of Applied Physics. 2000;**87**(9): 5167-5169. DOI: 10.1063/1.373283

[25] Hedgcock FT, Muir WB, Raudorf TW, Szmidt R. Magnetoresistance and magnetization in Cu: Fe alloys below 4.2°K. Physical Review Letters. 1968;**20**:457. DOI: 10.1103/PhysRevLett.20.457

[26] Zhang X, Pan T, Flood A, Chen Y, Zhang Y, Liou F. Investigation of copper/stainless steel multi-metallic materials fabricated by laser metal deposition. Materials Science and Engineering A. 2021;**811**:141071. DOI: 10.1016/j.msea.2021.141071

[27] Zhang H, Chen Y, Pan T, Cui W, Li L, Liou F. Joining of copper and stainless steel 304L using direct metal deposition. In: Proceedings of the 30th Annual International Solid Freeform Fabrication

Symposium; 12–14 August 2019; Austin, TX. Austin: The University of Texas at Austin; 2019. pp. 388-403

[28] Prasad HS, Brueckner F, Volpp J, Kaplan AFH. Laser metal deposition of copper on diverse metals using green laser sources. The International Journal of Advanced Manufacturing Technology. 2020;**107**:1559-1568. DOI: 10.1007/s00170-020-05117-z

[29] Bai Y, Zhang J, Zhao C, Li C, Wang H. Dual interfacial characterization and property in multi-material selective laser melting of 316L stainless steel and C52400 copper alloy. Materials Characterization. 2020;**167**:110489. DOI: 10.1016/j.matchar.2020.110489

[30] Osipovich KS, Gurianov DA, Chumaevsky AV. Influence of 3D-printing parameters on bimetallic products manufacturing process of Cu-Fe system. IOP Conference Series: Materials Science and Engineering. 2021;**1079**:042089. DOI: 0.1088/1757-899X/1079/4/042089

[31] Liu M, Kumar A, Bukkapatnam S, Kuttolamadom M. A review of the anomalies in directed energy deposition (DED) Processes & Potential Solutions - part quality & defects. Procedia Manufacturing. 2021;**53**:507-518. DOI: 10.1016/j.promfg.2021.06.093

[32] Qi H, Mazumder J, Ki H. Numerical simulation of heat transfer and fluid flow in coaxial laser cladding process for direct metal deposition. Journal of Applied Physics. 2006;**100**(2):024903. DOI: 10.1063/1.2209807

[33] Laeng J, Stewart J, Liou FW. Laser metal forming processes for rapid prototyping – A review. International Journal of Production Research. 2000; **38**(16):3973-3996. DOI: 10.1080/00207540050176111

[34] Grigoryants AG. The Fundamentals of a Laser Treatment of Materials. Moscow: Mashinostroenie Publishing; 1989. 301 p. ISBN: 5-217-00432-0

[35] Vedenov AA, Gladush GG. Physical Processes during Laser Treatment of Materials. Moscow: Energoatomisdat; 1985. p. 208

[36] Levchenko EB, Chernyakov AL. Instability of surface waves in a nonuniformly heated liquid. Zhournal Eksperimental'noy i Teoreticheskoy Fiziki. 1981;**51**(1):202-209

[37] Khosroshani EF, Heydari A, Mirzayi B, Shamkhali AN. A model to calculate concentration-dependent surface tension of binary systems. Fluid Phase Equilibria. 2016;**423**:34-42. DOI: 10.1016/j.fluid.2016.04.007

[38] Smith WG, Tiller WA, Rutter JW. Canadian Journal of Physics. 1955;**33**:723

[39] Chen YZ, Liu F, Yang GC, Xu XQ, Zhou YH. Rapid solidification of bulk undercooled hypoperitectic Fe–Cu alloy. Journal of Alloys and Compounds. 2007; **427**:L1-L5. DOI: 10.1016/j.jallcom.2006.03.012

[40] Wang CP, Liu XJ, Ohnuma I, Kainuma R, Ishida K. Thermodynamic database of the phase diagrams in Cu-Fe base ternary systems. Journal of Phase Equilibria and Diffusion. 2004;**25**: 320-328. DOI: 10.1007/s11669-004-0150-5

[41] Curiotto S, Greco R, Pryds NH, Johnson E, Battezzati L. The liquid metastable miscibility gap in Cu-based systems. Fluid Phase Equilibria. 2007; **256**:132-136. DOI: 10.1016/j.fluid.2006.10.003

[42] Chuang Y, Schmid R, Chang Y. Thermodynamic analysis of the iron-copper system I: The stable and metastable phase equilibria. Metallurgical Transactions A. 1984;**15**: 1921-1930. DOI: 10.1007/BF02664905

[43] Ma E, Atzmon M, Pinkerton FE. Thermodynamic and magnetic properties of metastable FexCu100−x solid solutions formed by mechanical alloying. Journal of Applied Physics. 1993;**74**:955. DOI: 10.1063/1.354837

[44] Raghavan V. Al-Cu-Fe (Aluminum-Copper-Iron). Journal of Phase Equilibria and Diffusion. 2005; **26**(1):59-64. DOI: 10.1361/15477030522509

[45] Raghavan V. Al-Cu-Fe (Aluminum-copper-iron). Journal of Phase Equilibria and Diffusion. 2010;**31**(5): 449-452. DOI: 10.1361/15477030522509

[46] Chen HL, Du Y, Xu H, Xiong W. Experimental investigation and thermodynamic modeling of the ternary Al-Cu-Fe system. Journal of Materials Research. 2009;**24**(10):3154-3164. DOI: 10.1557/jmr.2009.0376

[47] Svetlizky D, Das M, Zheng B, Vyatskikh AL, Bose S, Bandyopadhyay A, et al. Directed energy deposition (DED) additive manufacturing: Physical characteristics, defects, challenges and applications. Materials Today. 2021;**49**:271-295. DOI: 10.1016/j.mattod.2021.03.020

[48] Kou S. Solidification and liquation cracking issues in welding. JOM. 2003; **55**:37-42. DOI: 10.1007/s11837-003-0137-4

[49] Wei HL, Mukherjee T, Zhang W, Zuback JS, Knapp GL, De A, et al. Mechanistic models for additive manufacturing of metallic components. Progress in Materials Science. 2021;**116**: 100703. DOI: 10.1016/j.pmatsci.2020.100703

[50] Attalah MM, Jennings R, Wang X, Carter LN. Additive manufacturing of Ni-based superalloys: The outstanding issues. MRS Bulletin. 2016;**41**:758-764. DOI: 10.1557/mrs.2016.211

[51] Bi G, Gasser A. Restoration of Nickel-Base turbine blade knife-edges with controlled laser aided additive manufacturing. Physics Procedia. 2011; **12**:402-409. DOI: 10.1016/j.phpro.2011.03.051

[52] Ramirez AJ, Lippold JC. High temperature behavior of Ni-base weld metal part I. ductility and microstructural characterization. Materials Science and Engineering A. 2004;**380**:259-271. DOI: 10.1016/j.msea.2004.03.074

[53] Noecker FF, DuPont JN. Metallurgical investigation into ductility dip cracking in Ni-based alloys: Part I. Welding Journal. 2009;**88**(1):7s-20s

[54] Bahl S, Plotkowski A, Sisco K, Leonard DN, Allard LF, Michi RA, et al. Elevated temperature ductility dip in an additively manufactured Al-Cu-Ce alloy. Acta Materialia. 2021;**220**: 117285. DOI: 10.1016/j.actamat.2021.117285

[55] Wang XH, Liu SS, Zhao GL, Zhang M, Ying WL. In-situ formation ceramic particles reinforced Fe-based composite coatings produced by ultrasonic assisted laser melting deposition processing. Optics and Laser Technology. 2021;**136**:106746. DOI: 10.1016/j.optlastec.2020.106746

[56] Todaro CJ, Easton MA, Qui D, Zhang D, Bermingham MJ, Lui EW, et al. Grain structure control during metal 3D printing by high-intensity ultrasound. Nature Communications. 2020;**11**:142. DOI: 10.1038/s41467-019-13874-z

[57] Zhang D, Li Y, Wang H, Cong W. Ultrasonic vibration-assisted laser directed energy deposition in-situ synthesis of NiTi alloys: Effects on microstructure and mechanical properties. Journal of Manufacturing Processes. 2020;**60**:328-339. DOI: 10.1016/j.jmapro.2020.10.058

[58] Espacenet Patent Search. KR101820719B1 DED Metal Reinforcement Method of Direct Energy Deposition Using Ultrasonic Wave Excitation [Internet]. 2018. Available from: https://worldwide.espacenet.com/patent/search/family/058742114/publication/KR101820719B1?q=pn%3DKR101820719B1 [Accessed: 2021-11-12]

[59] Zhang Y, Guo Y, Chen Y, Kang L, Cao Y, Qi H, et al. Ultrasonic-assisted laser metal deposition of the Al 4047Alloy. Metals. 2019;**9**:1111. DOI: 10.3390/met9101111

[60] Yan S, Wu D, Huang Y, Liu N, Zhang Y, Niu F, et al. C fiber toughening Al2O3-ZrO2 eutectic via ultrasonic-assisted directed laser deposition. Materials Letters. 2019;**235**: 228-231. DOI: 10.1016/j.matlet.2018.10.047

[61] Espacenet Patent Search. CN109604603A Ultrasonic-Assisted Laser-Deposition Additive Manufacturing Method and Device [Internet]. 2020. Available from: https://worldwide.espacenet.com/patent/search/family/066021766/publication/CN109604603A?q=pn%3DCN109604603A [Accessed: 2021-11-12]

[62] Gorunov AI. Development of the science and technological basics of creating materials with enhanced physicomechanical and exploitational properties via the DED [thesis]. Saint Petersburg: Peter the Great St. Petersburg Polytechnic University; 2021

[63] Zhang K, Wang S, Liu W, Shang X. Characterization of stainless steel parts by laser metal deposition shaping. Materials and Design. 2014;**55**:104-119. DOI: 10.1016/j.matdes.2013.09.006

[64] Guo P, Zou B, Huang C, Gao H. Study on microstructure, mechanical properties and machinability of efficiently additive manufactured AISI 316L stainless steel by high-power direct laser deposition. Journal of Materials Processing Technology. 2017;**240**:12-22. DOI: 10.1016/j.jmatprotec.2016.09.005

[65] Azinpour E, Darabi R, Cesar de Sa J, Santos A, Hodek J, Dzugan J. Fracture analysis in directed energy deposition (DED) manufactured 316L stainless steel using a phase-field approach. Finite Elements in Analysis and Design. 2020;**177**:103417. DOI: 10.1016/j.finel.2020.103417

[66] Stepinski T, Wu P. Evaluation of ultrasonic attenuation and estimation of ultrasonic grain noise in copper. AIP Conference Proceedings. 1999;**497**:431. DOI: 10.1063/1.1303084

[67] Riedel E, Liepe M, Scharf S. Simulation of ultrasonic induced cavitation and acoustic streaming in liquid and solidifying aluminum. Metals. 2020;**10**:476. DOI: 10.3390/met10040476

[68] Suslick KS, Neis U. The chemical effects of ultra-sound. Scientific American. 1989;**260**:80-86

[69] Shishkovsky I, Saphronov V. Peculiarities of selective laser melting process for permalloy powder. Materials Letters. 2016;**171**:208-211. DOI: 10.1016/j.matlet.2016.02.099

[70] Wang J, Wang Y, Shi J, Su Y. Effect of external magnetic field on the microstructure of 316L stainless steel fabricated by directed energy deposition. In: Proceedings of the International Mechanical Engineering Congress & Exposition; 11–14 November 2019. Salt Lake City. Utah: ASME; 2019. pp. 866-870. DOI: 10.1115/IMECE2019-12122

[71] Du D, Dong A, Shu D, Wang D, Zhu G, Sun B, et al. Influence of static magnetic field on the microstructure of nickel-based Superalloy by laser-

directed energy deposition. Metallurgical and Materials Transactions A. 2020;**51**:3354-3359. DOI: 10.1007/s11661-020-05783-4

[72] Du D, Haley JC, Dong A, Fautrelle Y, Shu D, Zhu G, et al. Influence of static magnetic field on microstructure and mechanical behavior of selective laser melted AlSi10Mg alloy. Materials & Design. 2019;**181**:107923. DOI: 10.1016/j.matdes.2019.107923

[73] Filimonov AM, Rogozin OA, Dubinin ON, Kuzminova YO, Shibalova AA, Okulov IV, et al. Modification of mechanical properties in directed energy deposition by a static magnetic field: Experimental and theoretical analysis. Materials. 2021;**14**: 5190. DOI: 10.3390/ma14185190

[74] Smith PH, Murray JW, Jones DO, Segal J, Clare AT. Magnetically assisted directed energy deposition. Journal of Materials Processing Technology. 2021; **288**:116892. DOI: 10.1016/j. jmatprotec.2020.116892

[75] Smith PH, Murray JW, Jackson-Crisp A, Segal J, Clare AT. Magnetic manipulation in directed energy deposition using a programmable solenoid. Journal of Materials Processing Technology. 2022;**299**:117342. DOI: 10.1016/j.jmatprotec.2021.117342

[76] Dubinin ON, Chernodubov DA, Kuzminova YO, Shaysultanov DG, Akhatov IS, Stepanov ND, et al. Gradient soft magnetic materials produced by additive manufacturing from non-magnetic powders. Journal of Materials Processing Technology. 2022; **300**:117393. DOI: 10.1016/j. jmatprotec.2021.117393

[77] Du Plessis A, Macdonald E. Hot isostatic pressing in metal additive manufacturing: X-ray tomography reveals details of pore closure. Additive Manufacturing. 2020;**34**:101191. DOI: 10.1016/j.addma.2020.101191

[78] Molaei R, Fatemi A, Phan N. Significance of hot isostatic pressing (HIP) on multiaxial deformation and fatigue behaviours of additive manufactured Ti-6Al-4V including build orientation and surface roughness effects. International Journal of Fatigue. 2018;**117**:352-370. DOI: 10.1016/j. ijfatigue.2018.07.035

[79] Li P, Warner DH, Pegues JW, Roach MD, Shamsaei N, Phan N. Investigation of the mechanisms by which hot isostatic pressing improves the fatigue performance of powder bed fused Ti-6Al-4V. International Journal of Fatigue. 2019;**120**:342-352. DOI: 10.1016/j.ijfatigue.2018.10.015

[80] Tran TQ, Chinnappan A, Lee JKY, Loc NH, Tran LT, Wang G, et al. 3D printing of highly pure copper. Metals. 2019;**9**:756. DOI: 10.3390/met9070756

[81] Agarwala M, Bourell D, Beaman J, Marcus H, Barlow J. Post-processing of selective laser sintered metal parts. Rapid Prototyping Journal. 1995;**1**(2): 36-44. DOI: 10.1108/ 13552549510086853

[82] Buffington FS, Hirano K, Cohen M. Self diffusion in iron. Acta Metallurgica. 1961;**9**(5):434-439. DOI: 10.1016/ 0001-6160(61)90137-7

[83] Kuper A, Letaw H Jr, Slifkin L, Sonder E, Tomizuka CT. Self-diffusion in copper. Physical Review. 1955;**98**: 1870. DOI: 10.1103/PhysRev.96.1224

[84] Pavlov VA, Popov BV, Yakunin SN. Method of Hot Isostatic Pressing of Large Cylindrical Articles from Powders. Copyright Certificate [Internet]. 1989. Available from: https:// www.elibrary.ru/item.asp?id=40585617 [Accessed: 2021-12-20]

[85] Foundry Management & Technology. World's Largest Hot Isostatic Press Again [Internet]. 2008. Available from: https://www.found

rymag.com/issues-and-ideas/article/21925022/worlds-largest-hot-isostatic-press-again [Accessed: 2021-12-19]

[86] Quintus Technologies. Hot Isostatic Presses [Internet]. 2021. Available from: https://quintustechnologies.com/hot-isostatic-pressing/products/hot-isostatic-presses [Accessed: 2021-12-18]

[87] Becker TH, Kumar P, Ramamurty U. Fracture and fatigue in additively manufactured metals. Acta Materialia. 2021;**219**(33):117240. DOI: 10.1016/j.actamat.2021.117240

[88] Karakurt I, Ho KY, Ledford C, Gamzina D, Horn T, Luhmann NC, et al. Development of a magnetically driven abrasive polishing process for additively manufactured copper structures. Procedia Manufacturing. 2018;**26**: 798-805. DOI: 10.1016/j.promfg.2018.07.097

[89] Bai Y, Chaudhari A, Wang H. Investigation on the microstructure and machinability of ASTM A131 steel manufactured by directed energy deposition. Journal of Materials Processing Technology. 2020;**276**: 116410. DOI: 10.1016/j.jmatprotec.2019.116410